Energy Policy in the U.S.

Politics, Challenges, and Prospects for Change

PUBLIC ADMINISTRATION AND PUBLIC POLICY

A Comprehensive Publication Program

RECENTLY PUBLISHED BOOKS

Handbook of Administrative Reform: An International Perspective, edited by Jerri Killian and Niklas Eklund

Government Budget Forecasting: Theory and Practice, edited by Jinping Sun and Thomas D. Lynch

Handbook of Long-Term Care Administration and Policy, edited by Cynthia Massie Mara and Laura Katz Olson

Handbook of Employee Benefits and Administration, edited by Christopher G. Reddick and Jerrell D. Coggburn

Business Improvement Districts: Research, Theories, and Controversies, edited by Göktuğ Morçöl, Lorlene Hoyt, Jack W. Meek, and Ulf Zimmermann

International Handbook of Public Procurement, edited by Khi V. Thai

State and Local Pension Fund Management, Jun Peng

Contracting for Services in State and Local Government Agencies, William Sims Curry

Understanding Research Methods: A Guide for the Public and Nonprofit Manager, Donijo Robbins

Labor Relations in the Public Sector, Fourth Edition, Richard Kearney

Performance-Based Management Systems: Effective Implementation and Maintenance, Patria de Lancer Julnes

Handbook of Governmental Accounting, edited by Frederic B. Bogui

Bureaucracy and Administration, edited by Ali Farazmand

Science and Technology of Terrorism and Counterterrorism, Second Edition, edited by Tushar K. Ghosh, Mark A. Prelas, Dabir S. Viswanath, and Sudarshan K. Loyalka

Handbook of Public Information Systems, Third Edition, edited by Christopher M. Shea and G. David Garson

Public Administration in East Asia: Mainland China, Japan, South Korea, and Taiwan, edited by Evan M. Berman, M. Jae Moon, and Heungsuk Choi

Public Administration and Law: Third Edition, David H. Rosenbloom, Rosemary O'Leary, and Joshua Chanin

Governance Networks in Public Administration and Public Policy, Christopher Koliba, Jack W. Meek, and Asim Zia

Public Administration in Southeast Asia: Thailand, Philippines, Malaysia, Hong Kong and Macao, edited by Evan M. Berman

Available Electronically
PublicADMINISTRATION*netBASE*

Energy Policy in the U.S.

Politics, Challenges, and Prospects for Change

Laurance R. Geri
David E. McNabb

NEW YORK AND LONDON

First published 2011 by Taylor & Francis

Published 2019 by Routledge
52 Vanderbilt Avenue, New York, NY 10017
2 Park Square, Milton Park, Abingdon, Oxon OX14 4RN

Routledge is an imprint of the Taylor & Francis Group, an informa business

ISBN: 978-1-4398-4189-1 (hbk)
ISBN: 978-1-3150-9450-2 (ebk)

Typeset in AGaramondPro
by Apex CoVantage, LLC

For Mom and Rachel,
and
For Meghan, Michael, Sara, and Janet

Contents

PART 2 POLICIES FOR ENERGY TRANSITION

List of Boxes

List of Figures

List of Tables

Preface

This book is about the policies and policy-forming processes that our governments—federal, state, and local—have taken, are taking, and are proposing to take in their efforts to reach the elusive goal of energy independence. The object of the book is to bring to light some of the difficulties that constrain governments' efforts to forge energy policy—and to stick with it once it is formed.

The task of framing an energy policy for the country involves balancing the national and international concerns of conservation, economic growth, greater efficiency in the use of energy, and promoting invention and innovation in both existing and new energy resources. An additional goal of energy policy today is continual improvement in the security of domestic and international energy resources and distribution infrastructure.

For many Americans, true security for the nation and its vital energy supplies will occur only when we have eliminated our dependence upon other countries for our energy supplies. For many others, however, the goal of energy independence is an unrealistic dream—a chimera that obscures the fact that international trading in energy is always going to exist and, therefore, should be adjusted to result in greater advantage for the United States. Of course, that is the nature of public policy making; for every position there are one or more alternative proposed solutions.

Just as it is not proscriptive, neither is this book prescriptive; it does not point fingers, demand program revisions, or offer panaceas. Easy answers to the problems of energy and energy policy simply do not exist. There are just too many options and too many barriers, none of which can readily result in enough reliable, affordable, dependable, clean energy to erase our need to import a significant proportion of our energy supplies. Instead, this is a book that describes the efforts that clear-headed and farsighted men and women in government, industry, and private life are taking to address the interrelated issues of energy, climate warming, and public policy.

Why Worry about Energy?

This book has three goals. First, it recognizes that regardless of what some people would have you believe, there are *no* viable immediate or even short-term solutions to the problem of our dependence upon foreign oil. To the contrary: achieving energy independence may not be possible for 50 years or longer, if then. Second, the Obama administration's early published energy policies warn that the process will indeed require a commitment to long-range objectives and programs, many of which require significant investment in research before they can be implemented. Third, the U.S. Department of Energy uses the target date of 2030 in many of its studies. This apparently allows sufficient time for the department's research and development proposals and programs to come to fruition and to begin to make an impact on our reliance on foreign energy supplies. We believe this is too long a wait.

Just because there are no short-term solutions to the dependence on foreign sources of energy, this does not mean to suggest that we can afford to stop trying to get a better handle on our energy problems. There are many things that can and should be done now to ease the immediate problems and possibly contribute to longer-term solutions. We must remember, however, that there is no excuse for not doing all we can to conserve energy and to make better use of the energy resources we now have. Developing and sticking to a comprehensive energy policy is the first step in resolving our energy problems.

The Obama/Biden energy plan described in campaign material released in 2008 identified a long list of short-, medium-, and long-term energy and environment-related objectives and identified high-minded-sounding programs that the new administration would work to implement. Believing that all of these can be accomplished in the time periods suggested requires a big leap of faith. Early progress suggests that many of the energy goals and objectives may have to be shelved while the administration's attention remains focused on economic issues. A review of some prior energy policy initiatives is included in the next chapter in order to emphasize the difficulties that must be expected and overcome in forming an energy policy.

Acknowledgments

Several persons stand out to whom we individually and together owe a large debt of gratitude for their present and past help in getting this book written and into print. Foremost among these is Dr. Evan M. Berman, distinguished university professor at the National Chengchi University in Taipei, Taiwan, and editor-in-chief of the American Society for Public Administration (ASPA) book series in Public Administration and Public Policy (Taylor & Francis, publishers). Dr. Berman worked tirelessly with the authors to ensure that the final product would turn out to be the best possible book for the time and proposed audience.

We also want to thank the faculty and staff of The Evergreen State College Masters Program in Public Administration (MPA), as well as the MPA students during the 2009–2010 academic year who read and commented on portions of the manuscript. The students in Larry's summer 2010 Energy Policy course deserve particular thanks for reading the entire draft text and providing very helpful feedback. Thanks to MPA student Michael Davidson for providing many useful citations from the energy literature. Ken Conte, staff director for the Washington State House of Representatives, Office of Program Research, was unknowingly a big contributor to the book. Ken strongly suggested, in a program advisory board meeting several years ago, that someone from the MPA program needed to teach energy policy, and Larry took up the challenge.

We wish also to thank the faculty, staff, and administration of Pacific Lutheran University and The Evergreen State College for their generous support for this and other research efforts over the years. Pacific Lutheran University and its School of Business Administration, which continues to endorse the unofficial research professor activities of Dr. McNabb by providing access to facilities and staff. Without that support and encouragement, this book could not have been written.

About the Authors

Laurance (Larry) Geri is a member of the faculty of The Evergreen State College, where he teaches in the Masters Program in Public Administration (MPA). He completed his doctorate in public administration at the University of Southern California and holds a BA from the University of Washington in economics, and an MPA from The George Washington University. He was director of Evergreen's MPA program from 2002 to 2006. Prior to coming to Evergreen, he was a manager and internal consultant for an agency of the U.S. Department of Agriculture, responsible for oversight of the agency's programs in Latin America and Europe. Dr. Geri's academic interests include energy policy, nonprofit management, and international affairs; he has written articles and book chapters on a wide variety of public administration subjects. His e-mail address is: geril@evergreen.edu.

David E. McNabb is business administration professor emeritus at Pacific Lutheran University and currently a member of the adjunct faculty of Olympic College. He has been associated with higher education institutions in Latvia since its 1991 independence, serving as visiting professor at the Riga, Latvia campus of the Stockholm School of Economics (SSER), and a visiting lecturer at the Riga Technical University and the Turiba School of Business Administration in Riga. He has been a visiting professor at the Tacoma Campus of the University of Washington, Seattle University, The Evergreen State College, and the American University of Bulgaria. He has taught management, marketing, and public administration courses in Belgium, Germany, Italy, and the United Kingdom for the University of Maryland. Professor McNabb received his PhD in administration and marketing from Oregon State University, an MA in communications from the University of Washington, and a BA in language arts from California State University-Fullerton. His work experience has included positions as director of industrial and commercial development for the city of Fullerton, California, and director of communications for a caucus of the Washington State House of Representatives. He is the author of six books on such topics as research methods, public policy, and public and nonprofit management, and has published more than 90 conference papers and journal articles. His e-mail address is: mcnabbde@plu.edu.

Introduction

> Today we are hooked for better or worse, not only on coal but also on oil and gas and vulnerable to twin threats—uncertain supplies and the danger of potentially irreversible changes not just in regional but also in global climate induced by emissions of the products of those fuels, most notably the greenhouse agent carbon dioxide.
>
> **—Michael B. McElroy, 2010**

Why another book on energy? Energy policy is perhaps the most Janus-faced of all the policy arenas in American politics. One face is in shadow: when supplies of crude oil, gasoline, natural gas, coal, and electricity are abundant and cheap, energy policy is generally off the agenda. When supplies tighten, costs increase, or the effects of our profligate energy use become too great to bear, people clamor for relief. Suddenly the second face is in full sun, and discussion begins on policies that are ordinarily obscure and ignored.

As we write, there is a strengthening consensus that global climate change is occurring and that it is influenced by human activities, particularly the burning of fossil fuels such as oil, coal, and natural gas. The recognition that such anthropogenic climate change could have disastrous and irreversible effects on Earth's biosphere has again placed an uncomfortable spotlight on energy systems in the United States as well as the government policies that have influenced and supported those systems. In his 2010 book on energy, Harvard professor Michael McElroy added that in dealing with the twin threats resulting from fossil fuel use the United States must adopt an energy policy "in a world where the human population has never been greater, where distances have contracted, and where an ever-greater number of people aspire legitimately (and often forcefully) to share the benefits of the good life" (McElroy 2010, 71).

Our public policies in the energy arena and the social, economic, and environmental practices they support have led us to this point. In many respects these policies have been successful. Since the infamous energy crisis and oil embargo of the late 1970s, we have had generally stable supplies of the energy sources we need, at affordable prices. But again, there is a second, less benign face to the story. Our fossil-fueled civilization also generates high environmental costs to our air, water, and land, and not all share in its benefits. The United States continues to use nearly one-quarter of the globe's yearly oil supply (19 million barrels per day [mbd], of total global production of 87 mbd), while a quarter of the world's population, about 1.6 billion people still lack access to electricity (United Nations Development Program [UNDP] 2009).

Although mountains of books have been written about various aspects of energy, they tend to treat energy policy as an afterthought. Governments worldwide, including the U.S. government, intervene in energy markets in a variety of ways. They provide financial incentives and regulations that encourage production of preferred energy sources, limit consumption, subsidize use by individuals, or try to influence the behavior of firms and individuals to encourage efficiency and limit pollution. These policies are the details of a country's national energy strategy, its overall approach to obtaining the energy resources needed to sustain its infrastructure and civilization.

Our intent in this book is to focus on those policies: to provide an overview of the important energy policies in the United States, including their history, goals and objectives, methods of action—how, exactly, do they achieve their intended results—and consequences, both intended and unintended. With an improved understanding of these past and current policies, we believe that individuals and policy makers will be in a better position to create and implement new policies more appropriate to the goal of limiting the advance of global climate change.

What *Is* Energy?

One of the ironies of work in this policy arena is that there is no generally agreed upon definition of exactly what energy is! It is a slippery, indefinable concept grounded in the hard sciences of physics and chemistry. Thus, energy has been defined as "the capacity to do work," or (better) "the ability to transform a system" (Smil 2008, 13). We tend to associate the concept with particular sources of energy (oil, natural gas, uranium), which through various types of conversions (particularly burning or oxidation) produce secondary products, effects, or services that we value. The critical insight is that such sources have little intrinsic value; we value them because of what they can do for us, the services they can provide.

Americans treasure individual *mobility*—oil, fractured into gasoline, just happens to be the substance that best enables such transportation (although, as we shall see, at shockingly low levels of efficiency). We need energy to cook our meals, heat and light our homes and offices, and fuel the manufacturing and services that run our economy.

Public policy impacts all of these conversions, and thus helps to shape the choices available to us. These policies reflect decisions about what kind of society we want and what we are willing to do (or put up with) to have access to the services that energy provides. For example, we take gasoline-powered individual transport for granted. But behind each fill-up of your car lies a complex web of subsidies and regulations influencing petroleum production, refining, transport, car manufacturing, driving, and urban planning, as well as a degree of willful blindness on the part of the driving public (see Chapters 4 and 7 for more details). Each year around 40,000 people die in traffic accidents in the United States and gasoline explosions are surprisingly common.

The availability of cheap energy is part of the American self-image. It powers our big houses, cars, and SUVs, and any change in energy policy that meddles with that formula is likely to have a short half-life. Yet the risks associated with climate change are too great to make inaction acceptable.

Structure of the Energy Industry

To understand energy policy in the United States, it is necessary to be thoroughly conversant with the many different institutions and organizations that make up the energy industry. Regardless of the particular product or service, the energy industry is composed of four fundamental components; three are economic; one is political. The economic institutions include the firms involved in the production, shipping or transmission, and retail distribution of energy products. Governments constitute the political and regulatory arm of the industry.

The production methods of energy resources are remarkably similar. Oil, the single greatest source of energy in the United States, accounting for some 86 percent of all energy consumed, is pumped from underground pools by wells sunk on public and private land and from offshore platforms. Natural gas is also collected at wells, often as a by-product of petroleum production. Coal and its derivatives, such as oil sands and shale, are mined from either underground seams or extracted from surfaced deposits, again from both public and private lands.

For electricity, production means power generation. Electrical energy is generated by falling water in hydroelectric systems, by steam-powered turbines using fossil fuel, geothermal energy, by nuclear power, wind-power generators, or by gas turbines fired by natural gas, methane, or propane. A very small number of turbines are fueled by the combustion of biomass products.

Transmission is also similar for all three energy supplies. Electric power is moved over high-voltage transmission lines; gas is transported by large-volume pipelines; oil is pumped through pipelines or moved by truck, oceangoing ships, or river barges; coal is moved by truck, and increasingly by special, dedicated railcar from mines to where it is needed for industrial use and for generating electricity.

Distribution of energy supplies to end users is less similar than production and transmission systems. Electric power is distributed over low-voltage overhead or underground power lines to local public or private utilities that distribute power to homes and businesses, and provide power for street lighting and for a declining number of public transit systems. Gas and petroleum products are distributed via underground pipe networks to local collection and storage facilities (*tank farms*), and then distributed to local distribution points by over-the-road vehicles. Gas utilities distribute via underground pipes in ways similar to electricity utilities. Petroleum products, including heating oil, are distributed at company or independent gasoline outlets and distribution systems.

Governments play a large and varied role in the energy industry, one that ranges from leasing extraction rights on public lands to private companies to wholesale and retail price controls at the consumption end of the cycle. The federal government, the states, and some large local municipalities have diverse responsibilities in the production, distribution, and consumption of energy. First, they are large consumers of all energy products. Second, they control access to energy resources. Third, they set prices and mandate adherence to environmental regulations in all phases of energy production and distribution.

Stakeholders in the Energy Policy Network

This book examines elements of the energy policy question as it affects and is affected by the four interrelated institutions that make up the industry: *energy prime movers, energy industry shapers, energy users*, and *energy regulators*. The prime movers in the energy industry include all the public and private producers and distributors of energy resources. Industry shapers include the financial, research, social, and supporting organizations that facilitate the movement of energy from producers to users. Enron was one of the more disreputable organizations in the group.

Energy users include the consumer, commercial, industrial, and governmental organizations that purchase and consume energy in all its forms. Industry regulators include the international, federal, state, and local government agencies that regulate some or all aspects of energy supply and demand.

Together, these institutions make up the energy policy network (Rhodes 2006). The players in the network include the president, elected members of the House and Senate, congressional staff members, administrators in government agencies such as the Departments of Energy, the Interior, and Agriculture, members of citizen, environmental, and business interest groups, and research organizations such as the Rand Corporation. Some of these organizations—the insiders—have close relationships with elected and appointed government institutions, whereas others may be outsiders and restricted in their ability to participate in the policy framing process (Hajer and Laws 2006). Insiders are called upon to supply background information and regularly testify before legislative committees. They often help prepare proposed legislation, standards, and regulations. Government policy makers see these insiders as responsible in their requests, willing to compromise when necessary, and always ready to cooperate with agency planners and policy makers. Government needs these organizations in order to achieve its policy objectives.

The individual organizations within each of these institutions have diverse and often diametrically opposed goals and objectives compared with other organizations within and outside of the institutional class to which they belong. Within the context of this discussion, those goals and objectives can be expressed as policy-making *priorities* that make up their tool chest of potentially negotiable concepts. In addition to these priorities, each organization must find satisfactory answers to questions that are particularly relevant to their segment in the industry, their organization, and their individual constituents. Together, these priorities and questions function as the *position framework* within which policy negotiations will take place. Typical of the types of questions that no negotiator would be likely to forget to have answered during a policy negotiation are:

1. What's in this policy proposal for us?
2. How much is the policy going to disrupt the status quo?
3. Who does this policy hurt and who does it help?
4. What compromises or losses will the proposal require of my group?
5. Who is going to pay for the changes or required actions?
6. How much is this proposal going to cost our members?
7. What long-term benefits can we expect to offset those costs?
8. How long can we expect this policy direction to be in effect? Can we afford to ignore it?
9. How can we be sure that our position gets a fair hearing and is not ignored in the final policy?
10. How long will it be before we really run out of oil and gasoline, and what changes will we be required to make?
11. How can we best be prepared for the day when our members can no longer afford to pay for fuels?
12. How will environmental requirements hurt my group? Who will pay for our losses?
13. How will global warming really affect my members? What can we do to alleviate any damage?
14. When is the government going to get serious about requiring us to use alternative fuels?

These and other, related issues are high on the executive and legislative policy agendas. In addition, readers will have their own list of questions pertaining to issues particularly related to their organizations. While it is not possible in this book to answer each of these questions for each of the four energy industry institutions or for any one or two organizations or groups within any one institution, they are part of the concept framework within which policy considerations for the industry sectors contained in the discussion that follows this section of the book.

While the focus of this book is on energy policy and the policy process, it is important to remember that energy issues have great impact on the economic and quality of life of everyone. At the one end of the issue spectrum are propositions affecting stakeholders such as the families whose only source of heat for cooking are twigs and dried dung. Soot from their cooking fires is seen as contributing to the melting of glaciers in the Himalayas. A priority here is to provide families with more efficient cooking stoves (Rosenthal 2009).

At the other end of the issue spectrum are factors such as the burning of fossil fuels in industrial nations to generate electricity and power motor vehicles. This is considered by many as the greatest single cause of air pollution and contributor of greenhouse gases, which are the cause of global warming. Requiring more fuel-efficient motor vehicles is a priority at this level. Doing so results in higher costs to consumers. Another is requiring more ethanol as a motor fuel. This drives up the cost of food and cuts into already limited food supplies. Burning coal to produce electricity is a major cause of acid rain and air pollution. The list of costs and benefits goes on and on. And that's why a comprehensive energy policy is so important to the nation and the world.

What Actions Should We Take?

But what kinds of action are needed? Tim Flannery's otherwise excellent 2005 book on climate change, *The Weathermakers*, ends disappointingly with a "Climate Change Checklist" of eleven individual and household actions such as installing solar panels and buying energy efficient light bulbs (Flannery 2005, 316). But though individual action to change patterns of energy use will be necessary and helpful—it will not be *sufficient* to slow climate change. Most studies of energy and climate change are skeptical about the usefulness of ad hoc, individual changes in consumption to climate change action. It is difficult, although certainly not impossible, for individuals to change their energy habits appreciably in the short run.

Autos and homes are medium- to long-term investments, and are linked to career, job, and lifestyle choices that many Americans find difficult to change quickly. Although high gasoline prices spurred a surprising shift to mass transit in 2008, it is not clear how long that shift would last. Even more challenging is the fact that the impacts of climate change are long term, while the costs to prevent it are incurred now. And the impacts of our choices are difficult to see: the 19 pounds of carbon dioxide emitted when we burn a gallon of gasoline in our cars are invisible.

A major impediment to transformational change in the energy sector is that these policies will inevitably be framed by some stakeholders as a large and expensive dose of bitter medicine. The argument: a decreased reliance on fossil fuels will raise energy costs, lower our living standards, increase costs for businesses, and make U.S. products uncompetitive, wrecking our economy. This narrative links easily to the strong antitax sentiment in the United States and it is likely that any attempt to raise the price of carbon will be portrayed as a tax increase.

Further complicating matters is the wildly uneven tone of the debate on energy and energy futures. As Chapter 1 will discuss, the history of prediction in the energy arena is a sorry one. This should encourage authors on this topic to approach it with caution and humility, but alas, sensationalism attracts attention. So on one end of the continuum, there are jeremiads such as James Howard Kunstler's *The Long Emergency* (2005), anticipating the Decline of Civilization as We Know It, and at the other, techno-utopian descriptions such as William Tucker's *Terrestrial Energy: How Nuclear Energy Will Lead the Green Revolution and End America's Energy Odyssey* (2008).

The mainstream media have generally, in our view, tried to apply strong standards of journalism to the complex story of climate change and the role of energy, but the blogosphere, not surprisingly, flows from one extreme to the other. This can leave both the average citizen, and policy makers, bewildered by the cacophony of these *nattering nabobs of negawatts* (apologies to the late William Safire).

The challenge for change agents in the energy sector is to both rebut the gloomy predictions and present an appealing, yet realistic vision for a transformed energy sector in the United States. Fortunately, as Chapter 2 will discuss, the dire predictions are largely off the mark; most estimates of the costs of comprehensive strategies to reduce greenhouse gas emissions are around 1 percent of gross domestic product (GDP). And the effects of such strategies need not be dire. Other countries with a level of social and economic development equal to ours use considerably less energy, and despite our size and climate and autophilic ways, we can do so as well.

As will be discussed in Chapters 1 and 2, we believe that long-lasting changes to the energy sector must be systemic, not ad hoc. Our society cannot avoid the need to take collective action against the challenge of climate change, and that means changing energy policies that are now not only outmoded, but counterproductive. Such collective action runs against the grain of U.S. society's individualist self-image. It will also require confronting institutions (businesses, unions, regions) that have benefited from decades of substantial government support, in a country where our dominant energy policy strategy could be described as, "it's the supply, stupid." What policies do we need, and how do we enact them in such a complex and turbulent policy environment? This book will attempt to answer these questions.

Before launching into our overview of past and present U.S. energy policies, it is essential for the reader to have a basic understanding of the concepts and terminology used in energy analysis, and a grasp of the current patterns of energy use in the country. Since the most ubiquitous form of energy in our daily lives—electricity—is also one of the most poorly understood, we'll begin with a case study of the fascinating chain of events that puts the power of the electron at your disposal.

Purpose for the Book

Our purpose for writing this book was to provide a greater awareness of the policy decisions that America's elected and appointed officials must make if we are ever to successfully move the country away from its dangerous and costly dependence upon foreign energy sources—or even if trying to accomplish that goal is the right one for America. Despite the fact that our energy policy affects every aspect of the national economy and the daily lives of all Americans, our political leaders have still not been able to stick to one sustainable policy path; billions spent by one administration have been ignored by the next.

The country has been in an on-again, off-again drive for energy independence for more than forty years, and it looks as if we are no better off today than we were when the drive began in the 1970s. Achieving energy independence has been, in the words of former President Jimmy Carter, the "moral equivalent of war." This has not been a conventional war involving guns, ships, and aircraft, although at times violence clearly related to energy resources has broken out or is just under the very fragile surface calm. Rather, it is more akin to a quest which, at times, has included the overtones of a Crusade (Revkin, 2010).

While the focus of this book is on *energy* policy, crafting a sustainable energy policy cannot be achieved without including an analysis of *environmental* policy. On one hand, a major reliability and security-related energy policy proposal on the 2010 political agenda is to shift the millions of automobiles and trucks now powered by fossil fuels to electric power. This will require major expansion of the electricity generation, transmission, and distribution infrastructure. However, most electricity is produced in very large coal-burning generating plants.

The burning of coal to generate electric power is this country's greatest single cause of air pollution and contributor to greenhouse gases and global warming. Even so, there are groups who want to remove many existing hydroelectric dams, forbid construction of any new nuclear power–generating plants, and replace millions of existing vehicles with new electrically powered automobiles. What seems to be forgotten is the fact that the nation's electricity grid is in many locations already at maximum capacity, and its facilities are obsolete and badly in need of the investment of billions of dollars in new generation, transmission, and distribution infrastructure.

Why an Energy Policy Is Important

America needs a comprehensive energy policy for a number of reasons; in addition to resolving the confusion over energy that reemerges with every new administration and every swing in energy prices, two others stand out. First, it is a fact that the globe faces a growing shortage of nonrenewable energy resources (oil and gas, but not coal). People disagree over the length of time before we run out of affordable fossil fuels, but they agree that global demand for energy has been far outpacing new sources of supply. In the United States, this has increased our reliance upon a few foreign suppliers for fossil fuel supplies. Political leaders describe this dependence as not only representing a threat to the economy and our way of life, but also as a threat to our national security. Second, the continued burning of fossil fuels, particularly petroleum products and coal, are poisoning the environment and contributing significantly to global warming through the emission of carbon dioxide and other greenhouse gases, light-absorbing particulates (black carbon), and ozone-creating gases. A collaborative and sustainable approach to energy and environmental policy is clearly long overdue.

Energy and Economic Growth

Policy makers press for energy independence because economic growth requires a sufficient, stable supply of clean energy resources. When the nation is dependent upon other countries for much of its energy supply, the health of our economy and the nation's security are at risk. Thus, the energy policies now in effect and new models underway in Washington, DC and state capitals, affect the air we breathe, the food we consume, and power the businesses and industries that produce the safety and quality of life we have grown to expect. This is also why we need a comprehensive and coordinated energy policy, not just an energy bill here, an energy bill there, with little or no all-fuels integration. But, as this book will reveal, coming up with a comprehensive energy policy is not a simple task. It has not happened often. And, when it has, all too soon, policy makers have seen fit to change the direction and focus of that policy.

For most of the history of the United States we did not have a comprehensive, coordinated national energy policy. Until the late 1970s, energy policy was centered in a variety of government agencies. Moreover, until that historic decade, the problem was not having enough energy, but how to handle the energy surplus with which the nation was blessed. Abundant supplies of fossil fuels made it possible for America to become the world's most powerful industrialized nation.

For most of her modern history, America was a net exporter of oil and coal. At the same time, control of energy matters historically remained divided among the individual states and the federal government. Also, a large portion of our energy resources are found on public lands, which were controlled by the Department of the Interior and various other federal agencies, including the Department of Indian Affairs.

Energy Policy and Public Opinion

Public policy is the plan of action that guides a government or its agencies in actions dealing with issues of public concern (Cochran, Mayer, Carr, and Cayer 1996, 1). Public policy shapes—and is in turn shaped by—the laws enacted by federal, state, and local executive and legislative actors. However, in democratic societies, public policy is increasingly shaped by public opinion. Goodin, Rein, and Moran (2006) consider public opinion to be one of the key elements in policy. Unless the public supports a policy, it will not succeed, regardless of the political power held by its policy makers. These factors are elements in the analytic framework that is followed while we examine how energy policy is shaped and who does the shaping.

The greater the importance of a policy to the various stakeholders in an issue, the stronger is the relationship between public policy and public opinion. Actions by special interest groups and economic groups increase the potential threat for influencing or changing public policy (Burstein 2003). However, the right of all citizens, our companies, organizations, and our political institutions to be involved in forging energy policy too often makes achieving a policy goal problematic in the extreme. Without a commitment to resolution of the energy crisis and strong bipartisan leadership, a sustainable energy policy cannot emerge.

The national, and some cases international, associations of electricity, oil and gas, and coal firms are forums for addressing the big-issue policy questions faced by all energy industry participants working collectively and each industry independently. Associations and individual industries meet early each year to identify the more pressing issues they feel will have the greatest impact upon their operations. With little change seeming to result, those organizations and industries spend millions to ensure that legislation unfavorable to their interests does not get passed in Congress. Thus, the act of producing a comprehensive energy policy is exceedingly complex. Just a partial list of organizations with a stake in the nation's energy policy is displayed at the end of the book and reveals the extent of the fragmentation of the industry.

Structure of the Book

The book is organized into three parts with a preface, introduction, and a comprehensive bibliography. The first section includes six chapters that frame the energy policy problem by reviewing the history of energy policy in the United States, identifying the players in the policy-making drama, and bringing to light the costs and benefits and economic and political realities of the policy alternatives now competing for dominance. The second section includes five chapters that delve into some specific energy policy strategies and strategic factors that influence and shape energy policies. Four appendices are also included in the book. Appendix A provides a number of review and discussion questions for each chapter to facilitate the book's classroom use. Appendix B is a timeline of salient energy policy events that have occurred since the end of World War II. Appendix C is a list of energy policy related acronyms, while Appendix D is a brief glossary of important energy policy terms.

Part 1: The Challenges in Crafting U.S. Energy Policy

The six chapters in Part 1 are an introduction and overview of the economic and political realities that make up the "big picture" of the U.S. energy problem in the early years of the twenty-first century. By reviewing the many failures and few successes of a half-century of energy policy, the tremendous difficulties facing today's policy makers should come into clearer focus. In the process, this will serve to frame the energy question within the scope of proposed solutions with which most everyone is familiar.

Chapter 1, "The Political Realities of Energy Policy," explores the political and economic realities of establishing a comprehensive energy policy that will satisfy all groups with an interest in energy. One major roadblock in the establishment of a workable energy policy has been the lobbying efforts of powerful major energy pressure groups. A second problem has been the short memories of policy makers.

Chapter 2, "Energy Policy in Transition," looks at the transition in efforts of American presidents, legislators, and public and private individuals and groups to forge a sustainable energy policy that balances the requirements to meet current energy needs at affordable prices, the environmental concerns and climate dangers, and investments in long-term, secure, renewable energy resources. These efforts include negotiating and compromising on many subpolicies required by a comprehensive plan for ensuring the availability of an affordable, reliable, nonpolluting energy supply. The competing players in this game include groups promoting greater conservation, higher energy efficiency, greater investment in research and development in such nontraditional energy sources and hydrogen and nuclear fission, and those who believe that we must do all we can to increase exploration and exploitation of existing domestic fossil fuels.

Chapter 3, "The Art and Science of Crafting Public Policy," is a discussion of the processes by which the many stakeholders with an interest in energy policy attempt to influence those policies. The chapter begins with a brief overview of the structure of the energy industry; it then reviews the processes various stakeholders employ as they attempt to influence national energy policy.

Chapter 4, "The Long Search for a Sustainable Energy Policy," is a discussion of some of the policies U.S. governments have implemented since 1945 to solve current problems such as spikes in consumer prices for energy resources, weaning the nation from its commitment to foreign energy supplies, making major changes to the culture of automobile ownership, and increasing the use and cost effectiveness of alternative fuels and energy conservation. The goal of all of these policies and programs has been to come up with a sustainable supply of clean, efficient energy for this and future generations.

Chapter 5, "Difficulties in Achieving a Balanced Energy Policy," is a discussion on the on-again, off-again overriding goals that have driven energy policy since the 1970s: keeping energy affordable, protecting American producers, maintaining energy security, achieving energy independence, and others. The question is examined from four points of view: demand, supply, the environment, and national security.

Chapter 6, "What's on the Current Energy Policy Agenda?" reviews some of the salient issues involved in competing to achieve or maintain a high position on the energy policy setting agenda. This jockeying for position is a natural result of the ever-changing nature of the energy concerns of the polity. When prices for fuels skyrocket during an energy "crisis," for example, this results in a national commitment to conserve energy resources and penalize energy suppliers for high prices. Political leaders call for steps to eliminate our dependence upon foreign suppliers of fossil fuels by intensifying the application of scientific research for the development of alternative energy sources. When supply and demand again reaches equilibrium, energy concerns fade with a flooding of the market with an oversupply of fuels and steep drops in prices.

Conservation and efficiency are often promoted as the easiest and quickest ways to reduce dependence upon foreign energy sources. The chapter concludes with arguments supporting the policy position that conservation and efficiency, together or apart, should be the foundation of any final energy policy adopted.

Natural gas is the cleanest burning of all the fossil fuels. As such, it is enjoying increasing popularity for business, industry, peak-period electricity generation, and is also used to power some public transit vehicles. The chapter also looks at the proposed development of port and distribution facilities to off-load foreign liquefied and transport natural gas from foreign suppliers—clearly the antithesis of eliminating the country's dependence on foreign energy supplies.

Coal, in its various forms and degrees of energy efficiency, enjoys widespread distribution across the United States. The nation's coal policy in the twenty-first century goes beyond just the use of coal as a source of energy. Today, this policy must also focus on expanding existing research into producing chemical feed stocks and the environmental damage resulting from both the mining and burning of coal. The Department of Energy is providing millions of dollars for research on developing clean fuels from our extensive supplies of coal, coal shale, and coal sands. Energy conversion options for coal include goal gasification, producing liquid fuel from coal, and making efficient use of coal sands and coal shale. While these resources are, indeed, plentiful, no one can change the fact that they remain fossil fuels and their use does not reverse the damage to the environment. Finding a way to burn coal without greenhouse gas, black carbon, and fly ash emissions is a necessary goal if the nation is to meet its greenhouse gas reduction goals.

Part 2: Policies for Energy Transition

Chapter 7, "Crafting Policy with Subsidies and Regulations," is a focus on economic and regulatory policy options that together help shape national energy policy. These policies include very large investments in research development in alternative and renewable energy resources to wean America from its dependence on domestic and foreign supplies of fossil fuels, devote more effort into research and development on more efficient use of existing sources of energy, and encourage greater effort by Americans on programs of energy conservation.

Chapter 8, "Policies Shaped by Taxes and Market Mechanisms," examines the controversial view of relying on market mechanisms to facilitate the needed transition away from fossil fuels. Conservation and efficiency are indeed important and necessary elements in a comprehensive energy policy. Significant payoffs remain to be gained from efforts in these programs. However, it is clear that alone they can only slow, not eliminate, the nation's continuing reliance on traditional energy resources—coal, oil, and gas. Depending upon the political and economic philosophies of the administration in office, conservation and efficiency programs have been tempered by calls to balance efforts to achieve energy independence with proposals for eliminating greenhouse gases and protecting the environment.

Chapter 9, "International Cooperation on Energy Policy," highlights the importance of gaining and maintaining international cooperation and collaboration in developing a policy that promises the availability of a sustainable supply of clean, safe, and affordable energy for this and future generations.

Chapter 10, "Policies for a New Energy Future," reexamines the misdirected efforts of the renewed interest in a wide variety of problematic, expensive, and dangerous renewable and non-traditional energy resources. These include hydrogen, nuclear fusion, biomass (including algae and corn-based ethanol, among others), wave and tide motion, geothermal, photovoltaic, and wind-generating processes. Pressure is also growing to open the National Arctic Wildlife Refuge to oil drilling, along with opening more coastal offshore fields for oil production.

At various times, one or more of these sources has enjoyed great favor among policy makers—often resulting in a flood of federal research funds. Antinuclear opinion has long been against any expansion in the use of nuclear energy to generate electrical power until an acceptable plan is produced for dealing with nuclear waste and ensuring greater protection against nuclear accidents. There is, however, a renewed interest in nuclear power, particularly in nuclear fusion, which results in creation of more fuel instead of nuclear waste. The chapter looks at the major policy agenda items through two sets of filters: (1) scientific and technology, economics and finance, laws and regulations, and politics; and (2) physical, production, and environmental constraints.

Chapter 11, "Aftermath of the Gulf Oil Spill: Prospects for Policy Changes," looks at policy options in light of the world's largest maritime oil disaster to date. Options are examined including renewable and alternative fuels, particularly as they relate to the production of a steady supply of safe, efficient energy that is affordable and ceases to result in catastrophic damage to the environment. There is no hiding the fact that the United States will rely on liquid fossil fuels for many years to come. Thus, while a goal of U.S. policy between now and 2030 and beyond remains dedicated to finding a way or ways to greatly reduce, if not eliminate, the nation's dependence upon the Middle East and Venezuelan petroleum, there seems to be no viable short-term substitute on the horizon.

Key elements in the policy planning hopper must continue to include expanding spending on research and development of alternative energy sources, encouraging greater conservation in the use of existing energy resources, improving the efficiency of energy distribution and consumption, and changing the way Americans think about and use limited energy resources, both to ensure that existing supplies tide us over until a viable alternative energy source is available, and to cease polluting the global environment. We will talk in much greater detail about these and other key elements of energy policy in the chapters that follow.

1

THE CHALLENGES IN CRAFTING U.S. ENERGY POLICY

Chapter 1

The Political Realities of Energy Policy

> A small knowledge of human nature will convince us that, with the greatest part of mankind, interest is the governing principle; and that almost every man is more or less, under its influence. Motives of public virtue may for a time, or in particular circumstances, actuate men to the observance of conduct purely disinterested; but they are not of themselves sufficient to produce persevering conformity to the refined dictates and obligations of social duty.
>
> **—George Washington (quoted in Galston 2006 and Morgenthau 1978)**

Public policy is a deeply paradoxical subject; this is particularly true for the subfield of energy policy. Our attempts to improve outcomes in the energy sector reflect noble aspirations: to provide energy sources that are cheaper, more reliable, provided equitably to all, and generate less pollution. U.S. performance has been respectable along many indicators, including the reliability and cost of our most important energy sources, and significant decreases in air pollution. Yet the specific actions intended to bring about the benefits we desire—the public policies—have on occasion led to results that were the opposite of what was intended, or brought about deeply injurious and unintended effects.

There are some clear reasons for this. One, which we will explore later in this chapter, is the systemic nature of energy phenomena. We intervene in energy markets because we are dissatisfied (and often, rightly so) with the outcomes, but our interventions push systems out of balance, in ways we don't anticipate. Past attempts to set energy price ceilings, for example, often led to shortages, and even worse price pressures than existed initially.

Another challenge is that although energy use is fundamental to our civilization, most of us are buffered from its production, and its consumption is mostly indirect. We make few decisions *directly* to buy or consume it. Although most of us pay an electric bill each month, we don't just decide to buy more kilowatts because we like them. We want goods and services that we value: homes with heat, cooling in the summer, light, refrigeration, computers, vehicles that move us where we want to go.

The energy cost, which is for most of us minimal relative to our incomes, comes along as we decide we want more of these things. For example: 26 percent of U.S. households and 41 percent of households with annual incomes of more than $80,000 now have a second refrigerator. But most people care little about how these are powered as long as their power sources are reliable and not expensive. We don't buy coal, we buy electricity; we do buy gasoline, but it is *mobility* that we want. So although our lives are structured by energy-related products and decisions, indeed, as Nye (2003) suggests, energy creates the boundaries around our sense of possibilities, these decisions rarely take the energy implications into account. This has profound implications for policy making in this sector, as policy makers are forced to leverage their influence over other elements of the energy system.

Demystifying Energy Policy

Since one of our goals for this book is to demystify U.S. energy policy and policy making, we will begin with a brief case study that aims to explain a phenomenon that most of us take for granted, but few understand: electricity. In a 1978 study, Southern California Edison surveyed its customers and asked "Where does electricity come from?" The typical answer: "Out of the plug in the wall" (Sovacool 2006, 297). A more recent survey by the Harris Poll in June 2009 found only 9 percent of respondents identified themselves as "very knowledgeable" about sources of energy and energy efficiency, while another 50 percent self-identified as "fairly knowledgeable." But 41 percent admitted they were "not that knowledgeable" or "not at all knowledgeable" (Gstalder 2009).

Yet this poll did have good news. The responses suggest that those polled had a reasonable understanding of the relative levels of risk from various sources of electricity. When asked, "Do you believe the benefits of each source outweigh the risks, or do the risks outweigh the benefits?" renewable sources such as solar (82%), wind/turbine (78%), and hydroelectric (73%) had a high proportion of those surveyed respond that the benefits outweighed the risks. Nuclear energy was supported by 44% of respondents, while 34% felt its risks outweighed its benefits, and a substantial 22% were not sure. Only 36% believed the benefits from one of the country's prime energy sources—coal—outweighed its risks; whereas 42% responded that its costs outweighed its benefits, and again 22% were not sure.

This is an important finding. Coal remains, literally, a cornerstone in the U.S. energy economy, a fundamental primary source of energy. But it appears that widespread public discussion of its contributions to climate change and other forms of air pollution has influenced its public perception. To gain a better understanding of coal's importance and the problems associated with it, let's examine what it is, and how it is used to generate electricity. In the process, we will also introduce some important concepts and show how public policy is intimately interwoven throughout all facets of the energy sector.

An Example: Coal, from Mine to Furnace

In the United States, the journey of electricity to your wall socket usually starts with coal. As shown in Figure 1.1, in 2008, slightly under 50 percent of U.S. electricity was generated from coal, about 19 percent from nuclear energy, 20 percent from natural gas, and 7 percent from hydro (see Box 1.1).

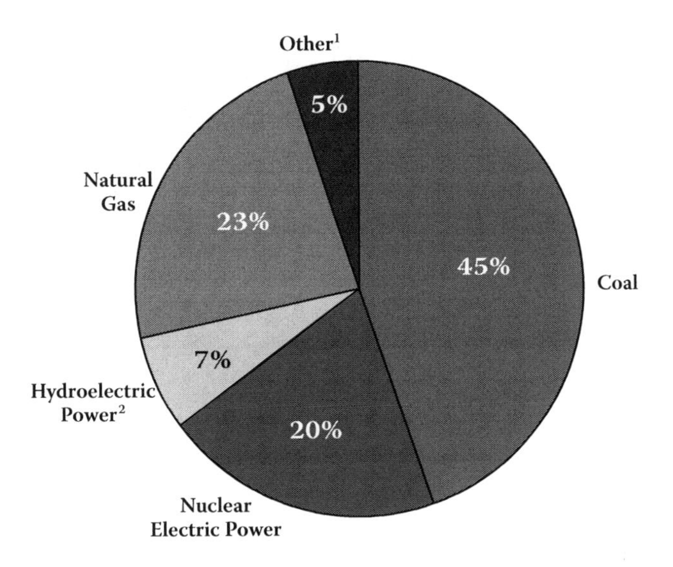

Figure 1.1 Net U.S. electricity production by source, 2008. (From EIA [U.S. Department of Energy, Energy Information Administration]. 2010a. *Electricity explained: Your guide to understanding energy.* Washington, DC: U.S. Department of Energy, Energy Information Administration, http://www.eia.doe.gov/energyexplained/print.cfm?page=electricity_in_the_united-states.pdf, accessed March 31, 2010.)

BOX 1.1 COAL MAKES A COMEBACK

Until the economic downturn that began in 2008, many of the nation's utilities were retreating to an earlier technology as a way out of this dilemma. Nearly 100 new coal-fired electric power generating plants were planned in 36 states. If all are completed, they would add something like 62-gigawatts of low-cost electricity to the nation's generating capacity. Illinois led the rush to coal with a total of ten new generating plants proposed. The retreat to coal is seen as the only way to keep electricity prices low while also adding to energy security by offering an alternative to foreign oil and gas.

Coal already produces about half of all the electricity generated in the country. However, coal-fired generators also pump mercury and greenhouse gases such as carbon dioxide, nitrogen oxide, and sulfur dioxide into the air. The new plants are estimated to add roughly one-tenth of 1 percent to the world's annual carbon-dioxide emissions. Environmental groups have filed suit to stop new construction. The United States, with more than 250 years worth of coal reserves, has been called the "Saudi Arabia of coal."

(From Clayton, M. 2004. The coal rush. *The Seattle Times*, February 27, A3.)

Why do we rely so much on coal? We do so in part because the United States has the world's largest coal reserves, roughly 28 percent of the current world total. There are several varieties of coal, including anthracite, bituminous, sub-bituminous, and lignite. All are sedimentary rocks produced over eons through the decay of Carboniferous Era plant matter under heat and pressure, and in the absence of oxygen. As with other so-called fossil fuels, coal is thus a preserved form of solar energy.

Most of what is now burned in U.S. power plants is black or dark brown sub-bituminous coal. The different types of coal vary in their energy density, or energy per unit of weight, often expressed in megajoules per kilogram or mj/kg. Sub-bituminous coal has an energy density of between 19 and 26 mj/kg, depending on the grade. By comparison, higher-quality bituminous coal has an energy density of 27 to 29 mj/kg; dry wood has an energy density of 16 mj/kg; crude oil and gasoline energy densities are about 45 mj/kg (Smil 1999; McElroy 2010).

A Good Fuel for Generating Electricity

The amount of energy packed into coal makes it a good choice as a fuel to generate electricity, but limits its usefulness for other energy services. You no longer see people loading coal into their steam engine cars, since gasoline has a much higher energy density, and can also be burned directly in an internal combustion engine.

U.S. coal deposits are located mostly in Appalachia, the Midwest, and the Intermountain West. Digging it up and transporting it to power plants is a carefully choreographed process of vast scale that begins with brute force: blowing off mountaintops in Appalachia or taking off 50 to 100 feet of dirt or overburden in Wyoming's Powder River surface mines.

Wyoming now leads the United States in coal production, with about 450 million tons per year—removed from some of the largest surface mines in the world. Although Wyoming coal has a lower energy density than Eastern bituminous coals (more must be burned to achieve the same power output), Powder River coal is prized for its relatively low sulfur content (Goodell 2006). The scale of the largest of the Powder River mines south of Gillette, Wyoming, is such that they are easily visible on satellite maps online. Given the advantages of coal as a fuel source for electricity, it is not surprising that its use has grown along with U.S. electricity output (Figure 1.2).

For a surface mine, the overburden is taken off in layers, loaded into enormous trucks, and transported to previously mined areas. The coal, often up to 100 feet thick in the largest mines, is

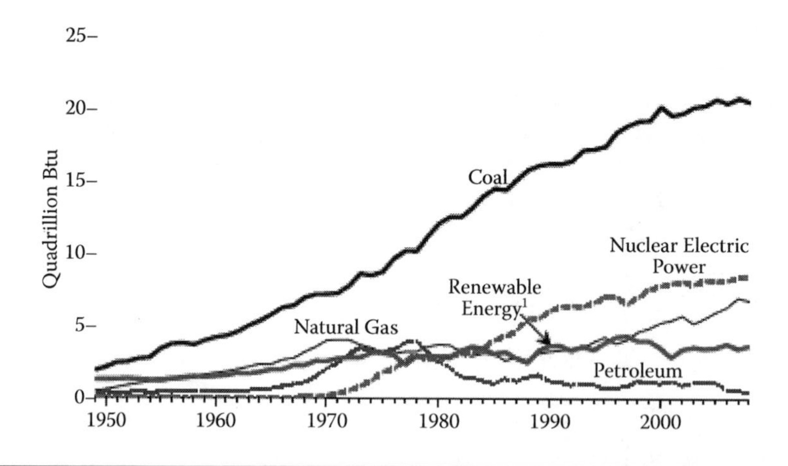

Figure 1.2 Fuel sources for the U.S. electric power sector, 1948–2008. (From EIA [U.S. Department of Energy, Energy Information Administration]. 2009g. *Annual energy review 2008*. Washington, DC: U.S. Department of Energy, http://people.virginia.edu/~gdc4k/phys111/fall09/important_documents/aer_2008.pdf, accessed December 29, 2009.)

thus exposed in seams. Explosives loosen the coal, which is then loaded into trucks and put into railcars for transport to power stations.

On average, coal is transported 500 miles to its final destination. Each railcar holds around 100 tons of coal, and often up to 100 cars are linked in a "unit train" that will travel directly to one of the 476 power plants in the United States (as of 2007) that burn coal as their primary energy source. The amount burned depends on the quality; a 500-megawatt coal plant operating at full capacity requires roughly one unit train of sub-bituminous coal (10,000 tons) every three days; a lignite plant of the same output may require one such train per day.

A Big Footprint

A typical coal-burning power plant, including the actual generation facility and related infrastructure (coal storage, rail siding, etc.) has a "footprint" of 1,000 or more acres. Most of these facilities are located away from population centers and are rarely seen by the public. Although this sounds like a large area, it is relatively small in relation to the power generated. Centralized utility systems powered by fossil fuels thus have a high power density, generating large quantities of power and then distributing them for a variety of end uses. Renewable means of generating power such as solar, wind, or even hydropower, have much lower power densities—they take up far more space per watt of power generated (Smil 2003). This may have significant implications for how the country organizes a renewable-based energy system.

After arrival, the coal is removed from the cars and placed in a holding area. It is usually washed, pulverized into powder, and pulled by conveyor belt into the facility, then air-blown into the furnace, where it is burned at 1500 to 1700°C. More advanced plants use a fluidized bed combustion process in which the coal is burned on a bed of air created by blowers. This allows more complete combustion at lower temperatures with fewer emissions. Coal is transformed into heat and a series of by-products, including ash, sulfur dioxide, nitrogen oxides, carbon dioxide, particulates, plus small but measurable amounts of mercury and uranium.

The Rankine Cycle Process

The burning or oxidation of coal is used to power the Rankine cycle. Pressurized water in pipes running through the furnace is boiled and becomes steam, which is superheated to 1,000°F and blasted into turbines. Using Faraday's principle of induction, the mechanical energy of the steam driving the turbine is converted into electricity, when the rotor on the turbine shaft (spinning at 3000 or 3600 revolutions per minute as required by the power grid), passes through the magnetic field of the stator (large magnet) surrounding it. This electricity is transformed to higher voltage and whisked away on transmission lines, eventually transformed to lower voltages at local substations, then distributed to households and businesses. At the plant, the superheated steam exiting the turbines completes the Rankine cycle as it is condensed into water, cooled, and cycled back to the furnace for reheating.

This somewhat simplified account provides a rough description of the process that generates almost half of the electrical energy in the United States; most of the rest is also generated in thermal power plants, but as noted, with natural gas or nuclear energy used to heat the steam that moves the turbine generators. Many aspects of this account are critical to the topic of this book: how the United States approaches the complex task of crafting its energy policy.

Understanding the Scale of Energy

To understand the scale of resources involved we will clarify how electrical energy is measured. First, the scale of energy production and consumption is vast. For example, consider electricity, fast becoming the favorite of energy policy makers. Electricity—the flow of electrons—is measured in watts; watts are the rate of electricity flow at a given moment (literally, joules per second). Power is the rate at which this energy is both generated and consumed—an important fact about the electricity system is that it must constantly equalize production and consumption.

At the household level, power is usually measured in kilowatt (1,000 watts) hours; the capacity of a power station is usually measured in megawatts or gigawatts (million or billion watts); the United States as a whole typically uses 4 terawatt hours (4,000 billion kilowatt hours) per year. Total global electricity use is about 16 terawatts per year. So, the United States with about 5 percent of global population uses about one-quarter of all electricity generated in the world.

Electricity keeps our lights, computers, space heaters, and flat screen televisions on. We must move huge quantities of our preferred primary energy sources to produce this electricity. We must then move that electricity through the country's 157,000 miles of high-voltage transmission lines. Our landscape is crisscrossed with these transmission lines because power generation usually takes place far from population centers.

As of 2007, the 476 coal power plants in the United States operated 1,470 coal-fired generators, producing a total of 2 billion megawatt hours of energy per year. In 2008 those plants consumed 1.042 *billion tons* of coal (EIA 2009f). How much coal is that? Enough to cover roughly 445,000 football fields with coal one foot deep; or 918 square miles of coal—a square about 30 miles per side.

Coal is only one primary source of energy used for generating electricity. We also use 7.2 trillion cubic feet (TCF) of natural gas per year to generate electricity in the United States, plus another 5.3 TCF for other residential uses—mainly space and water heating, plus cooking, and another 7.2 TCF for industrial uses (EIA 2010f). And we need 19.5 million barrels of oil per day—819 million gallons—out of a world total of 87 million barrels per day to power our transportation system and other needs.

These energy sources are generally not consumed at the point of extraction, so they must be moved, often repeatedly, with each stage of movement reflecting a loss of efficiency. In other words—it takes lots of energy to create energy. The more energy required per unit of energy produced, the lower the Energy Return on Energy Invested (EROEI). Second, the energy sector requires an intricate and enormous infrastructure, which in the United States is mostly privately owned.

The total capital stock in the U.S. energy industry has been estimated at $1.85 trillion as of 2008, with $946 billion in the electric utility sector alone (EIA 2008b). That is using historical cost figures; current cost estimates are twice that high. The infrastructure for our gasoline and diesel-dependent transport sector, including not only the ubiquitous corner service station with its invisible tanks, but tanker trucks, refineries, pipelines, and so on, is also valued in the trillions of dollars. The annual capital requirements for the entire U.S. energy sector are in the range of $300 to $400 billion per year. Would it be feasible for the U.S. government, and or state governments, to own, operate, and maintain this infrastructure? The short answer is yes, but whether that would be desirable is another matter.

Conversion Processes Needed

Energy processes such as this one almost always require conversion, from one state to another, in the process releasing heat or electrons. Even hydroelectricity, considered a "primary" source of electricity, converts the potential energy in a reservoir to kinetic energy as water flows down to and through a turbine.

During these conversion processes, consistent with the first law of thermodynamics, the total energy in the overall system remains constant. But the second law of thermodynamics mandates that the entropy, or disorder, in the system must increase. Once burned and turned into heat and its many by-products, the coal may not be magically re-created, and the process of conversion is inherently inefficient. Despite considerable research and investment, coal power plants have reached a ceiling of about 45 percent efficiency; most average 33 percent efficiency. Much of the energy contained in the original energy source is lost, and not turned into useful services.

This is reflected in Figure 1.3, the most recent in a series of charts produced by the Lawrence Livermore Laboratories of the U.S. Department of Energy that summarize the flows of energy through the U.S. economy (DOE 2009a); about 57 percent of the energy in the system is wasted overall as a result of inefficiencies in both generation and power transmission.

Why Worry about These Conversions?

Why are these conversions important? Why are such enormous quantities of primary sources of energy such as coal, oil, and natural gas required for our economy? Part of the answer lies in the relative inefficiency of these conversions. Most of the energy policy instruments we will summarize in Part 2 attempt to influence the supply or demand for these primary energy sources, for secondary sources such as electricity, or to limit their unintended, polluting effects. But another way forward is to create incentives to bring about improved technical efficiency so as to meet the demand for energy services with fewer inputs. This was one of the arguments of Avory Lovins in his landmark 1976 Foreign Affairs article, "Energy Strategy: The Road Not Taken?"

In addition to the inefficiency of these conversion processes, they create huge amounts of unwanted by-products. When the throughput of materials such as coal, natural gas, and oil is so vast, even a relatively small output of a pollutant can add up quickly. And some of these pollutants are produced in very large quantities. For example, for every ton of coal burned, *two* tons of carbon dioxide (CO_2) are released. Figure 1.4 compares gas, oil, and coal emissions.

In the United States, the Environmental Protection Agency (EPA) estimates that energy-related activities generated 86 percent of all greenhouse gas emissions in 2007. Electricity generation is responsible for 41 percent of all CO_2 emissions. Overall, coal-fired power is responsible for 33% of all energy-related CO_2 emissions (EPA 2009b). Because we are so dependent on coal-generated electricity, particularly in the southeastern and eastern United States, we become inured to its effects.

The infamous August 14, 2003, power blackout in Ontario and the eastern United States forced 508 generating units at 265 power plants in that area offline. A study of air quality in Maryland measured during the outage found dramatic *decreases* in sulfur dioxide, ozone, and particulates (Marufu et al. 2004).

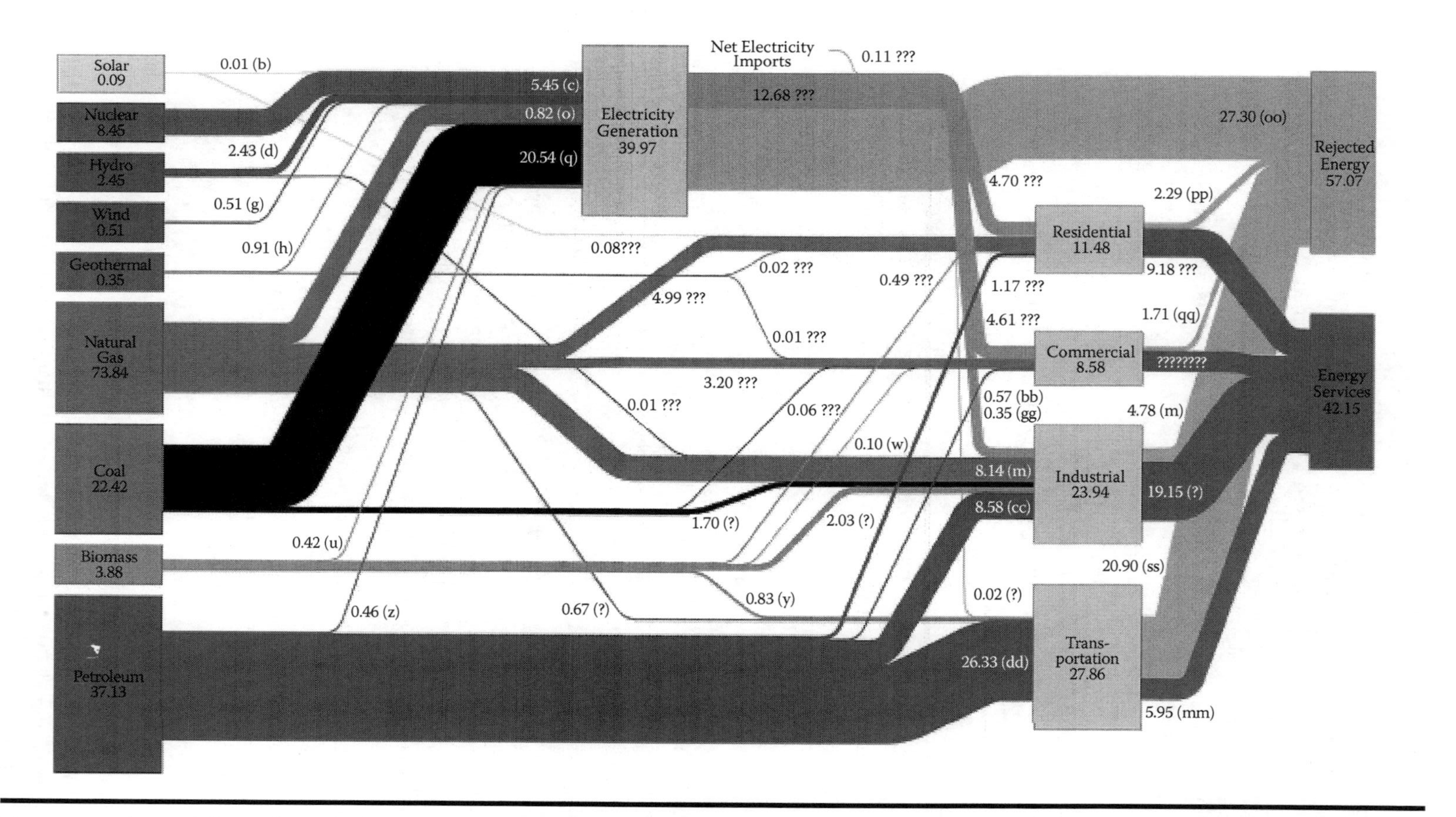

Figure 1.3 **Energy flows through the U.S. economy. (From LLNL [Lawrence Livermore National Laboratory]. 2009. *Estimated U.S. energy use in 2008*. U.S. Department of Energy, Lawrence Livermore National Laboratory, https://publicaffairs.llnl.gov/news/energy/energy.html, accessed December 8, 2009.)**

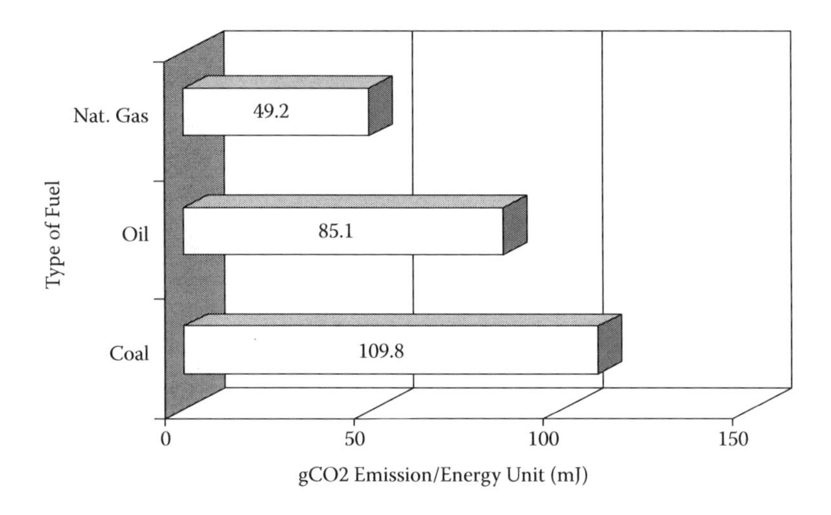

Figure 1.4 Comparisons of CO_2 emissions from coal, oil, and natural gas. (From Liner, C. 2010. CO_2-reactions. *Seismos*, http://seismosblog.blogspot.com/2010/04/CO_2-reactions.html, accessed August 28, 2010.)

The Economic Perspective

Energy systems may also be viewed from an economic perspective. Two tools used for economic analysis of energy—energy return on energy invested (EROEI) and product life-cycle cost—are described in Box 1.2.

Coal-fired power appears to be economically "efficient" in part because coal is relatively cheap as a source of fuel compared to the available alternatives, particularly natural gas. The long-run price of coal increased prior to the 2009 recession, thanks in part to increasing demand from China and other newly industrializing countries. Factoring in all other direct costs of production, electricity from coal cost on average 3 to 5 cents per kWh in the United States in 2008, less than the 4 to 8 cents per kWh for natural gas over the same period. Despite this price differential, most new electrical generating capacity added for the past two decades in the United States relies on natural gas.

These plants have much lower capital costs and gas is a cleaner fuel than coal that generates far lower carbon emissions (EIA 2009c). However, these prices omit a wide array of social and environmental effects of the electricity production process, literally from beginning to end. Some of these are: water pollution occurs in the wake of surface coal mining as pollutants work their way into the water table, and more water pollution occurs at the end of the production process as chemicals from coal ash leach into the ground. More seriously, Earth's atmosphere is a common-pool good, available to all, and the operators of our coal-fired power plants take oxygen from the air for burning the coal and return effluent to it. Due to our relatively recent recognition of human-caused climate change, the effects of carbon dioxide are now being acknowledged, but the other products have long been recognized as important contributors to air pollution and related diseases.

These externalities or spillovers are effects (both negative and positive, or alternatively, costs and benefits) from production and consumption processes that impact individuals and society, although those others had no impact on the decision, and the effects are not priced into the final good. They occur in part due to a property rights problem: no one owns the air, so from a production perspective, there is no incentive to preserve it. The energy sector is replete with such

BOX 1.2 TWO TOOLS FOR ENERGY ANALYSIS

How do decision makers, either in the public arena, such as members of Congress, or in the private sector, make decisions about energy projects, or products with energy implications? Two interrelated analytical approaches that have been somewhat useful are calculations of energy return on energy invested (EROEI) and of a product's life-cycle cost, using life-cycle cost analysis (LCA).

EROEI, also termed *energy return on investment*, is a calculation of the ratio between the energy extracted by a process and the energy used in the process of extraction or production (Cleveland 2008). Calculations of EROEI require careful accounting of the inputs required in extraction and production as well as the nature of the final product. Petroleum has a high energy density and typically generates a high EROEI, even when obtained in costly locations. We can also produce oil and gasoline from tar sands, but the high energy input required considerably lowers its return on investment.

A related process is the measurement of life-cycle cost. This can either emphasize the cost to the producer or consumer, or more broadly consider the total cost of the entire product system to the environment and society. What resources are required to produce and consume an item from beginning to end (European Environment Agency 1997)? Again, the calculations are detailed, are best done with computer software, and the results are sometimes counterintuitive. Which is "better," a flimsy plastic shopping bag that is made from oil and must be thrown away, or a nice, recyclable paper bag? On balance, the plastic bag uses less energy, in part because the greater weight and volume of paper bags requires costly shipping, usually via diesel trucks.

Application of both EROEI and LCA analysis is complicated by the need to set boundaries on each analysis. The choice of boundaries can make a difference; thorough calculation of the LCA of an automobile requires inclusion of all the inputs to the car, not just its assembly (Smil 2008). These concepts are also useful in considering a society's overall energy use. Fossil-fueled civilization has a high overall EROEI, which is likely to decline as we become more reliant on renewable sources (Cleveland 2008).

spillovers, both positive and negative, and they are related to both production and consumption of energy. Much of the public policy in the energy sector aims to reduce the external costs of energy production that are borne by society, and not by the producers of energy.

The Function of Energy Policy

Where does energy policy fit into this narrative? It lies behind each of the major steps, from the beginning at the surface coal mine, to the transport and storage of the raw material, its conversion to energy at the power plant, and distribution of the resulting power to our houses and businesses. As Table 1.1 shows, "energy" policy is somewhat of a misnomer: a wide variety of public policies influence the energy sector. Perhaps the most important energy laws have an environmental focus. They influence where mines and other energy facilities may be opened and how they operate, to protect vegetation, ground water and wildlife, as well as to limit the effluents that are allowed to enter the atmosphere.

Table 1.1 Federal and States of Wyoming and Washington Energy Statutes

Federal:
Surface Mining Control and Reclamation Act, P.L. 95-87 (requires remediation of mined lands).
Clean Air Act of 1980, as amended, 42 U.S.C. 85 (comprehensive statute providing the U.S. Environmental Protection Agency with authority to regulate sources of air pollution).
Staggers Rail Act of 1980, Public Law 96-448 (deregulation of the U.S. railroad industry).
Coal Excise Tax, Section 4121 of the Internal Revenue Code (to fund Black Lung disease programs).
State:
1973 Wyoming Environmental Quality Act (Statute §35-11-410): Wyoming's statute authorizing regulation of surface mines and providing for monitoring of compliance with remediation and other required actions).
Washington Clean Air Act, RCW 70.94 *et seq.* ("WCAA"), (Washington's clean air statute).
Washington: Energy Facilities - Site Location: Chapter 80.50 Revised Code of Washington (authorizes state control over siting of energy facilities).
Washington: Utilities and Transportation Commission regulation of utility tariffs, Title 80, RCW (authorizes regulation of utility rates).

Regulation of utility rates is a more direct form of energy policy. The U.S. Constitution created a federal system of government that attempts to establish boundaries for the roles and functions of the national and state governments. These boundaries, including the Supremacy Clause, which makes the Constitution the "supreme law of the land," and the Commerce Clause, which gives Congress the power to regulate commerce between the states, have been sorely tested by the need for environmental and energy policies with a national scope. States often rebel against federal action perceived as overly intrusive, but have also acted boldly when they believe the feds are being too timid, such as the state of California's consistent efforts to regulate carbon dioxide as a polluting greenhouse gas (Emerson 2002).

Federal laws and resulting regulations mandating efforts to remediate mines and to limit air and water pollution are mirrored by state statutes that at least match, or exceed, federal requirements. State-level regulators are usually the officials implementing policy on the ground, and it is often a challenging position. They are pinned between conflicting pressures to protect the environment, but also to say "yes" to energy projects that offer badly needed economic development and jobs to cash-strapped communities.

Unexpected Influence of Nonenergy Policies

Policy makers must also consider the unexpected influence of nonenergy policies. In this case, the Staggers Rail Act of 1980 almost eliminated federal government regulation of the railroad industry in the United States. It encouraged a series of mergers in the industry that led to increasing concentration, so that now the United States has only five major railways, and two in the western

United States that provide almost all coal haulage of the coal transported from Wyoming. It also enabled the remaining railroads operating in the intermountain west coal belt to raise their rates for hauling coal. Thus the generators that rely on steady supplies of coal face a restricted market for rail services for their essential input—in many cases they are effectively captive to a monopoly railroad service. As you would expect, this has led to a complex relationship between these two industries, with increasing lobbying for a return of federal regulation (Wilks 2006).

Despite the significant influence of the Staggers Act on the cost of shipping the country's most important source of electricity, it is not mentioned in a comprehensive analysis of U.S. energy policies created by the U.S. Government Accountability Office in 2005 (GAO 2005). It is hard to criticize the GAO, however, since the list already runs to over 160 different federal policies.

This case illustrates the complexity of the energy systems that provide the foundation for our fossil-fueled civilization. Technology, culture, politics, law, economics, business, and government interact to provide us with the electricity that powers much of our society. But this discussion of a key element of our energy supply leads to a different question: What is the source of Americans' seemingly insatiable demand for energy?

Why Does the United States Use so Much Energy?

The United States is often described as an "exceptional" country compared to the rest of the world—its citizens more individualistic, more religious, more suspicious of government than those of other countries. The term is particularly appropriate in the energy sector. We use more energy on an absolute basis than any other society on Earth. As shown in Figure 1.5, in 2006 with 4.6 percent of world population, the United States produced 15.1 percent of world energy, and consumed over 21 percent of world energy. The 6 percent differential between our domestic production and consumption was met by imports, largely of petroleum.

What primary energy sources do we rely on, and how do we use all this energy? Figure 1.6 shows the supply sources and demand sectors, and how as a society we have matched sources to their most appropriate uses. Overall, industrial uses gobble up 31 percent of all U.S. energy, closely

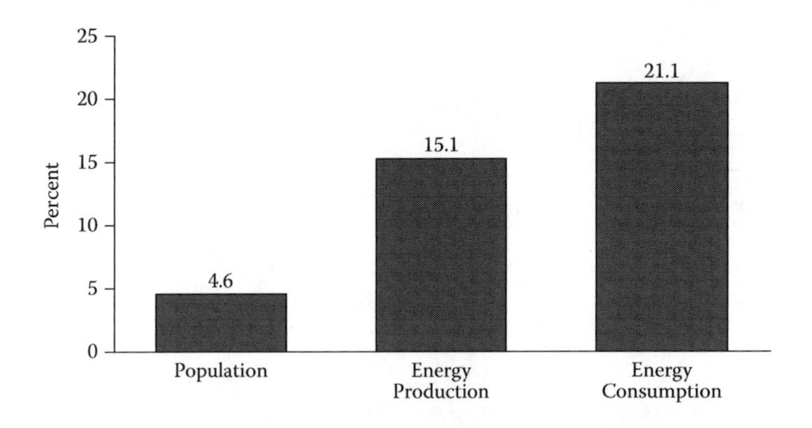

Figure 1.5 U.S. share of world energy consumption and production. (From EIA [U.S. Department of Energy, Energy Information Administration]. 2009g. *Annual energy review 2008*. Washington, DC: U.S. Department of Energy, http://people.virginia.edu/~gdc4k/phys111/fall09/important_documents/aer_2008.pdf, accessed December 29, 2009.)

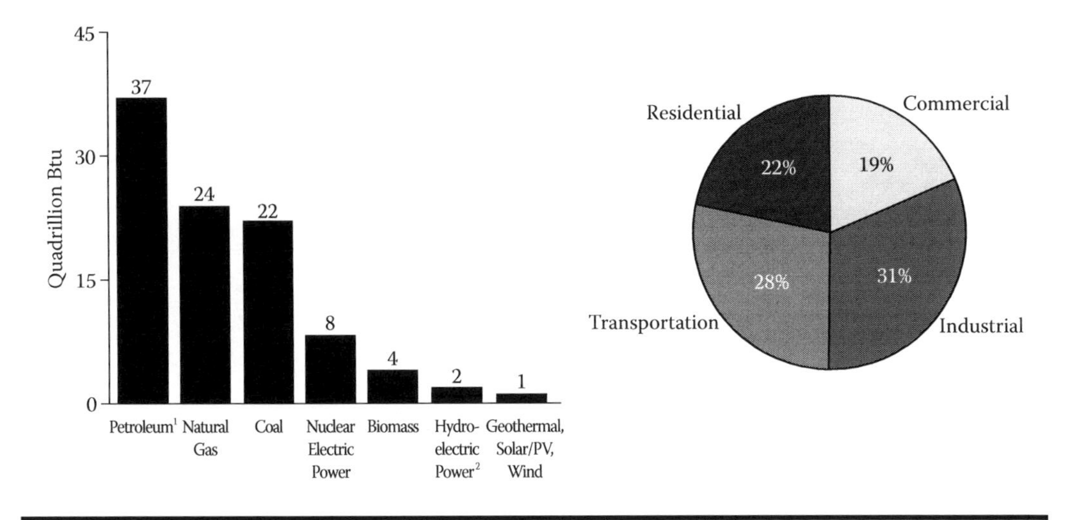

Figure 1.6 U.S. primary energy consumption by source and sector, 2008. (From EIA [U.S. Department of Energy, Energy Information Administration]. 2009g. *Annual energy review 2008.* Washington, DC: U.S. Department of Energy, http://people.virginia.edu/~gdc4k/phys111/fall09/important_documents/aer_2008.pdf, accessed December 29, 2009.)

followed by transportation (28 percent). Residential/household use and commercial use reflect 22 and 19 percent of the country's energy use, respectively.

Not shown in these figures are the high reliance of the transportation sector on demand for petroleum products; 95 percent of all oil use in the United States goes to transportation. Also, the residential and commercial combined sector uses 76 percent of the country's natural gas, primarily for space heating and cooking. As shown in Figures 1.7 and 1.8, industrial energy demand has been erratic since the 1970's oil shocks, while demand in the other three sectors (other than a slight dip for the early 1980's recession) has grown consistently since the late 1950s.

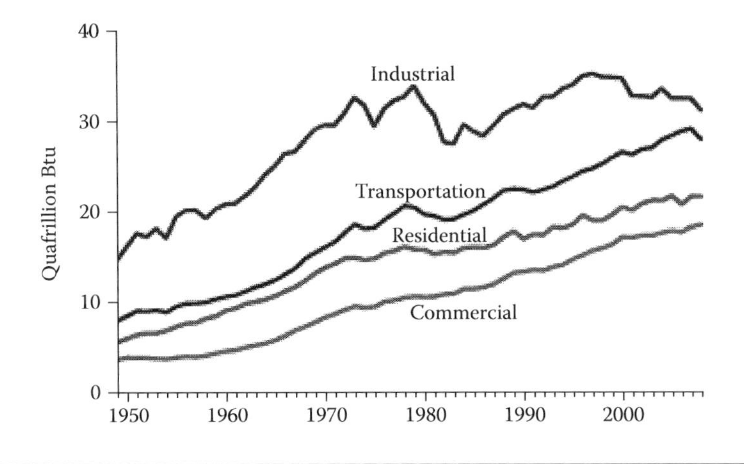

Figure 1.7 Total energy consumption by end use sector, 1949–2008. (From EIA [U.S. Department of Energy, Energy Information Administration]. 2009g. *Annual energy review 2008.* Washington, DC: U.S. Department of Energy, http://people.virginia.edu/~gdc4k/phys111/fall09/important_documents/aer_2008.pdf, accessed December 29, 2009.)

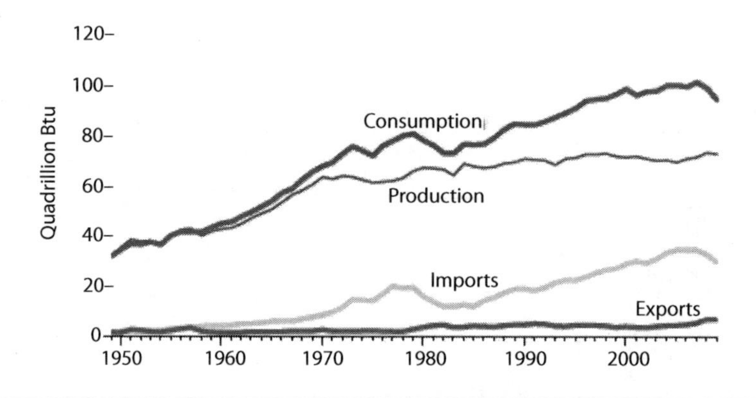

Figure 1.8 U.S. Primary Energy Overview. (From EIA, U.S. Department of Energy, Energy Information Administration). 2010c. Annual energy review 2009. Washington, DC: U.S. Department of Energy.

The United States also uses more energy per capita, and relative to the size of the economy (a measure termed *energy intensity*), than almost any other country. As shown in Figure 1.9, only Canada uses more energy on a per person basis than the United States. Although there are considerable methodological limitations to these data (countries are not entirely consistent in how they collect and present data on gross domestic product (GDP), and on energy production and use), they are accurate enough to allow a fair comparison between countries.

Factors Influencing U.S. Energy Use

The factors influencing Americans' huge appetite for energy have been explored by authors such as Mumford (1937), Nye (1998), and Smil (2003, 2006). The advance of civilization and the very concept of progress are linked to the ability of humans to make use of new and increasingly powerful sources of energy. Over time, energy use has advanced almost proportionately with economic growth. While this is true for the "industrialized" world as a whole, the United States differs in our resource endowment, climate, sheer size, culture, and other factors (Smil 2006). We have substantial supplies of oil, coal, natural gas, and uranium. It is easy to forget, now that we are reliant on imported oil, that the United States remains the world's third largest producer, behind only Saudi Arabia and Russia. Climate is critical, since colder climes requiring more space heating, and muggy locations made more livable through air conditioning both demand sizable amounts of energy for heating and cooling.

Sheer distance traveled also makes a difference. Americans travel farther by car, on average, than any other society, an average of 12,293 vehicle miles per year (2007 data), while the fuel economy of our vehicle fleet for both commercial and personal use has been stagnant for a generation (FHA 2008). We also travel by air more frequently than citizens of other countries. North Americans (including both the United States and Canada) generated 755 million passengers in 2008, one-third of global air passengers, with just over 5 percent of global population (ICAO 2009).

Americans also like single-family houses, which are 63 percent of the nation's housing stock as of 2007 (U.S. Census Bureau 2008). We have chosen to allocate some of the benefits of economic growth to even bigger houses. In 2008, the average new single-family house had 2,519 square feet.

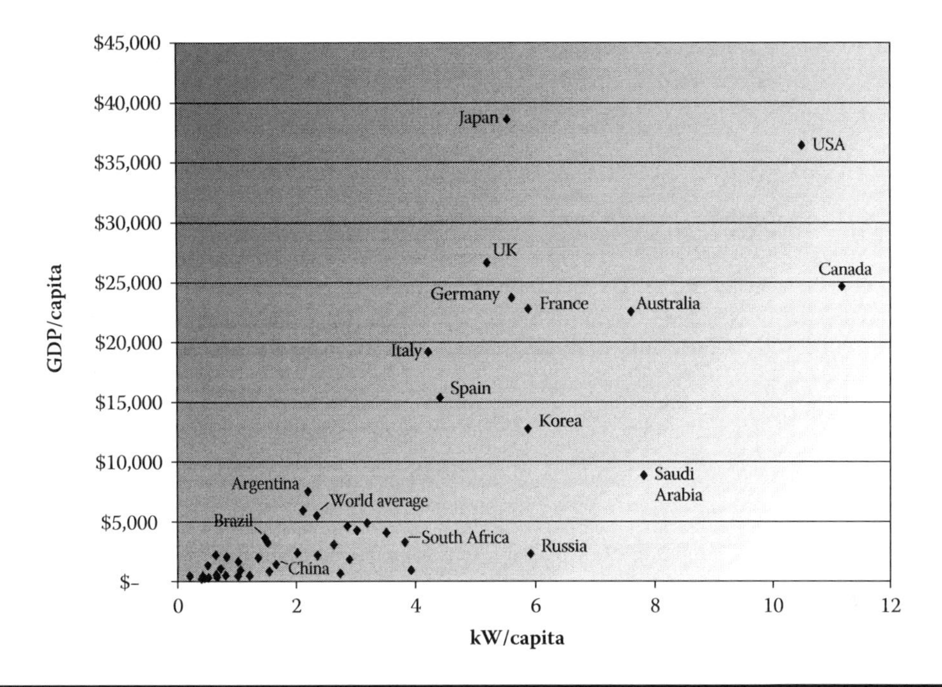

Figure 1.9 Energy consumption vs. GDP for selected countries. (From UNDP [United Nations Development Program]. 2009. *Decarbonizing growth: Some countries are doing better than others.* United Nations Development Program, Human Development Reports, http://hdr.undp.org/en/statistics/data/climatechange/growth/, accessed April 1, 2010.)

This was an increase of 43 percent (764 square feet) since 1978 (U.S. Census Bureau 2009) and double that of new houses in the 1950s.

Sources of Our Preferences

Where do these preferences come from? Nye traces the rise of the country's consumer culture, and our evolution into a wealthy society in which *wants* are manufactured by marketing and we are defined by what we own and what services we consume. Manufacturers invent increasingly sophisticated technology that creates ever-fancier products, to be bought when the previous model becomes unfashionable (not obsolete). This process relies on the easy availability of energy, and works most adroitly when it exploits strong cultural preferences, such as the American sense of restlessness and desire for space and freedom—epitomized by our infatuation with the automobile and preference for suburban lifestyles. Nye cautions us to avoid a simplistic technological determinism that makes advancing technology and its related effects seem inevitable, and thus avoids the reality of human agency: we make choices, they don't simply happen. But even he concludes that "electricity and the automobile transformed society" (1998, 182).

Most pertinent for this book is that public policy has been an important factor encouraging both the creation of the consumer society, and the use of energy. The path toward a consumerist society with high levels of energy use emerged in the United States during and after the Great Depression. As historian Alan Brinkley recounts, economic collapse inspired many proposals for the economic and social transformation of U.S. society, including corporatist and social democratic models (Brinkley 1995). The desire to bring about full employment during the postwar era moved the country toward a Keynesian model reliant on high consumer consumption, when necessary supported by government spending and policies to help regulate the business cycle.

President Franklin Roosevelt used public action to make electricity more broadly available and enhance economic development, through the creation of the Tennessee Valley Authority, rural electrification, and the construction of dams in the Northwest. Over time, public policy, working in concert with the private sector, was generally effective at providing reliable supplies of gasoline, heating oil, and electricity, often at costs that decreased over time. For example, the real (inflation-adjusted) cost of electricity declined on average in the United States between 1960 and 2007, despite a big jump in price in the mid-1970s and early 1980s. The journey of oil and gasoline prices over that period is much more complex, as we will discuss in Chapter 4. One thing is certain: oil and gasoline *decreased* in price in real terms for an almost twenty-year period beginning in 1981, before starting its run toward the breathtaking 2008 peak.

The good news about declining energy prices is that they save consumers money, which may then be spent on other goods and services. The bad news is that they encourage increased consumption. One of the paradoxes of energy policy is the so-called rebound effect: savings from investments in energy efficiency may be allocated to increased consumption of the same product (a direct rebound effect) or in more consumption of another energy-intensive good or service, an indirect effect (CRS [Congressional Research Service] 2001; Sorrell 2007). This is most evident in automobiles; when our new car has better gas mileage we are likely to drive more often and longer distances. This is a disturbing implication for investments in energy efficiency, as it implies that such investments will be less than fully efficient.

Encouraging Energy Trends

Several long-term U.S. energy trends are moderately encouraging: Although total energy use in the United States has tripled since 1950, the rate of increase has been much reduced since the 1970's oil shock, and close to zero since 2000. An emphasis on increasing energy efficiency has had a positive impact on our overall energy use. The United States has plateaued at around 99 quadrillion BTUs per year (EIA 2009c). Per-capita energy use has begun a gradual decline, and the overall energy intensity of the U.S. economy has decreased steadily since the 1950s. This has been termed the "decarbonization" of the economy and is due at least in part to offshoring of manufacturing to other countries.

The U.S. Energy Sector

The first section of this chapter provided an overview of energy use in the United States as well as important energy trends, definitions, and issues. To round out this chapter, we will consider energy from a structural or organizational perspective. We begin with the questions: How is the energy sector organized? What elements of our economy and society does it touch, and how critical is it in each instance? And how do these sectors interact?

The ubiquity of energy use in industrial society makes it worthwhile to view energy from a systems perspective. A system is a set of elements that interact in concert so as to form a whole. Organizational systems rely on inputs of resources, create outputs, and are oriented toward achieving a goal—they are teleological. They are sensitive to feedback loops and usually (but not always) tend toward maintaining an equilibrium or homeostasis. They also rely on subsystems that absorb inputs, and provide outputs important to the success of the system as a whole.

The systems model is helpful in considering energy phenomena because the principles of biology and physics are used to explain energy phenomena; the subdiscipline of *energetics* is also amenable to a systems approach. The essential input to the Earth's energy system is the 1,367 W/m^2 of sunlight that enters the atmosphere each day—only about 20 percent of which is absorbed at the surface. This is the original source of the stored energy that we use as we consume the fossil fuels oil, coal, and natural gas. These in turn have been essential inputs for our economic and social systems. Whether the Earth itself is a goal-seeking and self-regulating system, as suggested by James Lovelock's Gaia hypothesis, is a hotly debated question.

Energy in Contemporary Society

In envisioning the role of energy in contemporary society, it is also helpful to think of the interaction of multiple subsystems with an energy nexus intertwined. These include our economy, commerce, and industry; agriculture; the social/residential; transportation; political/government; and military subsectors. Although they can be described as discrete entities in their own right, these subsectors should not be considered in isolation, since each ultimately impacts the others. This section will provide a quick sketch of each of these elements of the U.S. energy system, and conclude with some thoughts on the nature and goals of this system as a whole.

Energy as a Business

The U.S. energy industry is a behemoth—one of what the U.S. Bureau of Labor Statistics terms a series of *supersectors*. Table 1.2 displays employment and revenue data on each of the subsectors within this supersector and lists key companies. These supersectors include:

- **Oil:** oil companies, refiners, fuel transport, gasoline sales
- **Natural gas:** natural gas extraction and processing, coal gas manufacture, and distribution and sales
- **Electric utilities:** electricity generation, high-voltage transmission, distribution and sales, nuclear power, energy services
- **Coal:** Mining, shipping, and distribution
- **Renewable energy:** wind power, solar power generation, alternative fuels
- **Energy markets:** NYNEX exchange, regional exchanges

The importance of this sector to the U.S. market system is difficult to understate. It fluctuates as a share of the overall economy, based primarily on movements in oil prices, but averaged 8.8 percent of U.S. GDP from 1970 to 2006, with a peak of 13.7 percent in 1981 and a low of 6 percent in 1998 and 1999 (EIA 2009c). Its importance is perhaps best measured in symbolic terms. The oil industry, in particular, is emblematic of the market system itself, with companies such as Exxon Mobil representing the best and worst of the free enterprise system. The fascinating history of oil and the oil business is explored in depth in Daniel Yergin's *The Prize* (Free Press, 2008). American distrust of the oil business can be traced back to John D. Rockefeller's Standard Oil monopoly.

Long after the breakup of this monopoly, the perception of cartel-like behavior in oil markets persists. The revenues and stock prices of these companies surge along with upward movements in oil and gasoline prices. Riding a recent oil price spike, Exxon Mobil set a record for profit by

Table 1.2 U.S. Employment and Revenues in Energy Supersectors, 2008

Sector	*Employment*	*Estimated Revenues ($US billions)*	*Typical Companies*
Petroleum and Coal Products	242,100 (2008)	1,370.5	Exxon Mobil, Chevron, BP, Occidental
Natural gas distribution industry	107,240 (2007)	92.131	XTO, Chesapeake Energy, Devon Energy, Apache Corp.
Electric Utilities	559,500 (2008)	365.4	So. Cal Edison, CenterPoint Energy, PG&E
Renewable Energy	Wind: 16,000 Solar: 7,600 Biomass: 9,600 (2007/08) Total secondary jobs estimate: 8.5 million	40.0	GWS Technologies, Dupont, GE, Westcorp Energy, Archer Daniels Midland

Sources: U.S. Census Bureau. 2008. *American community survey*, Table B25045. U.S. Census Bureau, http://www.census.gov/acs/www/index.html (accessed January 4, 2010).

a single company with profits in 2008 of $45.2 billion. This reflects a direct transfer of national wealth from consumers to stockholders in these companies. Big Oil's reputation has also been tarnished by its history of destructive environmental practices and willingness to support unsavory regimes that control substantial oil reserves.

The status of the oil majors—the largest U.S. oil companies—in 2010 has changed dramatically over the past 10 to 15 years, however. Concentration has reduced the number of true U.S. oil majors to three: Exxon Mobil, Chevron, and ConocoPhillips, while Organization of Oil Exporting Companies (OPEC) member countries and their national oil companies (owned primarily by their governments) now control about 88 percent of global reserves (EIA 2010g). Oil companies make tremendously large bets on exploration, and the U.S. mainland has now been mostly exploited; the continental shelf remains the last rich source of U.S. oil, but is off limits for the moment. So these remaining companies are facing increasingly limited options.

Public Utilities

Traditionally, public utilities, private but regulated companies that generate electricity, were perceived as a tad boring by the business world. Their role was to generate, transmit, and distribute electricity within a defined service area, as a monopoly business with prices and profits regulated by state public utility commissions and in part by the Federal Energy Regulatory Commission. An engineering mindset dominated the business and aimed to provide reliable service and steady, predictable profits. The industry was perceived as a safe, sedate cash cow sector; utility stocks were treated more like bonds that would regularly generate predictable dividends.

This began to change with the 1970's oil crises. A destabilized energy market resulted in a huge increase in electricity prices. At the same time, environmental concerns mandated significant investments in pollution control. Electricity price increases also helped create a new business category—energy services companies (ESCOs) that worked with industry and households to increase their energy efficiency. New technology and federal policy also created space for a renewable energy industry that has experienced a series of boom and bust cycles.

Need for a New Business Model

This period created the need for a significant change in the utility business model, which suddenly had to contend with both planning for future increases in demand, and how to sustain profits when expected to try to decrease use of their basic product. The 1978 Public Utility Regulatory Policy Act triggered a dramatic shift in the industry as it created market openings for new categories of private and often unregulated power generators, which stole away large industrial customers through lower rates. Since the 1970s, the utility industry has become riskier and more complex; the dramatic rise and fall of Enron illustrates the vulnerability of the sector to financial manipulation.

As these capsule descriptions suggest, the energy industry has evolved dramatically over the past three decades, in the process growing larger and more important to the economy. Not surprisingly, energy industries have sought to preserve their markets and profits through the political process. Possibly the most glaring example of this was the crafting of the infamous 2001 National Energy Strategy by Vice President Dick Cheney, with energy companies at the table but almost no other public input. Now, the energy sector faces an even greater test: the need to adapt to climate change and a world in which carbon—the unwanted by-product of the fossil fuel economy—will eventually be priced and its output increasingly limited.

Even with a new carbon economy, U.S. use of fossil fuels will remain large for generations, so this sector will remain significant. But the role of federal and state energy policy in encouraging a shift to renewable sources will increase the role of government.

Energy in Commerce and Industry

In 2008, the U.S. industrial and commercial sectors used over 50 trillion BTUs of total energy, about 50 percent of all energy consumed in the country that year. But the energy trends of these two sectors of the economy are diverging. Although the United States remains the world's largest manufacturing country, the global share of U.S. industrial activity has declined. With the contribution of investments in energy efficiency, this has lowered the amount of energy needed for the country's industrial sector by more than 11 percent since its 1997 peak. The drop in 2008 also reflects the onset of a serious economic recession that lowered energy use. The commercial sector, however, reflects the continued growth of services, including commercial real estate, with its demand for energy continuing to expand.

Since the 1970's oil crises, the sophistication of U.S. business toward energy use has increased considerably. For most large businesses, it is no longer just another manageable resource. Managing energy is now a cost center, and firms hire individuals and staffs that pay for themselves through reductions in energy use. This is now easier, thanks to the availability of technology to help manage energy use and costs.

Firms also must manage how they are perceived by the public at a time when being "green" is becoming mandatory for marketing success. Yet some aspects of business energy use are still in need of attention from a policy perspective. For example, buildings continue to be responsible for 40 percent of U.S. carbon emissions. The nation's reliance on petroleum as a transportation energy source receives much attention, but virtually all of it is focused on personal automobiles and light trucks; business vehicles, most using diesel, remain relatively inefficient and are ripe to be replaced by short-haul electric trucks.

Energy and the Consumer

Private citizens need energy because they need energy services. An analysis of household electricity demand in 2001 found that appliances are responsible for about 65 percent of electricity use, with refrigerators responsible for the most consumption (14 percent), with lighting (9 percent) in second place. The good news is that despite larger houses with their increased need for heating and cooling, and greater demand for electricity for electronics, the rate of growth in the residential use of energy has fallen dramatically. Since 2000, residential energy demand has increased by less than 1 percent. This is partly due to increased efficiency, replacing of old appliances with newer models that demand less power (thanks to the Energy Star program described in Chapter 4), and a shrinking of household size. Also, the effects of compact florescent lights, which typically use 75 percent less power than an incandescent bulb, are beginning to be felt.

We are also driving less; vehicle miles traveled were down by about 3.6 percent in 2008 compared to 2007. One of the energy surprises of 2008 was the realization that the demand for gasoline went down when gasoline reached the $4-per-gallon threshold. The price elasticity of demand for gasoline in the United States—the change in demand that results from an increase in price—has been shown to be low in the past, since there are few substitutes for driving for most Americans. But suddenly people were willing to carpool, work from home, or take mass transit. But prices of $4 per gallon of gasoline and $140 per barrel of oil are difficult to sustain politically in the United States. High fuel prices also strike fear in the OPEC states that now produce most of the world's oil. High prices change consumer behavior and encourage the development of sustainable substitutes that threatens their energy hegemony.

Help for Low-Income Families

Although the market system in the United States does a reasonable job of supplying energy to most people and families, it often fails for low-income households whose basic living circumstances differ dramatically from people and families with higher incomes. According to 2007 data from the Department of Energy, families below 150 percent of the poverty level ($30,975 for a family of four in 2007), are much more likely than people not in poverty to live in apartments than single-family homes; to not own their homes if they are in a single-family house; and to live in a much older home (HHS 2009). They spend less on energy for their households than families with higher incomes: an estimated $1,715 in 2007, compared to $2,132 for households not in poverty. However, energy purchases represent a much larger portion of their income: 9.9 percent compared to only 2.5 percent for households not in poverty.

Low-income families are also much less likely to own a vehicle. In 2001 (the latest data available), more than 20 percent of households with incomes of less than $25,000 per year did not own a vehicle, compared with about 2 percent of households with incomes higher than that threshold (BTS [Bureau of Transportation Statistics, U.S. Dept. of Transportation] 2003).

Energy and Agriculture

Although agriculture reflects only 1.2 percent of the $14 trillion U.S. economy as of 2008, and less than 1 percent of the labor force, the relative importance of this sector to our economy and society is significant. This country is more than self-sufficient in food, and agricultural goods remain an important category of U.S. exports. The country provides around 40 percent of world grain exports, for example. The evolution of U.S. agriculture toward larger but fewer farms has been accompanied by increasing sophistication in the use of energy, and energy use per unit of agricultural output has decreased steadily since the 1970's energy crisis (USDA [U.S. Department of Agriculture] 2006).

However, the big story in the agricultural sector can be summed up in one word: biofuels. There is no better example of the systemic, global nature of energy policy than the vast expansion of the country's ethanol production since 2001. Ethanol is alcohol generally produced in the United States through the industrial fermentation and distillation of corn, although any feedstock may be used that contains substantial sugars or starch, including sugar cane. U.S. policy has been to encourage ethanol production as a gasoline additive and substitute. Although pure ethanol has a significantly lower energy density than gasoline, in a 90/10 gas/ethanol blend, it helps lower emissions while only lowering the energy density of the fuel by about 3 percent.

The rapid increase in oil and gasoline prices beginning in 2006, plus changes in federal policy in the 2005 Energy Policy Act, encouraged a rapid increase in ethanol production in the United States. Production more than tripled in volume, from 2.8 gallons in 2002 to 9 billion in 2008. Corn production increased, from 9 billion bushels in 2002 to 13 billion in 2007. Prices doubled over roughly the same period, from $2 per bushel in 2005 to $4.20 in 2007. The proportion of the U.S. corn crop used for ethanol increased from 6 percent in the 1999/2000 growing season to 24 percent in 2007/2008 (USDA 2009).

To meet the increased demand, some farmers switched from growing white corn suitable for human consumption to field corn, which is used to feed cattle and to make ethanol. Demand for cropland and animal feed increased, along with the prices of other farm commodities, including meat, soybeans, dairy products, and wheat; these increases quickly led to significant increases in U.S. food prices. The U.S. Congressional Budget Office (CBO) estimated that the ethanol binge was responsible for 10 to 15 percent of the 5 percent increase in food prices from April 2007 to April 2008, and increased federal spending on food assistance programs (CBO [Congressional Budget Office)] 2009). Mexico was plunged into a crisis as white corn shortages led to a tripling or quadrupling of the price of tortillas, a cultural staple.

Tightening markets for energy, agricultural commodities, and fertilizer, plus low stocks of stored foods helped to drive up food prices worldwide. The World Bank estimated that these increases led to a 3 to 5 percent increase in poverty across the planet, as well as serious nutritional consequences as the urban poor switched to cheaper but less nutritious sources of food (World Bank 2008).

By early 2010 the boom in farm prices had been replaced by a classic bust. In early 2008 high corn prices made ethanol production uneconomic for smaller producers and scores of planned investments in ethanol plants were cancelled or finished plants left idle. The severe 2008 worldwide recession lowered demand for energy generally, including gasoline and ethanol. Ethanol producers, plus the USDA began lobbying for increasing the 10 percent "blend wall" of ethanol with gasoline to 15 percent. Food and fuel prices have moderated. We will return to our discussion of biofuels later in the book to provide a clearer description of the role of U.S. public policy in encouraging this cycle.

Conclusion: Energy out of Balance

This discussion of the U.S. energy system and its interdependent energy subsystems omits an important reality: it is out of balance. In 1955 the country crossed the threshold of 1 million barrels per day (mbd) of net oil imports, and oil imports have risen relentlessly ever since. As of 1996, net oil imports began to exceed domestic oil production; now we import around 57 percent of our daily oil needs, over 12 mbd (EIA 2009c). We do not produce enough oil to meet current demand and are reliant on the world oil market to make up the difference.

Our inability to meet our oil needs from domestic sources for decades has been one of the main irritants in U.S. energy policy. It also raises fundamental questions about our long-term supplies of not only oil, but other types of energy, and the extent to which we should seek to be self-sufficient. The recognition that global climate change is almost certainly linked to our vast appetite for energy is causing a reappraisal of our energy habits. The next chapter will consider these and other challenges and our need to begin an energy transition.

Chapter 2

Energy Policy in Transition

> Clearly, there are no easy choices facing future production and use of energy for most countries. … Only large reductions in global primary energy use, with all its difficulties of implementation, can meet the resource, environmental, economic and political problems that future energy use will face.
>
> **—Patrick Moriarty and Damon Honnery (2009)**

In the summer of 2009, one of the authors traveled to British Columbia's Okanogan Valley. The Okanogan has beautiful lakes and mountains and warm summer weather that make it an ideal vacation spot. Over the past two decades it has become famous for a new reason: the quantity and quality of its wines. Local winemakers were initially boosted by Canadian government support for removing previous grape varieties and replacing them with superior *vitis vinifera*. They also received an assist from a gradually changing climate. Warming winters have extended the region's wine growing area further south and made it possible to grow a wider variety of red wine grapes. Skillful vintners have taken advantage of the weather change; the valley is now one of the top wine-producing regions in North America.

But the same warming winters have had a devastating effect on British Columbia's (BC)'s forests. The mountain pine beetle has long been established in the vast pine forests of the province's central plateau, but its effects were relatively minor as long as extended cold winters kept its numbers low. Low temperatures of –35 to –40°C (about –31 to –40°F) for several consecutive days are needed to kill the various stages of the beetle. But warmer, dryer winters that don't reach these bone-chilling temperatures have allowed the beetle to survive and flourish in these forests, which have also been weakened by hotter and dryer summers. As of 2009, 35.8 million acres, or almost 56,000 square miles have been affected, with infested trees typically turning a rust color. Large areas of BC's interior forests are now colored an unearthly red. What's worse, as the dead and dying trees decay and emit carbon, BC's forests have become a net emitter of carbon dioxide. Although the infestation appears to have peaked, its effects on communities in the affected area and on the overall environment will likely linger for a generation. Similar forest pest infestations have caused extensive damage in Colorado, Alaska, Montana, and Idaho and the province of Alberta.

Clearly, the Earth's climate is changing. Average surface temperatures around the globe rose by 1.3°F over the last century, with varying effects, as the cases above suggest—sometimes beneficial,

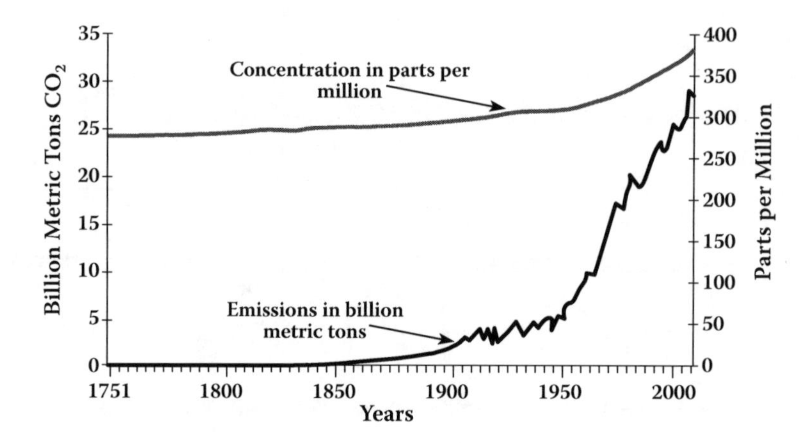

Figure 2.1 Carbon emissions and atmospheric concentrations, 1750–2000. (From Oak Ridge National Laboratory. 2004. Carbon emissions and atmospheric concentrations, 1750–2000; U.S. Department of Energy.)

sometimes disastrous. Are these cases evidence of global climate change? Individually, no. But the data suggesting that climate change is occurring, and that human activity is a primary cause, are now too overwhelming to ignore. Figure 2.1 displays the increase in carbon emissions from 1750 to 2000.

A French chemist, Joseph Fourier, first described the so-called *greenhouse effect* in 1824. About 28 percent of the sunlight reaching the Earth's surface is reflected as infrared radiation—heat. Gases in the atmosphere prevent a portion of that heat from simply going back into space, and re-radiate it back to the surface. Without an atmosphere, average global temperatures would be closer to –19°C, or 0°F; with it, temperatures are much higher—around 14°C or 59°F on average worldwide. Adding even relatively small amounts of certain gases and aerosols, including carbon dioxide and methane, causes the atmosphere to retain more of this radiation. In sufficient quantities—and we are now adding around 4.1 billion tons of CO_2 to the atmosphere yearly—the overall climate system retains more heat. Climate scientists term this process *radiative forcing.* The amount of CO_2 added would be much higher without the "carbon sinks," the world's forests and oceans, which absorb about half of human-caused emissions.

Early Climate Research

Scientists studying the Earth's climate began to speculate about the influence of development on the atmosphere and global temperatures as early as the 1890s, when Svante Arrhenius, a Swedish chemist, first suggested that human activities, including industrialization, could impact global temperatures. Over the course of the next 120 years, the steady accumulation of new data, combined with improved climate models, better instruments, and information technology, led to a series of critical conclusions (Figure 2.1).

One of the first was that the available data—ice core samples and other sources going back thousands of years—reveal a clear, consistent link between the proportion of carbon dioxide in the atmosphere and global temperature. Higher CO_2 levels did not always precede temperature changes, but the data suggest a complex feedback loop between the two. Another was that

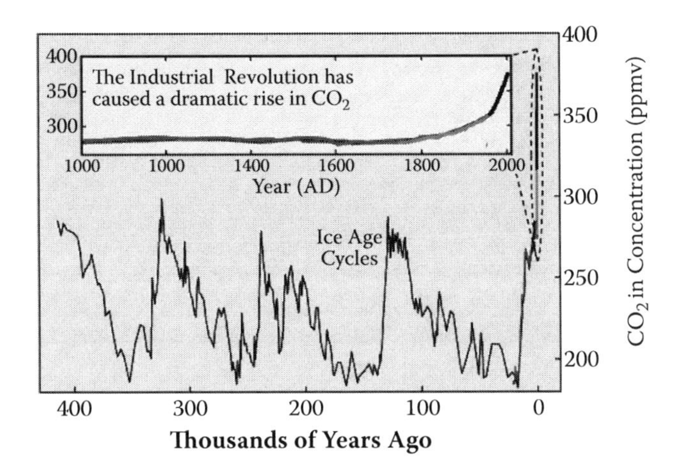

Figure 2.2 CO_2 variations and Ice Age cycles. (From Globalwarmingart.com. Gallery of Data Related to Carbon Dioxide. Global Warming Art, http://www.globalwarmingart.com/wiki/Carbon_Dioxide_Gallery, accessed February 5, 2010.)

atmospheric CO_2 began to increase at about the same time as the Industrial Revolution, just before 1800, and has steadily increased since then—from the 180 to 280 parts per million (ppm) typical of fluctuations around ice ages, to 387 ppm measured in December 2009 (Figure 2.2). In 1995, the United Nations (UN)-sponsored Intergovernmental Panel on Climate Change (IPCC) released the first report tentatively linking human activities—mainly the burning of fossil fuels—to observed changes in climate (UN 1995).

Today the consensus projection from the IPCC Fourth Assessment Report is that by the end of the twenty-first century, increasing concentrations of CO_2 and other greenhouse gases, in the absence of significant mitigation, will raise average global temperatures by 1.4 to 6.9°C, with the center of that range of estimates at about 4°C (UN 2007).

Predicted effects include increasing temperatures in many areas, with temperatures increasing disproportionately near the poles, increasing precipitation in already rainy areas, arid areas becoming even drier, and a decline in mountain snowpacks and glaciations as rising temperatures bring more rain to higher elevations, and less snow. Some regions may become cooler. Melting polar ice and glaciers will increase global sea levels, threatening coastlines and low-lying islands. And these changes in rainfall patterns and temperature will have substantial and uneven effects on agriculture and wildlife, as ecosystems adapt to changing conditions. The worst case scenarios suggest that rising temperatures have discontinuous effects. That is, once (now unknown) temperature thresholds are crossed, catastrophic effects such as the melting of the polar ice caps, extensive tropical forest fires, thawing of arctic permafrost or the shutdown of critical ocean currents such as the Gulf Stream could occur, with severe effects for climate and civilization.

The likely culprit? The burning of fossil fuels by consumers and industry directly and for generating electricity in developed countries. The IPCC estimates that fossil fuel combustion and cement manufacture are responsible for 75 percent of global CO_2 emissions since 1800; deforestation and agriculture account for most of the remaining 25 percent (UN 2007).

As of 2007, 60 percent of the various greenhouse gases in the atmosphere released since the mid-1700s (including methane, nitrous oxide, and fluorocarbons) had been released by developed countries. Hydrocarbons such as oil, natural gas, and coal are burned through combustion, using air as the primary source of oxygen; heat, water, carbon dioxide, and other impurities result, including nitrogen compounds such as nitrous oxide. Since 1995, the IPCC has become more direct in its statements about the link between burning hydrocarbons and climate change, stating in its 2007 report that, "Human induced warming of the climate system is widespread" (UN 2007).

The Threat of Climate Change

The policy system in the United States has yet to systematically respond to the threat of climate change, for reasons we will discuss in Chapter 4. However, there is a broad consensus among energy policy analysts that the U.S. energy system will require fundamental transformation in order to limit the rise in greenhouse gases and thus the overall warming of the Earth's climate.

But what form should that transformation take? And what factors influence our ability to gather data that would be useful to reach conclusions on how to proceed? This chapter will focus on several challenges that complicate the process of making judgments on these issues. One is our knowledge of current and anticipated supplies of oil and natural gas. U.S. production of oil peaked in 1970, and some analysts have concluded that we have already hit a global oil peak, representing having consumed roughly one-half of the world's recoverable oil. Although current estimates of natural gas reserves suggest this resource is more abundant, it is also a highly desirable fuel and its rate of use is increasing rapidly.

Related to these is the biggest single challenge to analysts of energy policy: the poor record of forecasting of energy-related phenomena. Our ability to peer into the future is breathtakingly limited, in part because of the complex interactions of economic systems, technology, demographics, and ecosystems. The remainder of this chapter will explore these issues and their interrelationships.

Forecasts, Energy, and Creating the Future

One of the greatest difficulties confronted by those responsible for complex energy systems is our inability to accurately predict the future. Historically, humans have shown only a limited capacity to anticipate what the future holds, and this is nowhere truer than in the energy sector. The record is so bad that Smil (2006) titled a book chapter on this subject: "Against Forecasting." Over the past 50 years, major unanticipated events—surprises—include the Arab oil embargo and subsequent quadrupling of crude oil prices, additional oil price fluctuations in 2001 and 2008, and the effects of nuclear power plant accidents at Three Mile Island, Pennsylvania, in 1979 and Chernobyl, Ukraine, in 1986.

Perhaps the most grievous forecasting error was the collective belief of 1970's energy planners that demand for electricity would continue to steadily increase at 5 to 7 percent a year—doubling roughly every 10 to 14 years (Craig, Gadgil, and Koomey 2002). This trend had been consistent for decades … why wouldn't it continue? But increases in electricity prices in the 1980s changed consumer behavior, as the price elasticity of demand for the product proved to be higher than expected. Households and businesses reduced their use of electricity and were also willing to invest in increased efficiency. The resulting projection was only half of the official government forecasts.

These failures—the use of trend extrapolation and failure to anticipate changing consumer behavior and other important variables—also resulted in some spectacular errors. These include the $6 billion Shoreham, Long Island, nuclear plant constructed to meet projected growth in demand that was based on poor planning and faulty beliefs about future demand; the plant has yet to generate any electricity for commercial distribution.

Another is the Washington Public Power Supply System (WPPSS) debacle. Northwest energy planners in the late 1960s foresaw continuing 7 percent growth in electricity demand, coupled with looming shortages and now new sources of supply to meet that demand. One planning document, the Hydro-Thermal Power Program of 1968, actually proposed the construction of *twenty* new nuclear power stations.

Gradually WPPSS, a municipal corporation composed of public utility systems in the region, developed a regional plan to construct five new nuclear generating plants. Funding for three of these plants was to be subsidized by the Bonneville Power Administration. Two additional plants were to be constructed by privately owned utilities—for a total of 5 nuclear plants. The plan unraveled in slow motion over the following 15 years.

Competing forecasts released by various parties began to confuse the public, which began to ask if all these costly plants were really needed. The City of Seattle initially signed on to support construction of plants 1, 2, and 3, but the decision on whether to support the final two plants, 4 and 5, was intensely debated. An alternative planning document released in 1976, *Energy 1990*, concluded that lower projected energy growth could be met by conservation. The City of Seattle and its municipal utility, Seattle City Light, opted not to participate—a prescient decision. Construction costs ballooned as interest rates rose and the complexity of the project overwhelmed its managers. Total costs of the project eventually were estimated at $24 billion in January 1982. Shortly afterward, construction on plants 4 and 5 was halted.

Of the five planned WPPSS plants, only one, plant 2 at Hanford, Washington (now called the Columbia Generating Station) was ever completed. Three of the plants, one at Hanford and two more at Satsop, Washington, were partially completed, and were eventually torn down and sold as scrap. Eventually WPPSS defaulted on $2.25 billion in bonds for plants 4 and 5. This is still the biggest municipal bond default in U.S. history. Northwest ratepayers are still paying for bonds associated with plants 1, 2, and 3—by one estimate, 15 percent of local electric bills (Pope 2008). One of the two additional plants, the Trojan Plant in Rainier, Oregon, along the Columbia River operated from 1976 to 1992.

Why Forecasts Are Necessary

There are valid reasons for us to want to peer into our energy future. As the WPPSS case above suggests, energy systems are very costly, often involve new technologies, and have long time horizons. In the electricity sector, the initial project cost for a power plant varies depending on the fuel source, but all require substantial up-front investments. Highly efficient combined-cycle natural gas facilities now cost upward of $1,000 per kW of output, or $400 million for a 400-MW facility. But that is relatively cheap; coal plants in the 500-MW output range now cost $1.5 billion and more.

New technology nuclear facilities are estimated to cost from $6 to $8 billion, although some estimates run as high as $10 to $20 billion. A critical question for nuclear power is whether standardization and experience will significantly lower these costs as more are built. All of these facilities generally have a useful life of 20 to 50 years. Commitments to invest funds of these magnitudes require a relatively high degree of certainty that they will provide an economic return.

There are other, systemic reasons for prudent, long-term thinking that are valid across all energy subsectors. We want to be able to meet future needs reliably, without either unnecessarily building excess capacity (particularly for electricity generation), or building too little capacity and either allocating shortages, or meeting short-term peak needs with high marginal cost sources. We want to have confidence that the primary sources of energy on which we rely will be available; if we anticipate that they will not be, we can make shifting to other alternatives a high priority, and allocate resources to that process. We need to have individuals with the right skills trained to operate these systems. And how will our energy needs link with the particular mix of power available within a defined region?

New Planning Tools

Energy planners were chastened by the failure to anticipate the oil embargo, and subsequent changes in the demand and supply of important energy sources. Eventually a new approach to energy planning was developed, termed *integrated resource planning* (IRP) that combined improved forecasting techniques, use of scenarios to clarify possible combinations of anticipated demand, and used a least-cost decision rule for selecting future sources. IRP also emphasizes alternatives to conventional generation, including demand-side management approaches and systemic improvements. The overall IRP approach was mandated for public utilities in the Energy Policy Act of 1992. Whether IRP has improved utility performance is an open question.

The more difficult question of whether energy analysts have improved their capacity to forecast future energy needs and events is still under debate. Winebrake and Sakva (2006) concluded that U.S. Department of Energy forecasts had not improved over a twenty-year period. Smil (2008) is likely correct that a better strategy is to formulate scenarios of energy use and their implications for society as a whole. Those may be used as starting points for debate about the type of future we want, and how to enact and implement policies that help us create that future. Box 2.1 is one small contribution to an explanation of why a solution to our widely understood energy problem has yet to presented, let alone adopted.

Confusion over Peak Oil

In his 2005 book *Beyond Oil*, geologist and author Kenneth Deffeyes predicted that the world oil peak would occur on November 24, 2005 (Deffeyes 2005). On this date global production of oil would reach its highest level, and half of available world reserves would have been exploited; Deffeyes later revised this estimate to December 16, 2005. Deffeyes had predicted earlier that 2005 would be the year of the global oil peak, but with no specific date. He missed that forecast as well (DOE/NETL 2007). The wide variation in peak oil forecasts displayed in Tables 2.1 through 2.3 reveal why energy policy makers encounter difficulty when picking a peak oil forecast upon which to base their planners. Figure 2.3 displays U.S. oil production, consumption, and net imports. U.S. consumption and imports are in temporary decline due to the 2008–2009 recession; both are expected to rise again after 2010. Meanwhile, U.S. domestic production is not likely to ever again meet demand.

These somewhat tongue-in-cheek predictions certainly were meant to add a touch of humor to the grim reality they represent: the belief by most observers that the world is fast running out of oil. These estimates can all be traced back to the approach pioneered by famous geologist M. King Hubbert, who predicted the 1970 peak of U.S. oil production. Hubbert also predicted that production would both rise and fall in a bell-shaped curve.

Deffeyes and Hubbert—and many others—have thus far missed all of their peak oil forecasts; the amount of available global reserves has continued to increase past 2005. However, the turmoil in the world oil markets in 2008 made peak oil look like more than speculation. A combination of factors, including increased demand from China and India, plus speculation on the NYNEX oil market, increased nominal prices to as high as $147 on July 11, 2005. Luckily for the global economy,

BOX 2.1 RARE TORTOISES STOP SOLAR ENERGY DEVELOPMENT

BrightSource Energy, an Oakland, California-based firm had its plans to build a 400,000-mirror facility to collect the sun's energy on six square miles of federally managed land in the Mohave Desert slowed because environmentalists want the land to remain a near-pristine home for rare plants and wildlife, including the rare desert tortoise. Federal and state biologists recommended that BrightSource purchase a nearby 12-acre site and relocate upward of two dozen of the rare tortoises. The plan would cost BrightSource an estimated $25 million and commit it to perpetual maintenance of the relocation site. The Sierra Club and other environmental protection organizations are fighting to have the plant moved to a location closer to Interstate 15, the freeway connecting Los Angeles and Las Vegas.

The solar generation plans call for building three solar power plants on the site, which would use the sun's heat to create steam. The steam would then be used to drive turbines that generate electricity. Fully developed, the desert site would include seven tall metal towers, water tanks, boilers and steam turbine generators, a natural gas pipeline, and buildings for maintenance and administration. The electricity would be transmitted over an existing nearby transmission grid. Each of the three plants would generate enough electricity to meet the needs of 142,000 homes.

The Bureau of Land Management (BLM) has received more than 150 applications for large solar power plants on land it manages in California, Arizona, Nevada, New Mexico, Colorado, and Utah. The BrightSource proposal, known as the Ivanpah project, was approved by U.S. Interior Secretary Ken Salazar on October 7, 2010, after the company reduced its size and agreed to a series of mitigation measures to protect the desert tortoises (BLM 2011).

(From Blood, M. R. 2010. Endangered tortoises snarl solar-energy plans. Associated Press, January 2.

those high prices didn't last. By December of that year, oil prices dropped to the $33 per barrel range; more important, yearly global oil production dropped to a point lower than it had been in 2005.

Crude oil supply and oil markets are apparently inscrutable, and will continue to be for some time in the future. Despite the oil industry's high technology forecasting tools, they simply can't say with accuracy how much oil remains in the ground. That oil is what geologists have termed *ultimately recoverable reserves*, or URR. Oil companies, both private and national, carefully gather data on their oil fields and the "proven reserves" they represent. In the words of the Energy Information Administration (EIA), these are the "estimated quantities that analysis of geologic and engineering data demonstrates with reasonable certainty are recoverable under existing economic and operating conditions" (see Appendix D).

Governments, private producers, and oil consulting firms systematically gather and analyze those data on a frequent basis. The IEA (2007, 2009), World Energy Council, British Petroleum (BP), Cambridge Energy Research Associates (CERA), and other groups produce yearly estimates. Yet, there is substantial disagreement about how much oil is left. The IEA's current estimate for proven oil reserves is 1.3 trillion barrels (about 40 years at current rates of consumption); another 2.2 trillion barrels are considered "ultimately recoverable," and 3 trillion additional barrels are estimated to be available from unconventional sources.

Table 2.1 Important Peak Oil Forecasts to 2010

Predictor	*Affiliation*	*Year*
Pickens, T. Boone	Oil and gas investor	2005
Deffeyes, K.	Retired Princeton professor and retired Shell geologist	December 2005
Westervelt, E. T.	U.S. Army Corps of Engineers	At hand
Bakhtiari, S.	Iranian National Oil Co. planner	Now
Herrera, R.	Retired BP geologist	Close or past
Groope, H.	Oil and gas expert and businessman	Very soon
Wrobel, S.	Investment fund manager	By 2010
Bentley, R.	University energy analyst	Around 2010
Campbell, C.	Retired Texaco and Amoco geologist	2010
Skrebowski, C.	Editor of *Petroleum Review*	2009 to 2011
Meling, L. M.	Statoil oil company geologist	A challenge around 2011

Source: DOE/NETL (U.S. Department of Energy/National Energy Technology Laboratory). 2007. *Peaking of world oil production: Recent forecasts.* Washington, DC: Department of Energy/National Energy Technology Laboratory, http://www.netl.doe.gov/energy-analyses/pubs/Peaking%20of%20World%20Oil%20Production%20-%20Recent%20Forecasts%20-%20NETL%20Re.pdf (accessed February 5, 2010).

Table 2.2 Important Peak Oil Forecasts to 2020 and Beyond

Predictor	*Affiliation*	*Year*
UBS	Brokerage; financial services	Mid to late 2020s
Strahan, A.	CERA energy consultants	Well after 2030
Esser, R. W.	CERA energy consultants	"Peak oil is garbage"
ExxonMobil	Oil company	No sign of peaking
Brown, J.	BP, oil company CEO	Impossible to predict
OPEC	Oil producers cartel	Deny peak oil theory

Source: DOE (U.S. Department of Energy). 2007. *Renewable portfolio standards in the states: Balancing goals and implementation strategies.* NREL/TP-670-41409. Golden, CO: National Renewable Energy Laboratory, http://www.nrel.gov/docs/fy08osti/41409.pdf (accessed June 1, 2010).

Table 2.3 Some Suggested GHG Stabilization Strategies

More efficient vehicles: increase fuel economy from 30 to 60 mpg (7.8 to 3.9 L/100 km) for 2 billion vehicles.	Reduce use of vehicles: improve urban design to reduce miles driven from 10,000 to 5,000 miles (16,000 to 8,000 km) per year for 2 billion vehicles.	Efficient buildings: reduce energy consumption by 25%.	Improve efficiency of coal plants from today's 40% to 60%.
Replace 1,400 GW of coal power plants with natural gas.	Capture and store carbon emitted from 800 GW of new coal plants.	Capture CO_2 from hydrogen plants generating hydrogen from coal or natural gas.	Capture and store carbon from coal to syn fuels conversion at 30 million barrels per day (4,800,000 m^3/d).
Displace 700 GW of coal power with nuclear.	Add 2 million 1-MW wind turbines (50 times current capacity).	Displace 700 GW of coal with 2,000 GW (peak) solar power (700 times current capacity).	Produce hydrogen fuel from 4 million 1-MW wind turbines.
Use biomass to make fuel to displace oil (100 times current capacity).	Stop deforestation and reestablish 300 million hectares of new tree plantations.	Conservation tillage; apply to all cropland (10 times current usage).	

Source: Pacala, S., and R. Socolow. 2004. Stabilization wedges: Solving the climate problem for the next 50 years with current technologies. *Science*, 305 (13): 968–972.

Lower Amounts Predicted

Pessimists, including Deffeyes and many others, predict much lower amounts of recoverable oil. They point to the declining average size of new discoveries—between 1990 and 2006 production was double that of new oil discoveries (*Economist* 2009a). There are also concerns about the quality of the data about the 300 or so giant oil fields worldwide that provide a high proportion of all oil production. Many of these fields are decades old and rely on injections of sizable amounts of water. Organization of Oil Exporting Companies (OPEC) member countries currently produce about 40 percent of daily world supply, but are believed to hold 70 percent of world reserves.

However, many of these countries, including Iran, Kuwait, Saudi Arabia, United Arab Emirates, and Venezuela, revised their stated reserves upward in the late 1980s, when production allocations were changed to reflect both production capacity and reserves. Given the lack of transparency of the processes used to generate these estimates, they are not entirely trusted. If output from large producing fields continues to decline (lowering reserves), and sizable new discoveries do not make up the difference, demand could quickly begin to exceed production. Even the relatively optimistic International Energy Agency (IEA) recently predicted a peak in oil production in 2020.

But the production of oil is influenced by economics and politics in addition to geology. Optimists are likely to cite the effect of tightening oil markets and higher prices on technological

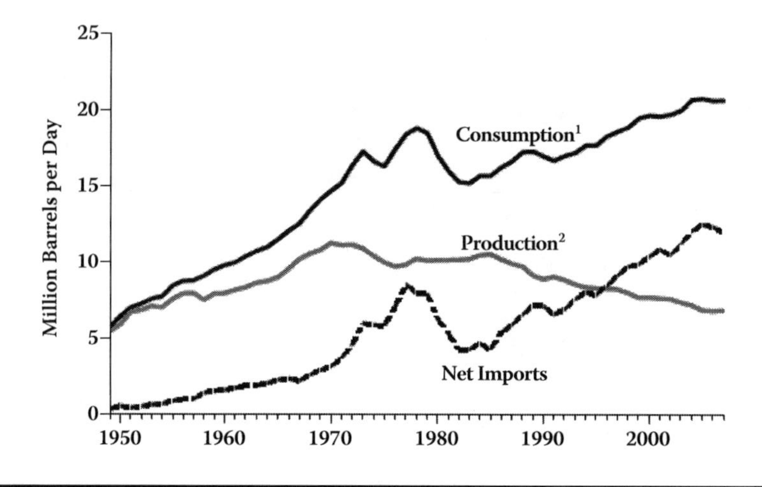

Figure 2.3 Petroleum consumption, production, and imports, 1950–2007. (From EIA [U.S. Department of Energy, Energy Information Administration]. 2010c. *Annual energy review 2009*. Washington, DC: U.S. Department of Energy, http://www.eia.doe.gov/cneaf/coal/page/acr/acr.pdf, accessed August 31, 2010.)

innovation, and willingness to recover oil in more costly environments, including deep water locations and in the Arctic (*Economist* 2009a). Moreover, higher prices make unconventional sources of oil, including tar sands, heavy oil, and even shale, economically viable. More optimistic scenarios suggest that increasing investments will stave off the year of peak oil production and lead to an extended period with a plateau in oil production at a relatively high level (Jackson 2009).

Transitioning from Peak Production

Does it matter whether peak oil arrived in 2005, or if it will arrive in 2015, 2020, 2030, or never, as some oil company executives would have us believe? In either case, we have to remember that the available window for beginning the transition to nonfossil fuels is relatively limited, considering that past energy transitions—such as the shift from wood, to coal, to oil and natural gas—have required decades.

Anxiety about the prospect of a future with limited supplies of oil is understandable given the general assumptions about the links between oil and economic growth, and its role as a ubiquitous fuel for transportation in a mobility-mad society. Past oil crises have slowed or halted growth in an earlier era; now, even more of humanity is reliant on fossil fuels. Growth in demand for oil is expected primarily from developing regions—as much as 90 percent of demand growth, by one estimate (McKinsey 2009). Tightening supplies could indeed result in hoarding, decreased mobility, impacts to agricultural output and productivity, and a squeeze on industry. Yet while oil is certainly important, it is not necessarily a "strategic" resource in the long run. Without succumbing to technological overoptimism it is possible to envision and enact a different kind of future.

The threat from human-caused climate change from continuing fossil fuel use creates an entirely different context for this transition than would have been the case in a more benign environment. Awareness of climate change will shape our attitudes to the transition, willingness to change energy related behaviors, and the range of energy options available to us. For example, the world has sizable reserves of unconventional oil. Yet these require considerable energy (primarily natural gas) and investment to drastically reduce their energy return on energy invested (EROEI). These processes also entail use of massive amounts of scarce water. Any hope of stabilizing atmospheric greenhouse gas (GHG) at any reasonable level will be lost if exploitation of unconventional oil becomes a primary strategy.

Alternative, "greener" scenarios include higher energy prices, use of oil to help create a renewable energy infrastructure, and extensive reliance on natural gas as a transition fuel. These need not have drastic limits on economic growth, as will be discussed in Chapter 6, especially since the long-term investments needed to create a new energy infrastructure will have strong impacts on employment.

Until quite recently there was considerable nervousness in U.S. energy markets about what were perceived as relatively limited domestic natural gas reserves. But improved horizontal drilling and hydraulic fracturing technologies have enabled extraction from previously untappable shale formations. Total estimated U.S. natural gas reserves rose from 1,532 trillion cubic feet in 2006 to 2,074 trillion cubic feet in 2008, an increase of 35 percent in just two years (Mouawad 2009). Now the IEA foresees a glut of U.S. gas to 2030 (IEA 2009). Yet this glut may disappear more quickly than anticipated if policy changes result in carbon prices that lead to a massive exodus from coal electricity production toward high-efficiency natural gas generation.

Climate Change: Challenges and Policy Goals

The Fourth Assessment Report released by the IPCC in 2007 is undoubtedly the dullest description of approaching apocalypse in print. But its multiple volumes describe in dense scientific prose (with an occasional detour into diplomacy-speak) the consensus of the world's top climate scientists that human actions are changing the biosphere, and that concerted, global effort is needed to avoid irrevocable damage.

Emissions of carbon to the atmosphere caused by fossil fuel combustion (plus cement manufacturing) totaled about 8.7 gigatons (GtC) in 2008. This is 41 percent higher than in 1990. An additional 1.5 GtC is released from land use practices, including deforestation and inefficient farming practices.

The annual growth rate of emissions has been estimated at 1.9 percent from 1959 to 1999. However, the rate of growth increased to an average of 3 percent per year from 2000 to 2008 (LeQuéré et al. 2009). Climate models suggest that limiting the atmospheric concentration of CO_2 to 450 ppm will keep the total increase in average global temperatures to 2°C above the preindustrial norm. This two-degree threshold is the temperature beyond which more serious systemic effects of climate change, such as the melting of the Greenland ice sheet and collapse of the Atlantic's Gulf Stream current, are believed more likely to occur. The challenge is that to meet this goal, the report concludes that total emissions for the entire twenty-first century must be limited to 490 gigatons of carbon, or about 5 gigatons per year (UN 2007).

At first glance, this doesn't look difficult—a 37 percent reduction, from 8 to 5 gigatons per year is all that is needed, right? Well—no. The IPCC models and scenarios are based on a set of interrelated variables: "[T]he main driving forces of future greenhouse gas trajectories will continue to be demographic change, social and economic development, and the rate and direction of technological change" (UN 2000, 5).

Each of these has a likely growth trajectory; thus, unless these phenomena can be delinked from their reliance on carbon, reductions to a 5-GtC level must be made from an escalating starting point. Global population, now about 6.7 billion, will increase to an estimated 9 billion

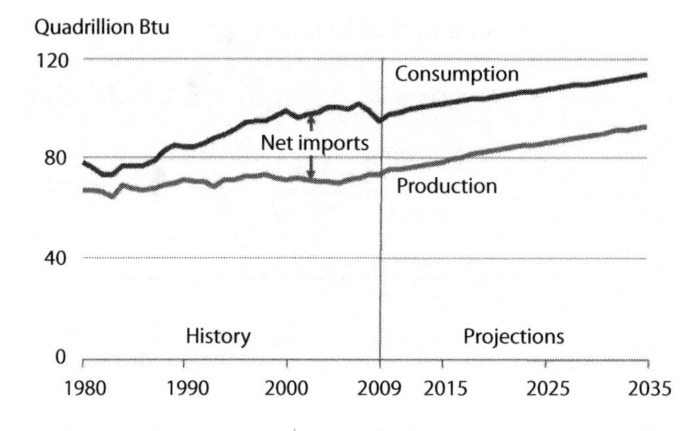

Figure 2.4 Total U.S. energy production and consumption, 1980–2035. (From U.S. Energy Information Administration, *Annual Energy Outlook 2011 Early Release Overview.* http://www.eia.gov/forecasts/aeo/pdf/0383er(2011).pdf, accessed January 24, 2011.)

by 2050, at a time when economic development and energy use are accelerating across much of the world. And despite the effects of globalization and growth, vast disparities remain between energy use in the Organisation for Economic Co-operation and Development (OECD) countries and the rest of the planet. As shown in Figure 2.3, CO_2 emissions in 2005 totaled 27 billion tons; on a per capita basis, the United States contributed over 19 tons per person, the European Union around 8 tons per person, China about 4—*and over half the planet around 1 ton per person*, for an overall average of about 4 tons per human that year. Reducing this inequity should be a basic goal of international development. Figure 2.4 displays estimates of continued growth in U.S. primary energy use. Figure 2.5 notes the sources of these CO_2 emissions by sector.

To meet the goal of limiting CO_2 concentrations to 450 ppm by 2050, overall emissions will need to be *reduced*, not just stabilized. This means per capita global emissions will need to be sta-

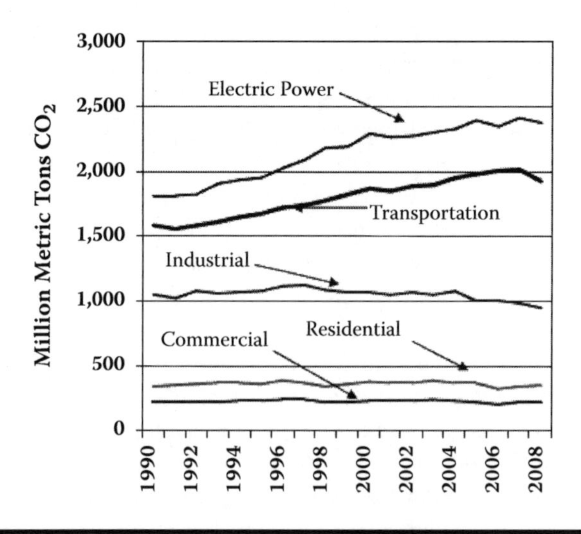

Figure 2.5 CO_2 emissions by sector fuel use. (From EIS 2009d.)

bilized at a low level—around 1 ton per person. Is such a shift in the U.S. consumption patterns that drive these emissions even possible?

Once in the atmosphere, carbon dioxide has a long, complex life span, one that lasts on average 125 years. So most analyses of climate change mitigation emphasize quick action to bring about a peak in human-caused CO_2 emissions, since even with a rapid response, concentrations of CO_2 are expected to plateau for an extended period, up to hundreds of years. And current computer models suggest that global temperatures could continue to rise despite a stabilized atmosphere.

High GHG Emissions Levels Continue

Since 2000, global greenhouse gas emissions have continued at a high level, although a decrease was likely beginning in 2008 due to the effects of the worldwide recession. The most recent analyses find parallels between the pre-2008 trends and the most extreme of the climate change possibilities presented in the IPCC's 2000 *Special Report on Emissions Scenarios* (UN 2000), the A1FI scenario, which envisions continuing high economic growth and social development, along with continuing reliance on fossil fuels (Schneider 2009). Should this trend continue post-recession, increases in global temperatures of 4 C by 2060 are possible (Betts et al. 2009). The science of climate change suggests that warming over the next 20 years or so will be driven by emissions that have already occurred (Trenberth 2010). Beyond that, our capacity to see the future depends on computer models that are improving, but will benefit from a gradually improving scientific understanding of the complex climate feedback effects.

The Challenge; the Response

The magnitude of this challenge to our societies and to political and social leaders is sobering. There are few similar examples of scientific findings—*unsettling* findings—that support a need for such an immediate, comprehensive, and costly policy response. One relevant example, the unlocking of the secrets of the atom, and the eventual development of atomic weapons and nuclear energy, is clearly troublesome. The Manhattan Project was a magnificent illustration of the capacity of governments, scientists, and citizens to mobilize in behalf of a mission, but it also exposed the limits of our ability to manage our own technological creations.

If we accept as likely true the scientific consensus that human activity is rapidly changing the climate, then action is essential. But what kind of action, and by whom? Applying a policy analysis perspective to these questions is helpful, as the range of possible societal responses is vast, and is driven by the framing of the problem. When framed as an inevitable end result of the growth model of industrial civilization, the response is to consider alternative models of civilization and human organization. If capitalism is driving the threat, then collectivized models become a possible solution. Pessimists suggest that the likelihood of coordinated action on mitigation of greenhouse gas emissions within and between nation states is low, and that we should instead focus on adaptation to a warming world.

A Supreme Challenge

We see the challenge as framed by the limitations of individual action and global action, and the benefits and limits of the capitalist nation-state as the primary form of organization of human affairs. While individual action is important and inspiring individuals can model changes in behavior and lifestyle for the rest of us, uncoordinated individual action will not be sufficient. Coordinated collective action is needed. Examples of existing energy collaboration and cooperation institutions and forums are described in Chapter 9.

The recent failure of the Copenhagen climate change conference again exposed the extreme difficulty of achieving coordinated global action. We do not have a functional world government, but a venue, the United Nations, for literally inter-national efforts to grapple with problems whose scope exceeds the boundaries of individual countries. To date, nation-states have been unwilling to give up sufficient sovereignty, power, and budgetary authority to make the United Nations functional as a means of global governance.

For the foreseeable future, the dominant model of economic production is capitalism, operating within nation-states with a variety of political systems, from authoritarian to generally democratic systems providing voter choice. The developed countries are governed by a variety of democratic systems that make elected officials accountable to citizens. Government action in such countries on challenges as monumental as climate change requires leadership; leaders must be willing and able to convince their citizens that action is needed. They must provide a vision of how to correct the problem and how the life of their citizens will be better as a result. This is the essence of public policy. So for at least the next few decades, the most important action on the climate change "problem" must be viewed through the lens of the public policy process in individual countries. The next chapter will examine this process in the United States.

Who Is Responsible?

The question of ultimate responsibility for the global warming problem is more vexing than it first appears. What degree of blame should the developed countries accept for emissions that occurred before the likely causes of climate change were understood? And what should be the metric used to measure country-by-country contributions? If the metric is the total CO_2 emissions from fossil fuels, then between 1890 and 2000, the OECD countries were responsible for about 60 percent of emissions. If, instead, global mean temperature in 2000 is the base indicator, all significant greenhouse gasses (CO_2, CH_4, N_2O, SF_6, HFCs, and PFCs) from all sources must be included. Over this same period, 1890–2000, the share of the OECD countries falls to 38 percent (Fuglestvedt and Romstad 2006).

Asia's contribution increases significantly due to methane emissions from rice farming. Also, thanks to its rapid economic growth, China has now caught or surpassed the United States as the world's largest emitter of carbon. Yet under any scenario, Figure 2.6 makes it clear that the United States is a significant contributor to the problem.

The very magnitude of this challenge presents a dilemma to policy makers. It seems to call for costly responses that require new visions of the future, new ways of thinking, and perhaps drastic changes of behavior that mean giving up cherished beliefs and habits. But citizens are already burdened, and the easiest response is inertia. Both policy makers and the public need to be reminded that our past experience with energy transitions is a largely positive one and that we need to avoid both underestimating the capacity of human ingenuity or overestimating the prospect of dreamy technological solutions. Hard choices will be required, but they need not bankrupt us.

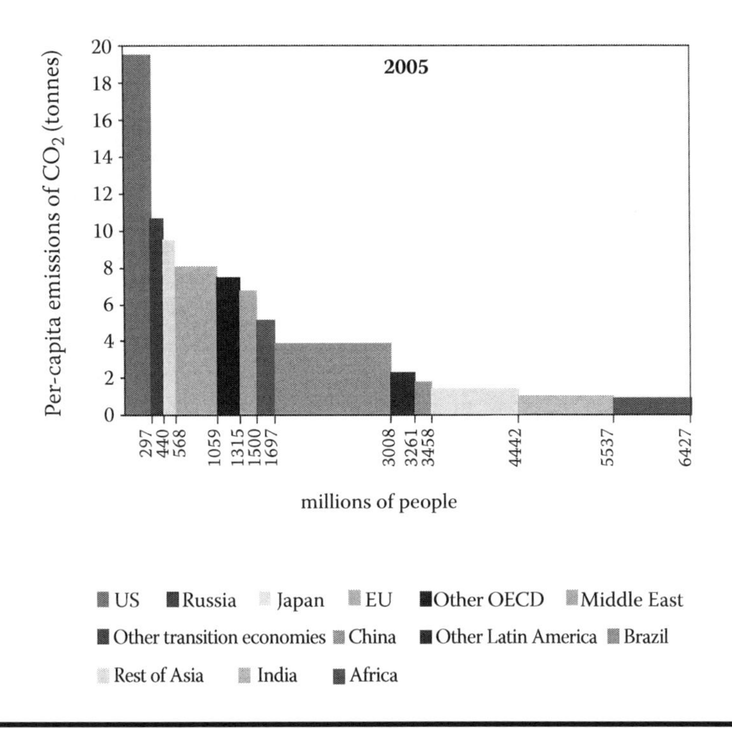

Figure 2.6 Per capita CO_2 emissions, 2005. (From IEA [International Energy Agency]. 2007. *World energy assessment 2007.* Paris: Organization for Economic Cooperation and Development.)

Thinking in Wedges

One of the basic tasks of the policy analyst is to help policy makers frame the magnitude of the problem they are considering and consider the range of feasible solutions. In the climate change arena, one of the most helpful models was provided by Pacala and Socolow (2004) through their concept of "stabilization wedges." Their assumptions: we need to limit total global carbon emissions to about 7 gigatons per year to meet a goal of a total atmospheric CO_2 concentration of 500±50 ppm by 2054. Since global emissions under the business as usual (BAU) scenario are likely to double by that time, a sensible policy goal is to eliminate that "stabilization triangle" between a flat emissions trajectory and the steeply rising BAU scenario. A total of seven "wedges" of 1 gigaton of carbon each will be needed of mitigation activities that begin at 0 and increase to a cumulative total of 25 GtC by 2054.

Pacala and Socolow suggest, "Wedges can be achieved from energy efficiency, from the decarbonization of the supply of electricity and fuels ... and from biological storage in forests and soils" (2004, 969). Their article suggests 15 possible wedges from which at least 7 would need to be selected. And if emissions continue at a higher than anticipated rate, additional wedges would be needed. A summary of their 15 potential wedges was presented in Table 2.3.

Thinking in wedges has become a common practice in the discussion of climate change mitigation strategies, but the approach has several limitations. The latest data suggest that to limit total warming to an additional 2 degrees, global carbon emissions need to be stabilized at 5 GtC per year to limit CO_2 concentrations to 450 ppm. This means a much more rapid implementation of the wedge strategies anticipated, and that additional wedges would be needed (Romm 2008).

The abstract for their paper begins, "Humanity already possesses the fundamental scientific, technical, and industrial know-how to solve the carbon and climate problem for the next half-century." This is a questionable assumption. As of early 2010 there is still (for example) no functioning example of a carbon capture and storage (CCS) system working effectively in a coal-fired power plant. Developing the technology for many of these mitigation strategies will require political will, capital, subsidization, and considerable investment in R&D. And despite such investments, some of their proposed mitigation strategies may not be technically feasible.

There is also a symbol noticeably absent from their paper: $. The wedge approach has been very effective at the important task of helping to define climate change as a difficult but solvable public problem. Expecting the authors to also attach cost estimates to their stabilization strategies, which is a separate and huge analytical task, is unfair. But considerations of cost are unavoidable in the policy process. Policy makers want to know the magnitude of costs that various policy alternatives will generate, who will bear them, and the implications of each alternative for the state of the public purse and for the state of the economy.

The Stern Review and Its Aftermath

The debate about the costs of climate change mitigation intensified with the release of *The Stern Review on the Economics of Climate Change* in October 2006. Nicholas Stern, an economist with the British Treasury, was assigned in 2005 the task of reviewing the economics of climate change by then Chancellor of the Exchequer Gordon Brown. The *Review* analyzed the evidence on climate change and the probable costs to global society, concluding that the BAU scenario would reduce global per capita consumption by between 5 and 20 percent, with the higher estimate more likely. It suggested that rapid action is needed to achieve stabilization of GHG at between 500 and 550 ppm CO_2 and estimated the cost of such action at around 1 percent of GDP per year (Stern et al. 2006).

The criticisms of the *Review* were mostly respectful, although some criticism focused on its relative one-sidedness (Nordhaus 2008). If you read it carefully, it can be easily perceived as an advocacy analysis. This reflects its emphasis on findings and studies that were consistent with its perspective, while omitting others more skeptical of the need for rapid and costly action on climate change mitigation. Still, the debate over the approach of the study and its recommendations has clarified challenges in the analysis of climate change impacts that continue to shape the policy debate today.

The Social Cost of Carbon

One of these challenges is the ideal carbon price, or using economist's lingo, the social cost of carbon, that policy makers should aim for in order to force the nonmarket effects of carbon emissions to be internalized into economic decisions. Effectively, this is a tax rate that should reflect our beliefs about the present value of damage caused by an additional ton of carbon emissions. In order to make quick headway on reducing CO_2 emissions, Stern et al. suggest a high carbon price, of about $85/ton of CO_2 or $314/ton of carbon (Stern et al. 2006).

Tol (2008a) found that of the many analyses of this issue, this figure is an outlier, up to ten times the preferred carbon price suggested by other analysts. This tax rate has immediate consequences. For example, Stern implies that the cost of gasoline in the United States should increase by over 90 cents per gallon. One proposal for a "starter" carbon tax of $37/ton carbon would only raise prices by about 11 cents. To rightly emphasize that although the *Review*'s figure is an outlier, it is not necessarily wrong—even if correct for the wrong reasons.

Why the Discrepancy?

Why the wide discrepancy? Climate change is to a significant degree a problem of intergenerational equity (Dasgupta et al. 2006). Climate change mitigation imposes costs on individuals and society today, for benefits that accrue primarily to future generations. Economic models that attempt to assess the value of actions today that generate streams of benefits (or avoided costs) in the future, typically discount those benefits and costs, using the *pure rate of time preference.*

If we are concerned about the possibility of significant harm resulting from a course of action, even if that occurs 50 or more years in the future, we will choose a very low discount rate to make clear that significant costs today are worthwhile to raise the likelihood of avoiding that harm. But those scarce resources could also be allocated to investments—both public and private—with a high return and make future societies much better off.

If we are somewhat more concerned about this lost opportunity and less concerned about future damages, we will choose a higher discount rate that—in this case—forces us to do less emission reduction today, but probably leaves future generations richer and better able to adapt to a warmer world.

Discounting is important in part because of its power. Even at relatively low discount rates, large damages 100 years or more from now have a drastically lower present value. For example: at a 5 percent annual interest rate, $1,000 of damages incurred 100 years from now is worth $7.60 today.

The Ethics Question

For many observers, this is an ethical issue: we have a responsibility to bequeath to future generations an environment that we have not further degraded. Padilla (2004, 536) suggests:

> The problem is to deal with uncontrolled emissions growth in ways that recognize the rights of future generations so that their ecological and socio-economic system does not deteriorate further in relation to the one we enjoy. This is not to 'give' anything to future generations, but rather to stop taking away something to which, from the sustainable development perspective, they are entitled.

The Equity Question

There is no guarantee that allocating more resources to growth now will generate higher levels of overall wealth in the future, if the value of natural systems is taken into account. This is also an equity issue since climate change impacts are projected to have more extreme impacts on poorer, weaker regions. And although growth creates wealth and makes people and societies wealthier, we could not use that new wealth to, in effect, "buy" a new ecosystem if our growth has degraded it beyond repair.

The Carbon Price Debate

The ongoing debate about carbon prices and discount rates is heavily influenced by the country's views about the uncertainty of many climate change phenomena. And there are many uncertainties.

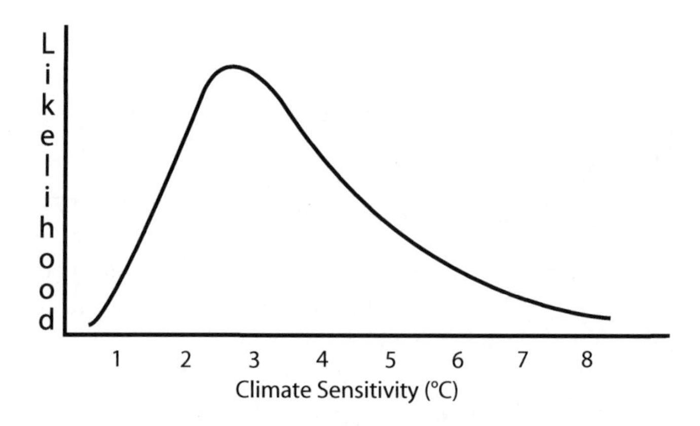

Figure 2.7 The "fat tail" of climate change costs. (Adapted from IPCC physical science report, 2007.)

Some of the questions in the minds of policy makers are: How rapidly will temperatures rise? What are the temperature thresholds at which catastrophic effects, including the possible collapse of the ocean "conveyor belt" or thermohaline circulation? What damage will result from melting of the permafrost and release of vast quantities of methane gas occur? Do we have any understanding of the feedback loops between the Earth's albedo (surface reflectivity, which decreases as warming temperatures melt sea ice), ice formation, and water vapor that occur at higher temperatures? Do we even have a way to predict what will happen once global temperatures rise by more than 4°C? And what are the implications of these climate changes for civilization?

Weitzman (2009) argues that the probability distribution of predicted effects of climate change does not look like a normal curve with tails on each side quickly approaching zero, but instead has a "fat tail," as shown in Figure 2.7. There is an uncomfortable probability of impacts that have a very high cost. Our knowledge of this uncertainty suggests that we should mitigate that risk through more decisive action than the results of other cost–benefit studies suggest.

The consensus estimate of the effects of climate change is about a 3°C increase in temperature. But the right tail of the distribution suggests a significant risk of a much greater increase, with correspondingly more serious impacts.

Arcane Issues with Global Ramifications

These debates over seemingly arcane issues have very serious real-world ramifications. The benefit–cost analyses and integrated assessment models (IAMs) that combine (mostly) marginal cost economics with scientific analysis to estimate best carbon prices and the timing of implementation generally suggest that the best policy mix will start with relatively low carbon prices, and increase them over time (see Nordhaus 2008 for a clear example). These models acknowledge that considerable damage from climate change will occur, but their assumptions (including positive rates of time preference) lead to the result that the costs of mitigation would exceed the value of avoided environmental and social damage. In other words, from this perspective, high-cost greenhouse gas mitigation strategies are not economically justified. We are better off encouraging economic growth through the burning of fossil fuels, and using the wealth we create to both adapt to climate change and to compensate those harmed by the continued use of fossil fuels.

This analytical approach and vocabulary is the *lingua franca* of economists and is common in the policy arena, but leads to results that are deeply troubling from an ethical perspective. Measuring "costs" and "benefits" using monetary metrics is an inevitably reductionist exercise that omits critical but difficult-to-measure phenomena. In the case of climate change, we can only hazily estimate the value of the oceans, land ecosystems, and biodiversity that may be irrevocably harmed by a warming climate. The "go for growth" option also elides the moral question: should we have the right to make this choice? Many authors have expressed similar reservations about the use of cost–benefit analysis to analyze value-laden policy choices in the health and environmental arenas.

In a useful series of papers, Tol (2008a, 2008b) and Weitzman (2009) carefully analyzed portions of the economic literature on climate change; their conclusions help to clarify the dimensions of the climate change problem and the appropriate range of policy choices. Tol performed a meta-analysis of 211 studies that used a variety of economic approaches to estimate a preferred social cost of carbon (2008a). He concluded that even conservative assumptions lead to a positive social cost of carbon. In other words, there is wide agreement that creating a carbon price is economically justified in response to the threat of climate change.

How High to Price Carbon?

Tol's (2008b, 439) extensive literature review of studies estimating the economic impacts of climate change reaches a skeptical conclusion regarding the usefulness of cost–benefit analysis to answer that question: "the policy suggested by cost-benefit analysis—emission reduction, but not enough to stabilize emissions let alone concentrations—is intuitively wrong. It cannot be the case that the best policy is to let the world get warmer and warmer and warmer still."

Weitzman's concern about the "fat tails" of climate change and the challenge of uncertainty suggests that stronger policy measures are appropriate now. This includes more rapid action on the potential threat of climate change, and attempts to reduce uncertainty about the effects of extreme climate change through increased research. Tol and Weitzman also suggest that it would be wise to pour research resources into geo-engineering, the use of technology to remove emissions from the atmosphere, since it may be essential to have a quick-to-deploy method of reducing global temperatures.

Conclusion: Climate Implications for Energy Policy

Crafting public policy by definition requires action in the face of a future that is literally unpredictable. The policy analysis literature distinguishes between risk and uncertainty, with risks being those future events whose probability we can estimate with some confidence. Policy analysts attempt to help policy makers by carefully examining those probabilities and placing boundaries around the extent of the problem and the likely contribution of public action to resolving it. Uncertainty, however, suggests a limited capacity to even make an educated guess about the probability of some event or problem.

One of the ironies of the climate change problem is that uncertainty about possible catastrophic effects is in itself a strong rationale for immediate action. Since roughly 75 percent of greenhouse gas emissions in the United States are energy related with power plants, factories, and transportation the largest emitters, we have no choice but to accelerate the transformation of our energy systems to lower their dependence on high-carbon sources. As our earlier discussion suggests, an important and interrelated issue is the inequitable distribution of energy, both within the United States and across the globe. National action is essential but will not be sufficient. The next chapter will examine the public policy environment in the United States, which shapes what problems are recognized as worthy of resolution, and the range of possible responses.

Public policy is a reflection of particular attitudes, intentions, and actions of policy makers that affect society; examples include industrial policy, education policy, welfare policy, public safety policy, environmental policy, energy policy, water policy, and the like. The results of public policy on the energy industry are exemplified in the character of the laws enacted, regulatory actions taken, court decisions handed down, and the behaviors and attitudes expressed by legislatures and the public on energy industry operations and issues. These policies seldom if ever remain permanent; they are always subject to change.

Chapter 3

The Art and Science of Crafting Public Policy

> Politics and policy making is mostly a matter of persuasion. Decide, choose, legislate as they will, policy makers must cater people with them, if their determinations are to have the full force of policy. … To make policy in a way that makes it stick, policy makers cannot merely issue edicts. They need to persuade the people who must follow their edicts if those are to become general public practice.
>
> **—Robert E. Goodin, Martin Rein, and Michael Moran (2006)**

On Presidents Day 2010, Washington State's capitol complex in Olympia was a busy place. Two groups of citizens took advantage of their right to assemble to gather on the steps of the neoclassical Legislative Building and make known their deep concerns about the state's plans to cope with its $2.8 billion 2011 budget deficit. The first group, organized by antitax forces affiliated with the so-called Tea Party movement, aired their opposition to proposed tax increases and threatened protax legislators with electoral Armageddon. A group with the opposite message then took over the same location, arguing for tax increases to avoid deep cuts in the social and health services programs and education. Legislators caught in this electoral pincer faced a dilemma: how to eliminate the deficit, as required by the state constitution, without alienating a wide swath of the state's voters. This brief example illustrates the difficulties of policy making in a democracy.

Policy Making in Action

The public policy process is the expression of the political system in action. Citizens and other stakeholders experience frustrations, perceive needs, and seek satisfaction through collective action and the formal political system, aiming to bend that system in their direction. Change in democratic systems is rarely easy, however. Like many institutions, political systems suffer from inertia; there is no political equivalent to the Second Law of Thermodynamics. Once changed, policies tend to

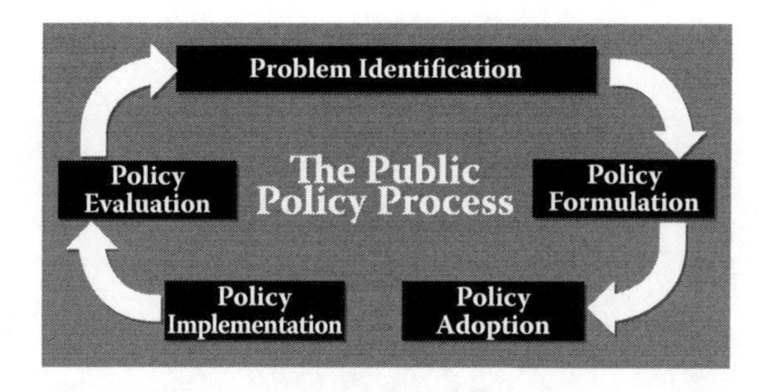

Figure 3.1 The stages model of the public policy process. (From Mott 2011.)

stay changed, often until elected politicians sense voters stampeding in the opposite direction or new conditions force the system to adapt.

Beyond this broad description, the public policy process analyst faces a formidable task. There is no satisfactory and consistently applicable explanation of the nature of the political system itself, or of the policy process in particular. For every evocative metaphor or model crafted to explain how policy is made, there are cases that exemplify it and cases that contradict it. This conundrum leaves those that wish to bring about change in a quandary; it implies that no single strategy is likely to be effective consistently as a means of accomplishing your policy goals.

The most common depiction of the public policy process in the United States is the *stages* model or heuristic, as shown in Figure 3.1. This begins with agenda setting for the appropriate level of government, and proceeds through formulation of the policy, a legislative decision to proceed with policy change, concurrence by the executive branch, implementation of the new policy, and evaluation of the resulting program. Part of the appeal of this model is both its simplicity and the fact that it mirrors the rational model of policy analysis. In this rational model, policy makers acknowledge a problem, seek a range of possible solutions, select a decision rule for choosing between those solutions, select the "best" solution, implement it, and then evaluate the results.

A model, of course, is only a representation of reality, not reality itself. These interrelated models provide a superficially helpful approach to thinking about an often opaque and complex process. Applications of the stages model often take the form of the story of "One Big Bill" as it winds its way through a dynamic legislature (or dies in the process) and is ultimately implemented by public agencies. Consider what this omits. How, exactly, do issues of concern land on the agenda? How do policies make the leap between stages, and what happens within each stage? How is power conceptualized in the model? Sabatier (2007) and others argue that the model obscures more than it clarifies.

The most glaring weakness of the stages approach is the lack of clarity about its context. It seems to assume a loop model of democracy and a pluralist system of competing groups of political actors. In the loop model (see Miller and Fox 2007), voters self-aware of their policy preferences vote for candidates that offer the best mix of policy proposals, coalitions of winning candidates pass laws enacting those preferences, and in the next electoral cycle, voters evaluate their performance.

This is a problem, for as Miller and Fox argue, "As an acceptable model of governance, orthodoxy [the loop model] is dead." Of course, the loop model is not at all dead. In the United States, we have a federal constitutional republic, a model of government that retains power in the hands of citizens but relies on elected representatives at the national, state, and local levels to make collective decisions. Whether that political model is still working acceptably is a deeper question. Since the 2008 national elections, there are both encouraging and discouraging signs. But to gain a deeper understanding of the policy process, a much broader consideration of our political community and political culture are needed. We need answers to such questions as these: Is this system capable of taking on the challenge of climate change and a transformed energy sector? Is it still a model of political community that citizens want to live in? What changes are possible?

The Evolution of U.S. Political Culture

The dominant political tradition in the United States is classical liberalism. Best described by English political theorist John Locke in his *Second Treatise on Government*, the liberal tradition emphasizes the primacy of the individual as a person with a set of natural rights, to life, liberty, and property. Men, Locke argued, lived in a free, but anarchic state of nature. Without a strong government that freedom was constantly endangered, and "the enjoyment of the property he has in this state is very unsafe, very unsecure."

Locke envisioned a state in which individuals could work together in a sphere outside of government, which he termed *civil society*, and through a limited government that would protect individual liberty and individual rights, and encourage civility between citizens. Ideally, this would protect individuals from both the anarchic dangers of the state of nature, and the very real dangers of an overly strong monarchic government.

In the economic sphere, Adam Smith described in his *The Wealth of Nations* a similar, limited state that would protect individuals in their role as economic self-maximizers, pursuing their self-interests in a competitive marketplace. Smith countered criticism that an economy and society based on such a rule would lead to chaos with his famous metaphor of the "invisible hand," arguing that free markets would work to produce a rich variety and amount of goods that would meet society's needs. Smith was a more nuanced thinker than many of his later admirers—including the members of the so-called Austrian School of economics—will admit: he acknowledged the need for government action under a variety of circumstances.

The psychologist Carl Jung suggested that people display a variety of dominant characteristics, but also a have a Shadow—those characteristics that are repressed, unconscious, and often denied. In the political realm, the liberal model also has a shadow: the human need for community, a collective approach to life whose advocates view with alarm the implications of a society based on the atomized, self-aggrandizing individual. Such a communitarian model is stronger than the notion of civil society suggested by Locke. From this perspective, the basic societal unit is the group, not the individual. Decisions about the goals of society and dominant values would be based on a collective, more egalitarian approach that sought to advance the interests of the collective, and also sought to provide more protection to individuals.

This tension between individualism and collectivism is a basic fault line in many political cultures and has dominated our political evolution. There are other important variables useful in considering the evolution of a political culture, including respect for tradition versus the role of science, and unity and centralization versus fragmentation, which we will discuss below. But in the United States, tension over the basic unit of society, and our obligations to those individuals and communities, continues to influence the design of our institutions and basic policy debates. Box 3.1 illustrates this conflict.

BOX 3.1 NIMBY CONFLICT OVER A BIOMASS DEVELOPMENT

Mason County, Washington, has always been a lumber town. Oakland Bay, located at the end of a narrow southwestern arm of Puget Sound, has for many years depended upon private timberlands and nearby national forests for its livelihood. At one time, sawmills were lined up side by side on the waterfront. Fir, hemlock, and cedar in seemingly endless quantities came out of the local forests to be cut into the lumber that helped build the West. Today, just one mill and a tugboat towing log rafts to Tacoma are all that remain of that not so distant past. The nearby woods are still logged, however, and tons of wood waste is still left behind, which must be openly burned each winter. The local area produces an estimated 1,430,000 dry tons of woody biomass from forest harvesting, mill by-products, and thinning each year. Mason County has one of the highest rates of unemployment in the state. That's why people's reaction to an announcement that a new biomass-burning power plant was to be built that would bring something like 750 construction jobs and 200 permanent jobs to the community came as surprise to country administrators.

Biomass is one of the renewable energy processes the Department of Energy strongly supports; among others are hydropower, wind, and geothermal energy. The proposed $250 million, 60-megawatt biomass power plant is one of several that are planned across the United States by Adage, a Maryland-based joint venture between North Carolina's Duke Energy and Areva, the giant French atomic power service company. The plant fits neatly into the national plan requiring all utilities to get at least 15 percent of their electricity from renewable sources by 2020.

Local environmentalists objected vehemently to the proposal. One spokesperson at a public hearing pointed out that the plant, which will consume some about 604,000 tons of wood waste annually, will annually emit an estimated 240 tons of nitrous oxide, 149.4 tons of sulfur dioxide, and 548,480 tons of carbon dioxide, along with large amounts of fly ash. A mailed flyer produced by John Deere, the company that manufactures the forest waste harvesters, notes that the facility will meet or exceed all state and federal regulations for emissions, will use air rather than water for facility cooling, and that controlled burning of forest and mill residue rather than open-air burning will improve the region's air quality.

Project opponents started a recall petition drive against county commissioners; the drive was inconclusive at the time of this writing.

(From K. Moore, *Mason County Journal*, 2010, and John Deere Biopower Facts, 2010.)

A New Conception of the State

The United States enjoyed an unprecedented era of growth between the late 1940s and 1973. During this period a new conception of the state began to develop in the United States that reflected a more inclusive political system and a more expansive role for government in attacking economic inequality. Groups traditionally excluded from the political process, particularly women and African Americans, demanded representation and civil rights, and won significant victories through the Civil Rights Act of 1964 and the Voting Rights Act of 1965. President Lyndon Johnson, aided by unusually large Democratic majorities in Congress in the mid-1960s, pushed through a series of Great Society programs in education, health, transportation, and the environment. The most important of these were health care programs for seniors (Medicare) and low-income people and children (Medicaid).

The American social model relies on strong economic growth to drive an expanding sense of economic and social opportunity. This model hit the rocks following the 1973 Arab Oil Embargo, for reasons that are still not entirely clear. Economic productivity decreased, and the long-run growth in consumption dropped in half, from around 3 percent per year during the period 1950 to 1973 to around half that afterward. The resulting economic turmoil set the stage for the so-called Reagan Revolution.

President Ronald Reagan defined government as the problem and sought to reduce federal expenditures, deregulate business, privatize where possible, and pull back the size and scope of the state. He was successful at having a cowed Democratic Congress cut taxes. However, his relative failure at reducing expenditures led to massive deficits that persisted until tax increases under the Clinton Administration and the dot-com boom of the late 1990s. This "neoliberal" approach, combined with U.S. openness to trade in a smaller, more globalized world, helped to reverse the gains in economic equality of the early postwar period and brought a new era of increasing inequality.

Emerging Themes

Several themes emerge from this summary of the evolving role of the federal government in the country's economic and social life. The U.S. political culture has emphasized individual initiative and a smaller government, and has limited the role of public policy. This is also evident when the U.S. public sector is compared with other "developed" countries, most of which have a significantly larger government sector. The United States, with roughly 32 percent of gross domestic product (GDP) devoted to government, has a much smaller public sector than most European countries, where it is common to devote 40 to 50 percent of national GDP to government. Government expenditures in the United States have expanded since the 1960s, and their composition has changed, as shown in Table 3.1. In 1960, military and international expenditures totaled 9.9 percent of federal gross domestic product, while entitlement programs represented only 4.7 percent. These percentages have shifted dramatically, with defense and international reflecting only 4.5 percent of GDP and entitlements increased to 12.8 percent of GDP in 2008.

Congress and Its Influence on Policy

Congress no longer commands the respect that it once did. It has evolved into a more partisan body with weaker leadership, making it more difficult to pass major legislation. The Senate is now a collection of individuals with ties to their party caucus but generally weak commitment to the institution itself and its august history as the "World's Greatest Deliberative Body." In

Table 3.1 Federal Government Expenditures by Major Category, as Percent of GDP

Year	*Total Gov*	*Defense and International*	*Entitlements*	*Other Federal*	*Interest on the Debt*
1960	26.2	9.9	4.7	1.9	1.5
2008	33.0	4.5	12.8	1.8	1.8

Source: U.S. Office of Management and Budget

recent years the filibuster has become a standard mechanism for obstructing the agenda of the party in power, rather than a means of preventing the passage of particularly egregious legislation. Senators regularly place "holds" on nominees for judicial and administrative posts, often as an exercise in political pique over ideological or policy differences, not over concern about the candidate's qualifications.

The gradual decline of the House began as Democrats held a majority in the House from 1954 to 1995. During the early years of that period, the institution was under the grip of both seniority rule and conservative Democrats from the South. The post-Watergate elections brought elections of more progressive members, who subsequently pushed through reforms to strengthen ethics standards and weaken the seniority system. The election of a Republican majority in 1994 moved the House into a more combative zone under Speaker Newt Gingrich, who sought to expand Republican control of the institution. Although Republicans held the House until 2006, their legislative accomplishments were meager until the election of George W. Bush as president in 2000. The 1994 reform of welfare was a major exception, when center-right President Bill Clinton was willing to endorse a bill that ended welfare as an entitlement.

The Battle for Control

The separation of powers in the Constitution created a continuing, structural battle between the president and Congress for control of the federal government. Through much of the nineteenth century, Congress dominated, with the exception of the Civil War presidency of Abraham Lincoln. The Progressive Era began the decisive shift toward a strong president, which was solidified with the Depression and the rise of the administrative state, which required a strong manager in chief. A hallmark of the postwar period was the rise of the national security state, with the Cold War against the Soviet Union from 1949 to 1993, major wars in Vietnam, Iraq, and Afghanistan, and the 9/11 terror attacks all contributing to the perception that a strong leader was needed to protect the United States from a chaotic world. Although the Congress on occasion attempted to restrain the presidency through, for example, use of committees to provide oversight over government operations, budget reforms and the War Powers Act of 1973, these proved ineffective at blunting the long-term trend toward a strong executive.

The Bush II era may prove to be the pinnacle of presidential power in the early decades of the twenty-first century. Bush showed considerable skill during his two terms at pushing through his policy agenda of tax cuts, wars in Afghanistan and Iraq, expansion of Medicare to include prescription drugs, a bill revising the bankruptcy code, education reform, creation of a Department of Homeland Security, and two major energy bills in 2005 and 2007. It is a curious record that reflects no particular ideology except enhancing presidential power and the interests of corporate elites. On balance, the Bush Era considerably expanded the scope and reach of the federal government, while undermining its fiscal soundness through tax cuts that mainly benefited a select few. In contrast, the Obama Administration consistently has faced strong resistance from Congress, even before the election of a Republican House of Representatives in November 2010. An intense battle over the direction of public policy will continue.

Pluralism, Elites, Triangles, and Networks

Producing an energy policy that meets security and environmental criteria begins with answering these questions: Who has decision-making power in this system of government? How does the political system acknowledge needs and concerns, and respond to them? To whom is the government more likely to respond? These questions are broader than the narrower workings of the policy process, but are also essential to it. The huge scale of the United States and its states and limited role of the citizen in a representative democracy leaves the individual citizen capable, in most cases, of little impact on important policy processes at either the state or national levels.

Beginning in the 1950s, political scientist Robert Dahl and others argued for a pluralistic or group model of political decision making. That model emphasized the competition between groups of relatively equal power in the political arena, and the policy outcomes that result from the bargaining between these groups. The pluralist model largely ignores the differences in resources and power between groups and the reality that large economic entities, especially corporations, have a considerable advantage in the policy bargaining process. Their ability to control investment and employment gives their requests more weight in Washington and the state capitals, although it may not be enough when voters feel betrayed by the corporate sector, as they have in the aftermath of the financial crash of 2008.

The Role of Technical Experts in Policy Making

A related issue in policy making is the role of elites and experts. The complexity of government action in a legalistic and highly technological society limits most citizens from acquiring the knowledge needed to become expert in policy making. Mastering the details of even one issue area is a significant achievement. This works to entrench the influence of experts within legislatures, bureaucracies, and in the burgeoning think-tank sector—elites with education and the resources to gather data and present it so as to support a preferred position.

Shifting Models of Government Power

In the 1970s and 1980s, the dominant metaphor for the relationship among Congress, the bureaucracy, and interest groups was that of the Iron Triangle, as shown in Figure 3.2. This model acknowledged the role of interest groups in policy development and decisions and suggested that the type of linkages among these three parties influenced policy outcomes.

The key to this set of relationships was the recognition that the staff persons in federal agencies were not simply passive recipients of congressional policy judgments, but instead were active constructors of their own policy reality. Government workers are responsive to members of Congress, but they also aim to shape perceptions of their performance and needs, and to equate what benefits them as being in the public interest.

The triangle model also suggests that boundaries between the private interest groups and public employees may blur; this is encouraged by the cycling of employees between the three corners of the triangle. The most powerful such triangle is perhaps that in the broad arena of national defense, which provides a powerful floor under all federal expenditures. It is the rare member of Congress who can afford to be perceived as weak on defense.

A New Approach: The Policy Network

Since 2000, the metaphor of the policy network has gained growing acceptance among analysts. The Iron Triangle metaphorically looks like a closed system with unitary actors at each corner. The network metaphor, on the other hand, suggests a profusion of different individuals and groups,

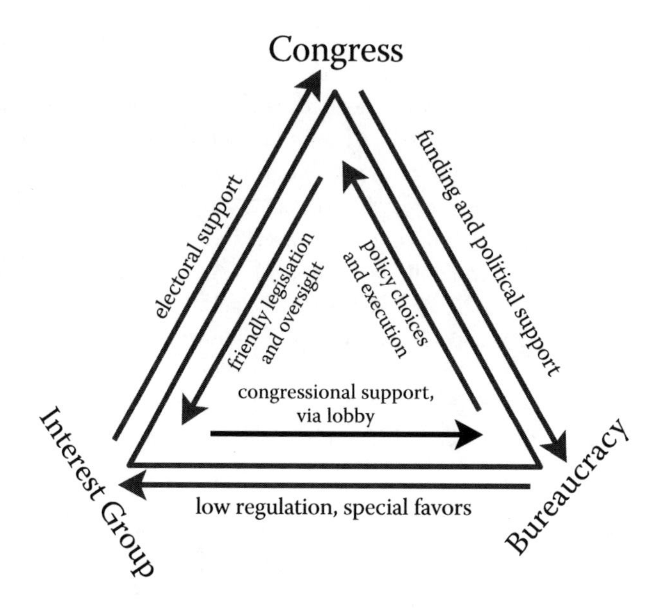

Figure 3.2 The Iron Triangle among Congress, the bureaucracy, and interest groups. (From Saylor Foundation 2010.)

each of which attempts to influence the nature of the ongoing conversation within the network. All actors have an interest in the outcome, monitor the actions of the other nodes in the network, and seek to join with other actors to advance the cause of their particular stake in the issue. Adam and Kriesi (2007) point out that this way of conceptualizing a policy space creates an opportunity to *manage* the network. It is not a passive collection of actors, or groups of actors, but a set of entities open to persuasion and coordinated action.

Power in the network may be concentrated or diffused. Concentrated power may delay or encourage change, depending on the policy stance of the powerful actors. A recent example is the status of costly weapons systems, which have long been viewed as immune to reduction or elimination as their production is distributed widely across the country and creates thousands of jobs. But Defense Secretary Robert Gates, with the support of President Obama, supports outright cancellation of additional procurement of the C-17 cargo plane and F-22 fighter.

Two Additional Policy Models

The quest to provide useful models of the policy process has led scholars in several additional directions. Two of these are particularly useful for the consideration of energy policy: the multiple streams framework, and the application of social construction to policy design. These models are useful in that they illuminate different aspects of the policy process, and provide frameworks that will be helpful as we consider the history of energy policy in the United States in later chapters.

The Multiple Streams Framework

Straightforward descriptions of the policy process give the appearance of logic and rationality. Yet John Kingdon (2003) in a classic study found in interviews with policy experts and politicians that the reality they described was ambiguous and anarchic; the policy arena was thus filled with uncertainty. To make sense of this puzzle, Kingdon deftly borrowed a model termed the "garbage can" from the organizational theorists Cohen, March, and Olsen (1972) that attempted to make sense of this chaos. Their critical insight was that organizations and decision processes were not orderly but could best be described as "organized anarchies."

Problems are separate from solutions. Participants in the policy arena come and go; they are often cagey about revealing their preferences. Many solutions are proposed during negotiations; most are discarded, but they do not disappear. All wait in the "can" for the right situation. Kingdon simplified the model for the policy arena, suggesting that three streams of problems, policies/solutions, and politics/elected officials/public sentiment flowed independently, until a policy window opened that enabled a policy entrepreneur to bring the streams together, resulting in enactment of a policy. This model is particularly helpful at conceptualizing how some issues are suddenly on the agenda and ripe for action, and many others are not.

The vocabulary of the multiple streams model is now part of political discourse. We can see policy entrepreneurs at work and are more likely to be able to spot when a policy window opens, as it did for health care reform following the 2008 election. An open question in the energy policy arena is whether, as we write in early 2010, the policy window is truly open for transformational climate change legislation.

Policy Design and Social Construction

Where the Kingdon model emphasizes agenda setting, the social construction perspective starts with policy designs—policies that have been crafted, implemented, and that have had an impact on their policy setting and on the target groups they are intended to benefit. Much of the literature in public administration describes these as *interventions* or *programs*, but they are much more powerful than those flat terms suggest. Policy designs influence both the target groups intended to be helped, and the institutions and culture as a whole. This influence occurs both through the "objective" impacts of the policies as intended and through the rhetoric used to describe these target groups.

Some groups are identified as more worthy of support than others. These groups are given both more benefits and positive messages about their impact on society. This in turn encourages their participation in the political process, in part to protect those gains. In the energy world, no group has benefited more from this process than the oil industry. The oil industry is typically characterized as entrepreneurial, dynamic, and risk taking. Counterintuitively, this industry as a whole receives massive yet largely hidden public subsidies.

For individuals unfortunate enough to be in a negatively constructed target group, particularly low-income people and families receiving public benefits, the drumbeat of negative messages about their basic unworthiness is a significant deterrent to their willingness to participate in the political process. Ingram, Schneider, and deLeon (2007) suggest that these messages convey who belongs to which group, whose interests are more important, what kind of "game" politics is, and whether the person has a place at the negotiating table. Their model identifies several categories of groups, the most important of which are: (1) the *Advantaged*, groups with high political power and positive social constructions; (2) *Contender* groups with high power but negative constructions; (3) *Dependents*, groups that are weak politically but viewed as deserving; and (4) *Deviants*, groups that are typically blamed for the social ills of society and face strong sanctions for their transgressions.

Perhaps the most intriguing of these groups are the Contenders, including corporations, who adroitly use their power to gain benefits that are often hidden from the public and receive little scrutiny. They are also often relieved of burdens, such as environmental regulations, that may be fought in court or priced into their products. Many of the laws that provide the legal framework of U.S. energy policy include provisions for tax expenditures and subsidies that receive less attention than funded programs. They reflect revenues foregone by the U.S. Treasury. Few members of the public have a clue about the magnitude of these lost revenues and the corporations that receive them have no incentive to publicize them.

The model suggests that policy makers craft and perpetuate these social constructions because they believe the public will support them. As President Obama has found with the massive bailout of the financial sector in 2009, the public is not sympathetic to policy designs that benefit the "scoundrels" believed to be responsible for a big mess.

Constructions Are Not Fixed

Yet social constructions are by definition not fixed. Much of the power of this model emanates from the capacity of politicians and pundits to change the social construction of target groups, for better or worse. For example, although scientists are generally considered Advantaged, they are vulnerable. The recent *Climategate* furor was seized by opponents of climate change legislation to depict pro-warming scientists as unethical and self-interested, only aiming to gain access to big streams of grant funds by hyping up the risks of climate change. This is where the social construction model blends with the policy "framing" perspective advocated by Goffman (1974), Schön and Rein (1994), Lakoff (2008), and others.

The System: Weakened, but Still Functioning—for Now

In their classic analysis of the Clinton administration's doomed health care reform effort, *The System*, Broder and Johnson (1996) describe a fractured political system with a divided, distrustful, and easily manipulated electorate, a political class that is alternatively cynical and public spirited, and a media sector that failed in its basic task, to fairly inform public debate on the issues. Although the forces against reform "won" that battle, the victory was pyrrhic: it undermined public trust in institutions of all kinds, and as they note (Broder & Johnson 1996, 639), "A thoroughly cynical society, deeply distrustful of its institutions and leaders and the reliability information it receives, is a society in peril of breaking apart."

Framing the Energy Policy Issue

How did the U.S. political system perform on energy issues in 2010? In 2009 and 2010, health care legislation was the signature issue; substantive reform of health care in the United States was the object of considerable executive and legislative attention during most of the first two years of the Obama administration. As a result, energy issues were pushed off center stage, with little public or legislative interest paid to the UN's 2009 Copenhagen environmental conference. Despite President Obama's personal diplomacy efforts, little progress toward controlling greenhouse gas emissions was achieved at the conference.

The 111th Congress acknowledged the seriousness of the health care issue and invested vast amounts of time, energy, and political capital in the task of crafting sensible legislation that would, over time, limit the vast disparities in the U.S. health care system. The bad news is that almost all of those supporting health care legislation were Democrats. Republicans refused to budge on the issue.

The political divide in the United States continues to widen, and partisanship has taken on a bitter edge that does not bode well for the capacity of the political system to take on the challenges of energy use and climate change. With the United States and world economy mired in a continuing economic crisis producing high unemployment and a weakened financial sector, the last years of the first decade of the twenty-first century is obviously a very difficult time to pass potentially costly legislation. The public also shows signs of disgust: recent polls found between two-thirds and three-fourths of the public are unhappy with the performance of Congress. Public trust in government is at an all-time low.

Worrisome Trends

Several trends in the broader political system in the intervening years since the collapse of President Bill Clinton's health care proposal give cause for concern. Winning election to our primary policy-making body, the U.S. Congress, is now so costly as to require virtually nonstop campaigning to pay for polling and television spots. The average cost of a House campaign in 2008 was $1.1 million, while Senate campaigns averaged $6.5 million. The fundraising required drives out capable people who are not personally wealthy or who do not have the stomach for the never-ending "ask."

Despite a succession of scandals, lobbying of Congress and the government as a whole is at an all-time high, with $3.47 billion spent on lobbying in 2009, according to the Center for Responsive Politics. The perception, right or wrong, is that members of Congress are for sale, trading positions and votes in exchange for campaign contributions. The endless parade of "earmarks," the narrow clauses attached to larger bills that are intended to benefit particular constituencies and the complexity and backroom deals needed to grease major legislation, seems to confirm the public's suspicions.

In discussing how difficult it has become to pass legislation in the public interest, Galston (2006) noted the explosion in the numbers of registered special interest associations and their lobbyists in Washington, DC, since 1955: the number of associations grew from under 5,000 to more than 20,000; membership in the Society of Association Executives grew from less than 2,500 to nearly 25,000; the number of lobbyists registered with the U.S. Senate grew from 3,000 to more than 10,000; and the number of lawyers in Washington grew from 12,000 to 76,000.

Partisanship politics is deeply entrenched in the United States. The Senate is now hostage to the filibuster, with the number of cloture votes required having more than doubled since the late 1980s. Redistricting has produced for each party in the House "safe" seats that are more likely to appeal to party extremists. The parties themselves continue to grow wealthier and stronger. With the rise of highly partisan political blogs and the need to stay exposed in the never-ending news cycle, many senators appear as political singletons, concerned first for their own reelection chances, and less for the institution. Both parties seek to exploit highly divisive cultural issues such as abortion, immigration, and gay marriage.

Noisy, news-grabbing partisan events are staged by the parties and interest groups to look like the natural mobilization of grassroots opposition, in a phenomenon called *astroturfing*. The end result is fewer members in the political center who can speak each other's language and seek common ground, and an attack mentality that aims to score quick political points, but ignores the long-term damage to Congress as an institution, and the country (Seib 2010).

Visible Lack of Consensus

This lack of consensus is visible in the constant, vicious battle to frame the nature of debate on public issues. *Frames* are "organizing principles that are socially shared and persistent over time, that work symbolically to meaningfully structure the social world" (Reece 2001, 5). At the broadest level, linguist and political analyst George Lakoff (2006) argues that there is a fundamental split between the conservative and progressive/liberal frames.

Conservatives see the world as a dangerous and competitive place in which a family-based, strict father is needed to emphasize discipline and self-reliance. Government should have the limited role of protecting the family from harm. The progressive frame is a nurturing model that emphasizes the equal contributions of women and men. From this frame there is a need for a more communitarian model that uses collective action to help individuals who can't make it on their own, largely because the system is stacked against them.

Conservatives have been far more adept than progressives at framing important political issues over the past 50 years. They have an easier task because the U.S. political system is rooted in a classic liberal model that emphasizes limited government. But examples abound of their framing skill. Who could possibly oppose such calls for action as: "American families deserve tax relief"; "A greedy government takes too much of your money and delivers too little in return"; and "Provide relief to hard-working Americans impoverished by taxes"?

The progressive frame—that we all benefit from the infrastructure and services provided by government and should be willing to support it—is a harder sell. Another recent example is voter registration. Efforts to ease voter registration are decried by one party as "opening the door to fraud," while the other party wants to "eliminate roadblocks to participation in our democratic system." Even the word *liberal*, twisted from its original meaning, has become a negative one-word frame, while the corresponding term *conservative* lacks the same bite in political discourse.

A More Divisive Nation

Related to this framing problem is the fact that there is some evidence that Americans themselves have begun to be less accommodating to political and religious diversity (Bishop 2008). There are signs that they are seeking refuge in media (radio talk shows, newspapers, and magazines), living circumstances (gated communities and ideologically concentrated neighborhoods), and faith communities that reinforce a consistent set of ideological messages, and make them less open to political compromise. In such circumstances, Bishop suggests, attempts to find common ground through dialogue with those who hold opposing views may instead lead to even more polarization. Like-minded, relatively homogenous subgroups can sharpen their differences in discussion with each other, frustrating attempts at consensus.

Subtext of the Framing Issue

The subtext of the framing issue is its acknowledgment that members of the political class manipulate voters/citizens, both through careful framing and other means. This is one of the many criticisms of the loop model (Miller and Fox 2007). Voters' wants are manipulated through the use of television and the news media. Local television news focuses on crime and sensational stories that attract viewers and in turn, advertisers. Also, citizens do not put much effort into their civic responsibility and have weak knowledge of the issues and institutions.

This rings an alarm bell for those concerned about energy and climate change, since the argument for policy change rests on a relatively complex causal argument involving energy choices, scientific findings, future risks, and the need for somewhat costly actions now to avoid possible future calamity. It is inevitable that the great clash over climate change and energy legislation will be over how the issue is framed. As Lakoff (2006) dryly notes, "when the frames don't fit the facts, the frames are kept, but the facts ignored."

A Glimmer of Optimism

We are somewhat more optimistic about the overall capacity of voters to broadly provide signals about their preferences to the political system. We are reassured by such work as Carpini and Keeter (1997), who concluded after an exhaustive analysis of voter knowledge and involvement that on balance, citizens do a reasonable job of this. Still, they found the political system problematic in many areas. For example, they found "systematic differences in political knowledge." People of color and low-income persons generally were found to be less knowledgeable and underinvolved, while wealthier and better-educated individuals and groups with power and resources were more knowledgeable and able to obtain access to the system and make their preferences known. Our energy policy is a clear reflection of the distinct differences in wealth and power between various widely divided groups.

Hard to Generate Citizen Involvement

The U.S. political and policy system simply is not designed to enable much direct citizen involvement. Will it be possible to bring about transformation in the energy sector within a political system so closed to individual citizens? Asked to share the burden of costly policies and lifestyle changes without sufficient input, many citizens may resist. An early 2009 poll on energy and environmental issues found broad support for renewable energy development, and willingness to change transportation behavior, but strong opposition to tax proposals—including a large rise in gasoline taxes or a "floor" gasoline price, which would quickly raise the cost of driving (Bittle, Rochkind, and Ott 2009).

Needed: A New Way of Thinking about Policy

A new way of thinking about energy policy is needed; one that is more grounded in people's concerns and lived experience. One approach is suggested by postmodern political theorists, who have concluded that the "modern" political and economic project supporting corporate-led economic growth through a distant government of experts and elected representatives has failed. Instead, the role of the expert is recast as a skilled facilitator of citizens, to help them engage in political discourse.

Hajer (2003, 89), for example, suggests that policy proposals from political leaders and others "make people aware of what they are attached to," and through dialogue with others—preferably those with a range of beliefs and opinions—may build a "sense of collective identity." Such conversations provide a forum, through which political communities are created and nourished, and people may reflect both on their vision of a preferred future and the policies needed to create it. Hajer provides examples of such processes in the Netherlands. King and Stivers provide similar examples in their 1998 book, *Government Is Us*.

Occurring at the Local Level

These processes occur almost exclusively at the local level. The federal government's approach to public involvement requires extensive, mostly one-way communication from citizens to the agencies about their opinion of policy proposals, but does not create a forum for respectful and thoughtful dialogue. Recent attempts to do so (including the absurdist health care policy forums held in mid-2009 that devolved into shouting matches) are fine examples of the challenge of creating respectful forums and seem to confirm Bishop's pessimistic analysis. A condition that appears necessary for effective dialogue is for individuals to have relationships across a wide swath of society, not only people from one's own "in-group" (Bishop 2008).

The knowledge that your opponent today may be part of your coalition tomorrow tempers legislators' willingness to label opponents in ways that make future cooperation impossible. We would be wise to nurture institutions and forums that encourage the formation and maintenance of such cross-cutting relationships at the local level.

The U.S. Policy Space: Today and the Near Tomorrow

Our emphasis in this chapter has been on the historical evolution of governance and policy in the United States. But it is useful to consider the situation in the broad, national policy arena at this moment, and longer-term trends that are likely to have an impact over the next 10 to 20 years on energy and climate change legislation.

Although recent polls of U.S. residents continue to show low levels of regard for the Congress, arguably the 111th Congress has been one of the most productive in the last several decades (Ornstein 2009). The vast American Recovery and Reinvestment Act of 2009 economic stimulus bill alone was an impressive legislative achievement that included tax cuts, aid to state governments, investments in transportation and health information technology, and an important array of energy provisions.

Despite these accomplishments, the Great Recession has not ended and President Obama's rhetorical skills did not win over a middle class worried about job losses and angry about his administration's support for bailouts of bankers. The slow-motion process of crafting a health care bill was also done with considerable transparency, and the public did not like what it saw. Trading benefits for particular states and legislative districts to gain votes is a standard legislative practice, but left many Americans feeling that the price of reform was too high.

A common metaphor used to describe the environment for public policy making is that of the policy *space*. How much room is available for new policies to be crafted and implemented, in the political system as a whole, or particular issue areas? Many factors influence the dimensions of this space. The power of the metaphor lies in its implicit suggestion that there are hidden factors exerting unexpected influence, which if identified, become more open to active engagement and opening. The previous discussion suggests that public attitudes about political processes are one of those factors. But there are several others that are likely to impact the capacity of the political and economic system to respond to many challenges, including those in the energy arena. We will examine a few of these below.

Meeting the Needs of an Aging Society

The United States is an aging society. We aren't alone; the entire world is aging. The United Nations predicts that the median age worldwide will increase from 29 to 38 by 2050, and that

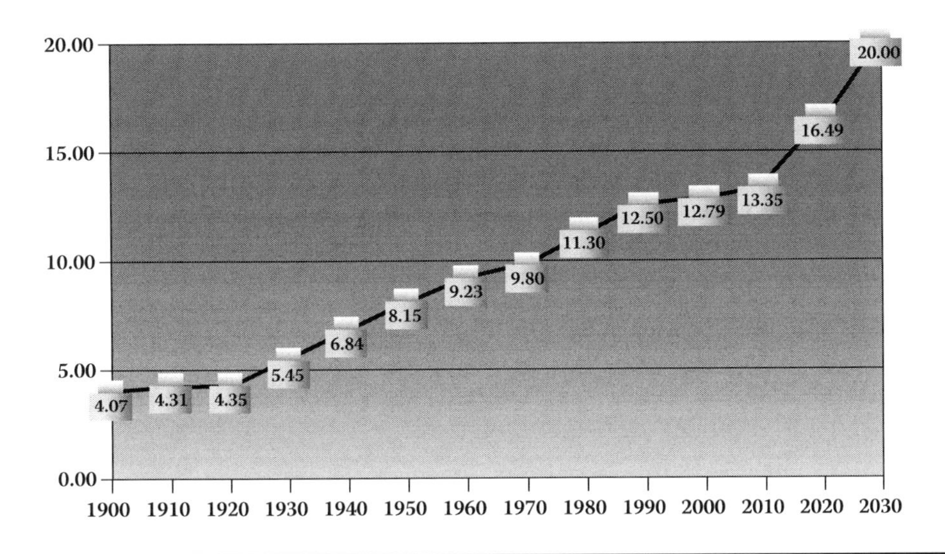

Figure 3.3 Trend in the percent of persons over age 65 in the United States, 1900–2030. (From Ageworks.com. 2010. *Demographics of an aging population: Module 2*. Ageworks.com, http://www.ageworks.com/course_demo/200/module2/module2.htm, accessed June 26, 2010.)

by 2050, 22 percent of the world population will be age 60 or over, compared to 11 percent now. In the United States, the baby boom generation is living longer and having fewer children. The number of people aged 65 years or older is expected to increase from the 40 million (12% of the U.S. population) to nearly 70 million (20% of the population) in 2030 (Ageworks.com 2010). Figure 3.3 displays the over 65 trend in the United States.

This increase in the senior population will place substantial additional burdens on the Social Security, Medicare, and Medicaid systems that currently cost about 10 percent of the U.S. GDP, or about 44 percent of the overall federal budget. Older people have more health problems.

Providing care to older people without resources or family is very costly. Part of the challenge to Social Security is the change in the dependency ratio, the number of dependents (children plus retired people) in society relative to the number of wage earners. Currently this is about 3.3 workers per beneficiary, but is expected to be about 2-to-1 by about 2045. The Social Security program is a "pay as you go" system that is projected to be financially sound until 2037; after that, Social Security revenues will cover about three-fourths of paid benefits.

Medicare is in much worse condition, with the main hospital insurance trust fund due to be exhausted in 2017. Substantial amounts of general revenues will be needed to cover Medicare costs even prior to that date. The relatively good news is that the overall cost of managing the rise in Social Security costs over the next few decades will be about 1 percent of U.S. GDP per year—a significant sum but not a backbreaker. But health care spending is projected by the Congressional Budget Office to increase from 16 percent of GDP to 37 percent by 2050 under current policy, a vast increase that would force a retrenchment of other federal government programs.

Structural Deficits and Entitlements

Overall, the current forecast is for a continuing structural deficit for the United States in future years of around 5 percent of GDP, a chronic gap between federal revenues and expenditures. The federal budget deficit increased to $1.4 trillion in 2009 and is estimated at $1.6 trillion in 2010, around 11 percent of GDP. After a brief surplus in the late 1990s, the government first slipped into deficit spending thanks to past tax cuts and increased domestic spending. Additional tax cuts, economic stimulus and other expenditures, and decreased receipts due to the Great Recession dug the hole much deeper. An immediate result: a substantial increase in the cost of debt service to the U.S. government, which will likely hit over $700 billion per year by 2015. And state and local governments are sharing the pain, with decreased tax revenues and increased expenditures on social services likely to tighten budgets for the next several years.

The implications of the country's changed financial picture for the U.S. policy space are profound. Deficits of such magnitudes are not sustainable without driving down the value of the dollar and endangering the country's economic stability. There is no guarantee that buyers in the global bond market will continue to finance such large deficits.

Political capital, political will, and presidential attention will be needed to pass the combination of tax increases, policy changes, and spending cuts needed to stabilize the budget. There will be less political capital left over to devote to other policy challenges. Although political capital is not a fixed asset, it tends to wane as administrations age. The most important of the policy changes is the reform of the U.S. health care system. As economist Henry Aaron (2009) argues, reforming the system to lower its long-run costs, while improving access to care and health outcomes, is the essential policy step required. But as we have already seen, this is a very, very difficult task.

Overall, funds for direct expenditures or tax expenditures and subsidies will be much more difficult to come by. In the energy arena, this will encourage consideration of policy instruments that use indirect means such as regulation to accomplish policy goals, or the political bravery required to increase the taxes and fees needed to finance new programs.

A Shifting U.S. Economy

Since 1995, the U.S. economy has swung through two boom and bust cycles. First was the 1995–2000 dot-com boom helped by investment in the Internet, followed by a recession deepened by the effects of the 9/11/2001 terrorist attack. Economic growth picked up in the United States in 2003, thanks to a housing boom pumped up by low interest rates and "innovative" financial strategies. With rising house prices, people felt wealthier, spent more, and saved less. U.S. savings rates actually became negative as households spent and borrowed more than they earned. Then, in 2008 the party ended. Now the country must figure out how to create the conditions for more sustainable growth, as households are saving more and spending less in the recession.

What will replace the roaring housing and financial sectors that supported the recent boom and the federal stimulus spending that helped the economy at the depths of the recession? That is anything but clear. A more balanced macroeconomic situation, with higher savings rates and a lower U.S. dollar, could encourage increased investments, exports, and support for manufacturing. In part due to this uncertainty over future sources of growth, growth forecasts for the upcoming decade are generally around 2.5 percent per year, and unemployment rates are expected to remain high.

With a relatively weak economy and high unemployment, policy makers will be exquisitely sensitive to the influence of policy proposals on the economy and on the federal budget. Much of the debate over climate change legislation will focus on how to frame its economic impacts. One side will emphasize its cost to both households and the economy, and the dangers of slowing the economy and raising unemployment; the other will tout the potential for a transformed energy sector to generate U.S. jobs through new energy investments, while minimizing its cost to households and the overall economy. How receptive will the public be to these arguments? Are policy makers right to be concerned that the United States could emulate Japan, whose tepid response to its 1980's combined housing and banking bust led to 20 years of economic stagnation? So far the U.S. policy response to its financial meltdown has been strong.

Conclusion: A Rough Terrain Ahead

The making of public policy in the United States as in any country, is usually not a pleasant spectator sport. This overview of the U.S. political system and its policy-making processes could lead one to wonder how any proposal makes it through such daunting terrain. But as Bismarck reminds us, "Politics is the art of the possible." Legislators are subject to persuasion by their peers and leaders, and make difficult judgments between their personal preferences and the needs and wants of their constituents. The expanding girth of the U.S. Code reminds us that political agreement is surprisingly possible.

The recent history of U.S. energy policy is a complex story, and not easy to characterize—some view it as farce, others as tragedy. But it has shaped the America we know today in numerous ways, and having an understanding of the goals, means, and accomplishments of these policies is essential to crafting policies appropriate for a new energy era.

Chapter 4

The Long Search for a Sustainable Energy Policy

> The most common form of human stupidity is forgetting what one is trying to do.
>
> **—Friedrich Nietzsche (quoted in G. Allison 2006)**

> Today's policy options are a product of policy choices made previously ... sometimes decades previously. ... Those earlier choices may have both a constraining, or "lock in" effect and an opportunity-enhancing effect.
>
> **—Eugene Bardach (2008)**

Although a wide variety of laws pertaining to one or more specific types of energy resources have been passed approximately every two years since the nation began to be more than casually concerned over the energy question, only three comprehensive energy policy acts have been passed since the energy crises of the 1970s: the Energy Policy Acts of 1992, 2005, and 2007. The Obama administration was expected to push for passage of a fourth policy act early after taking office until it got bogged down in efforts to revitalize the economy, but may still do so during the administration's term in office.

What does America want from an energy policy? The country's energy policy started out as a way to protect domestic producers and keep prices affordable (Farris and Sampson 1973). After the 1970's energy crises, it shifted into a program to ensure a steady, reliable, supply of oil, together with a return to energy independence. The goal of the 1970's energy policy was to never again let foreign suppliers—particularly members of the Organization of Oil Exporting Companies (OPEC)—hold the county hostage. One arm of this policy was investment in research into alternative energy and synthetic oil. As petroleum prices dropped after 1978, energy policy again shifted, this time to promoting hydrocarbon exploration, including drilling in Alaska's North Slope, and deregulation of the energy industry. In the 1990s, damage to the environment and global warming caused by the burning of fossil fuels took their place in the policy-making arena alongside renewed interest in renewable energy sources.

As the twenty-first century began, the attention of policy makers had turned to security and correcting the problems that resulted from deregulation, the terrorist attacks of 9/11, and dealing first with widespread electricity blackouts, power shortages caused by droughts, and then with economic problems. By 2007, speculation in oil resulted in rapid increases in fuel prices; energy independence again took center stage. The election of Barack Obama in 2008 brought another shift in energy policy, this time focusing heavily on research and development into alternative and renewable energy sources. This chapter looks at the history of energy policy actions and legislation since the end of World War II.

The History of U.S. Energy Policy since 1945

The history of U.S. energy policy from 1945 into the first decade of the twenty-first century can be conveniently grouped into several distinct time periods (Blackford 1988; Blackford and Kerr 1990; Clark 1990; Davis 1982; Munasinghe and Meier 1993; McNabb 2005; Thompson 1932; Vietor 1984):

1. 1945 to 1970: Managing an energy surplus
2. 1971 to 1980: Coping with energy shortages
3. 1981 to 1990: Deregulating the industry
4. 1991 to 1999: Calls for a comprehensive energy policy
5. 2000 to 2002: Linking energy and national security
6. 2003 to 2007: A comprehensive energy policy finally passes
7. 2008 to 2009: A renewed call for energy independence
8. 2010 and beyond: Energy efficiency, conservation, and the environment

1945 to 1970: Managing an Energy Surplus

In the post–World War II period of abundance and strong economic growth, the U.S. economy nearly completed its shift from a reliance on coal to the use of liquid and gas fuels. As a result, the share of the domestic energy market served by coal dropped from 43 to 22 percent. Petroleum products from domestic and foreign sources were plentiful and the price declined throughout the 1950s. Vietor described this period as one of "super-excess capacity." Research into synthetic fuels based on coal that began during the war was dropped. Huge reserves of natural gas together with demand that was only beginning to grow brought the price lower than it cost to produce and transmit. Congress did not see the need to produce an energy policy.

From 1945 until the early 1970s, America enjoyed near complete energy self-sufficiency. The United States was one of the world's largest exporters of oil; the use of coal for heating and for industrial production had been taken over by cleaner and easier-to-handle oil, and increasingly, cheap, abundant natural gas. A large quantity of American coal was, however, still exported throughout Asia and Europe, where production costs were higher or supplies tighter. A selection of some energy-related developments that occurred during this period is displayed in the timeline in Appendix B.

Natural gas was of particular interest in those regions of the country where gas was a convenient by-product of petroleum production—the American Southwest and California. It was still used largely for consumer heating and cooking or if no market existed, it was simply flared off as an inconvenient waste product.

Throughout this period America was not alone; most nations of the world did not bother with an energy policy. Supplies were everywhere thought to be endless and prices cheap for everyone; oil cost as little as $1.50 to $6 a barrel and gasoline sold for fewer than 20 cents a gallon. Five gallons or more also brought the buyer a free drinking glass or some similar premium. It seemed as if there was a gasoline station on every corner and all were engaged in a price war.

1971 to 1980: Coping with Energy Shortages

During the second period, supply and demand of all fuels remained more or less in equilibrium, with only gas prices increasing slightly. This was largely because the public policy of the time focused on protecting U.S. producers from foreign competition and price stabilization. The nation's supplies of fossil fuels seemed endless; the market was awash with cheap oil and gas and still cheaper coal. Stabilization resulted in the implementation of strict import quotas on foreign oil.

The nation's attention to energy policy questions underwent a series of shocks beginning in the 1970s when Middle Eastern suppliers first cut off oil supplies to the United States during the Arab-Israeli War, and dramatically raised the price of crude oil. During this time, the strict federal price controls on wellhead prices, which remained from the previous period when energy was plentiful, had resulted in a shortage of domestic natural gas. An even more disruptive oil shock occurred in 1978 when the oil-producing cartel OPEC again cut supplies. They demanded a larger share of production proceeds and again raised the price of oil. During both of the 1970's oil shocks, supplies of motor fuels were particularly curtailed; long lines appeared at fuel stations across the country, with most consumers limited to 10 gallons at a time. A selected list of energy-related highlights for the period is displayed in the timeline in Appendix B.

Events at the beginning of this period were a signal that things were about to change, and when they did, they would never be the same; the country's reserves of cheap energy—particularly oil—that seemed endless did, indeed, have an end. America was not out of oil and gas, but domestic supplies alone could no longer fuel the expanding economy and unprecedented growth in transportation; imports were necessary to meet the shortages.

The United States had plenty of coal, but its use declined steadily, replaced by plentiful and government price-controlled oil and gas. The future of this fossil fuel changed dramatically with the 1973 oil crisis, however. Suddenly, industrial firms and electricity power generators found almost overnight that the energy world would never be the same. They needed a ready, dependable, and relatively inexhaustible supply of cheap fuel. Coal filled that bill. By 1974 the U.S. Department of the Interior had completed its survey of the country's coal supplies and was ready to publish its findings. A detailed summary of its recommendations are presented in Table 4.1.

The gap between the domestic supplies of oil and gas and demand from the rapidly growing economy continued to widen every year. Energy shortages allowed the OPEC to raise the price of oil 1,700 percent (in 1984 dollars) during the period. Prices for natural gas and coal also rose, but far less than oil. The low price of oil reversed its decline in share of the energy market; coal would never be king again, but it was plentiful and cheap.

Energy policy in the United States began a new chapter during the administration of President Jimmy Carter (1977–1981). The major energy legislation passed during the Carter administration was the National Energy Act (NEA). Three important pieces of this energy legislation signed by President Carter in 1978 focused on natural gas: (1) the Power Plant and Industrial Fuel Use Act

Table 4.1 Selected Early Policy Recommendations for Coal

Policy Area	*Recommendations*
General Industry Recommendations	• Keep coal prices attractive enough to encourage production. • Require that all new power generating plants be built to use coal either exclusively or as a fall-back fuel (except for plants using coal gas or other fuels from fossil fuels). • Government should clarify its policies on land leasing, natural gas pricing, and other factors, including air quality. • Energy policies on expansion of energy resources should consider national, social, economic, and environmental goals and the long-term economic strength of subregions. • The U.S. Bureau of Mines should fund centers of excellence for coal mining, preparation, and mine reclamation research.
Environmental Recommendations	• Effective steps should be taken toward achieving greater compatibility between increasing energy requirements and environmental standards. • Congress should amend the Clean Air Act to determine where regulations could be relaxed.
Resource Recommendations	• Expedite leasing for mining on federal lands; encourage prompt development of new leases. • Obtain more information of the environmental impact of development of western coal resources before leasing. • Study water availability and conservation at power plants, conversion plants, and mines. • Develop coal in areas least sensitive to environmental damage.
Research and Development Recommendations	• Accelerate R&D on the reliability of stack gas scrubbers, sulfur removal before burning, surface mine reclamation, coal conversion technologies, geological research, new mining technologies, and technology to reduce emissions of nitrous oxide, sulfur dioxide, and particulates.
Mining Equipment and Supplies Recommendations	• Determine whether the Defense Production Act of 1950 gives authority to allocate raw materials to manufacturers of mining equipment to achieve the goals of independence. If not, Congress should enact such legislation.
Manpower Recommendations	• Fund research grants in mineral science, technology, and mining engineering at selected universities. • Fund development of new training techniques for coal miners.
Health and Safety Recommendations	• Periodic reviews of mine health and safety legislation for possible amendment of Coal Mine Health and Safety Act where technology indicates higher productivity or operating savings can increase supplies without sacrificing safety.

Table 4.1 Selected Early Policy Recommendations for Coal (Continued)

Policy Area	*Recommendations*
Transportation Recommendations	• Studies and surveys of short- and long-term constraints on coal shipments by rail and/or water and slurry pipeline. • Ensure rail car availability by railroads. • Require all commodity rail shipments to be at full rates for capital accumulation for equipment needs.

Source: DOI (U.S. Department of the Interior). 1974. *Federal energy administration project independence blueprint. Final task force report on coal.* Washington, DC: U.S. Department of the Interior. Report No. FE 1.18:C63.

(FUA); (2) the Natural Gas Policy Act (NGPA); and (3) the Public Utility Regulatory Policies Act (PURPA). Except for PURPA, these important gas (see Table 4.1) laws faded in importance after the era of deregulation set in under President Ronald Reagan.

Carter's chief accomplishment was establishment of the cabinet-level Department of Energy. Included in the Department of Energy Organization Act of 1977, the Federal Power Commission (FPC) was abolished and replaced with Federal Energy Regulatory Commission (FERC) with essentially the same areas of authority. FERC's main task was to begin setting wellhead prices for all natural gas, and to come up with a formula for pricing various categories of natural gas according to its source (Table 4.2).

1981 to 1990: Deregulating the Industry

The 1980s saw a shift from the highly regulated energy policy years and the energy shortages of the 1970s to deregulation, reliable supplies of cheap energy, and reliance upon market forces to control supply and demand. Ronald Reagan became president on January 20, 1981 and a little over a year later, he signed Executive Order 12287, removing federal control of crude oil and refined petroleum products. On May 24, 1982, he proposed legislation eliminating the Department of Energy (DOE) and transferring most of its responsibilities to the Department of Commerce. However, Congress failed to act on the proposal and the DOE was allowed to continue to function.

For the rest of the decade, the United States enjoyed plentiful supplies of energy at affordable prices. While research on alternative energy sources continued during this period, research tended to focus on known sources, such as "clean coal." Congress did pass several important nuclear-related bills, among which was the Nuclear Waste Policy Act of 1982, the nation's first comprehensive nuclear waste legislation, and in 1987, approved an amendment designating Yucca Mountain, Nevada, as the only site for the first high-level nuclear waste repository.

Throughout most of this decade, energy of all types was in high supply and prices remained low. In reaction to high oil prices during the 1970s, Americans accelerated energy conservation fuel efficiency; oil demand dropped from 18.8 million barrels per day (mbd) in 1978 to 15.2 million mbd in 1982. Prices of crude oil remained low until Iraq's invasion of Kuwait in 1990 cut off 4.3 mbd from the world supply. Prices that had averaged $16 a barrel in July 1990 were more than $28 a barrel by August and $36 a barrel in September 1990.

Table 4.2 President Carter–Era Natural Gas Acts

Legislation	*Objective*	*Status*
Power Plant and Industrial Fuel Use Act (FUA)	Passed during the Carter administration in response to concerns over energy security. FUA forbade construction of electricity generating plants using oil or natural gas as a primary fuel, encouraging use of coal, nuclear energy, and other alternative fuels.	Rising supplies and lower prices for oil and gas during the 1980s resulted in repeal of FUA in 1987, again allowing use of oil and gas for power generation.
Natural Gas Policy Act (NGPA)	From 1938 to 1978, the federal government regulated only interstate shipment of natural gas. NGPA gave regulatory control over intrastate gas as well by setting ceilings on wellhead sales of gas.	Declining gas prices and secure supplies resulted in 1985 removal of all gas price ceilings.
Public Utility Regulatory Policies Act (PURPA)	Passed as part of the National Energy Act, this bill required the states to implement utility conservation programs and create special markets for co-generators and small producers who meet certain standards, including the requirement that states set the prices and quantities of power that the utilities must buy from such facilities.	Amended in 2005 to include greater emphasis on use of renewable energy sources to generate electricity. Bill is expiring as 1980's contracts expire.

Sources: EIA 2005. *Annual Energy Outlook 2005, Natural Gas Supply and Disposition.* U.S. Department of Energy. http://www.eia.doe.gov/oiafaeo/pdf/aeicftab_13.pdf (accessed January 11, 2011). NaturalGas.org. 2010. *History of Regulation.* Natural Gas Supply Association. http://www.naturalgas.org/regulation/history.asp (accessed January 12, 2011).

1991 to 1999: Calls for a Comprehensive Energy Policy

The decade of the 1990s coincided with strong growth in the U.S. economy, partially fueled by the growth in the electronics and computer industries. But by the end of the decade, another energy crisis occurred—this one brought on by electricity shortages brought on by an exceptionally hot summer after a dry winter and smaller than normal show pack in the West. California was particularly hard hit. Rapidly increasing prices of energy futures resulted in a bubble in energy trading. It was learned later that the shortages were largely artificial, caused by shutting down production facilities in order to raise prices. The shortages were linked to the unethical energy trading practices of the Enron Corporation. Selected highlights of the decade are displayed in Appendix B.

During the 1990s, federal energy policy makers had determined that it was necessary to split the traditionally integrated functions of electricity providers—generation, transmission, and distribution—into separate functions. The deregulation of this arm of the industry was promoted as a way to expose cross-subsidies and inefficiencies in the industry, thereby increasing efficiency and lowering consumer prices. It was believed that competition among power generators would lead to lower prices for all customers. Restructuring was designed to introduce open market competition only in electricity generation, with transmission and distribution services still subject to varying levels of regulation. By 1998, nonutilities were responsible for 11 percent of the total generation in the nation and were contributing 406 billion kilowatt-hours (kWh) to the electric system (Smeloff and Asmus 1997).

An intensive effort to craft a national energy policy by the first Bush Administration led to the passage of the Energy Policy Act of 1992 (EPAct). One aspect of EPAct was to further press wholesale deregulation by opening up transmission access to nonutilities. In return, regulated utilities were permitted to build new merchant plants outside their service territories.

Other landmarks in restructuring the energy industry were Regulatory Orders 888 and 889 issued by the FERC. Both orders were issued in 1996 and were designed to pave the way for increased participation by nonutilities and promote wholesale competition by eliminating local energy monopoly control over transmission. The combined effect of these orders required public utilities that controlled transmission to develop open access, nondiscriminatory transmission tariffs and to provide existing and potential users with equal access to transmission information. These orders also began the process of unbundling existing energy functions by separating transmission of electricity as a stand-alone service from generation and distribution. The opening of access to transmission lines was a significant step in restructuring the energy industry.

By 2000, almost half of the states were pursuing some form of restructuring. However, several recent events have cooled the enthusiasm for abandoning the traditional heavily regulated and integrated production-distribution system. Foremost among these was the California electricity crisis.

Since its first commercial production in the United States, monitoring supply and pricing of electricity was considered to be a responsibility of the individual states. This policy continued despite the fact that power was regularly being wheeled across state and the Canadian and Mexican borders. More problematic was that power was being moved over what was proving to be an inadequately coordinated and aging transmission grid. However, deregulation and restructuring was forced upon state governments by policies made in Washington, DC. It was a recipe for disaster that, when it came, surprised everyone by its intensity.

2000 to 2002: Linking Energy and National Security

Appendix B also displays selected highlights of the energy policy actions and legislation that occurred during the years 2001 through 2004, the years of the G. W. Bush administration. This period is marked by a number of catastrophic events, all of which have some relation to energy.

The first was the terrorist attacks of September 11, 2001. The destruction of the World Trade Center Towers in New York and attack on the Pentagon by aircraft highjacked by Middle Eastern terrorists brought home the dangers of our relying upon the Middle East for such a large proportion of the nation's supply of oil. The United States–led invasions of Iraq and Afghanistan that followed not long after reinforced the problems associated with that reliance.

The public was just getting over the electricity crisis when gasoline and natural gas prices began to surge in 2007 and 2008. That increase was accredited to growing demand for energy in the fast-growing countries of Asia—particularly by China, India, and South Korea. From an average U.S. price at the pump of around $1.15 per gallon of regular in 2006, in some locations (including most of California) prices grew to close to more than $4.50 a gallon by early 2008. In the past, energy bills addressing portions of the problem have been passed in what seemed to be bits and pieces, often to benefit one portion of the industry at the expense of others.

War, Politics, and Energy Security

The ties between energy and national security are deep and strong—and becoming more so as the world moves ever closer to the time when oil is beyond peak supply. Energy, and particularly petroleum, has become a strategic commodity. But what does it mean for a commodity to be *strategic*? Is it worth going to war over? Does it mean we perceive it to be especially rare and valuable for national competitiveness? Or do we think of its supplies in a zero-sum context (i.e., your control of a new source of supply is my loss), so that countries believe they should engage in complex games to "win" long-term access to reliable supplies?

Petroleum has been a highly sought after commodity ever since the first commercially drilled well struck oil in Pennsylvania in 1859. Its unique properties, including its high energy density, capacity to be fractured into many distillates suitable for specialized applications, and relative ease of transport, have made it particularly valuable for modern industrial societies. Oil remains the fundamental energy source for transportation worldwide for both civilian and military applications. Oil provides the energy for about 95 percent of the transportation in the United States. The fact that vast fortunes have also been associated with discovery and development of oil—for both individuals and nations—is also a factor in its global desirability.

Over the last few decades, other fuels and minerals have become increasingly strategically important as well. Natural gas, for example, is important because of its suitability for residential uses and ease of use as a relatively clean fuel for generating electric power. Other minerals are also potentially important strategically, including uranium for nuclear energy and weapons to lithium for batteries, tellurium and indium for thin film solar cells, and neodymium for wind turbine gear boxes (Jacobson and Delucci 2009). But oil is still at or near the top of the list of strategically important commodities.

Energy security in its simplest form is "the availability of sufficient supplies at affordable prices" (Yergin 2006). Of course, both "sufficient" and "affordable" are open to interpretation. What is sufficient now may not be in the future for a growing population and economy. And, who wants to pay any more for a product than they have to? This meaning of sufficient and affordable is from the perspective of a consuming country. Producing countries have different priorities; they are particularly interested in maintaining stable demand at a reasonable price, and in consuming countries having the ability to pay for their oil in a stable currency (Rosner 2010). How should states aim to accomplish their energy security goals? The classic advice is to develop diversified sources of supply that avoid overreliance on a small number of sources, as well as a resilient energy supply system able to respond to disruptions (through storage and reserves) as well as transparency of information (Yergin 2006).

But this leaves many questions unanswered. One is: who should be responsible for accomplishing these goals? One approach is to rely on firms and the market, particularly in the case of oil with its relatively developed markets, to respond to supply shortages. Prices increase, demand falls, firms draw down their stocks and longer-term invest in exploration in response to the higher prices. Eventually a new equilibrium is found. Of course, this rosy scenario leaves out the economic impact on the consuming countries, which face a bigger oil import bill—one that must be paid in hard currency, generally dollars. The higher oil price pulls resources away from other uses and can cause a fall in aggregate demand, which can tip an economy into recession. Whether oil price and supply shocks have helped to cause recessions is one of the huge areas of dispute in this literature. And further, consumers are also voters, and let political leaders know of their displeasure at the rising cost of a basic commodity, increasing pressure on them to do *something*.

A second approach is to rely on government to take a more active role in the management of energy supplies. One tool is the use of strategic petroleum reserves (SPR) such as the SPR of the United States, and similar national reserves coordinated by the International Energy Agency (IEA). In addition, governments can be called on to exercise regulatory actions to temporarily ease requirements that can cause supply restrictions. A third approach invokes price and administrative controls rather than relying on market allocation of fuels. These are reactive, short-term actions taken in response to external supply shocks. The vast literature on this topic, on balance, favors a strategy of limited governmental intervention that avoids use of price controls.

But what about the longer-term, energy supply "game"? If fossil fuel supplies are limited, and the growth and future prosperity of your country are tied to their easy availability, can national leaders rely on the market to make supplies available, or use other means?

Japan, China, and the United States provide an intriguing contrast. Following the Arab Oil Embargo, Japan aggressively invested in efficiency, passed higher prices along to consumers, and increased its reliance on nuclear energy. It is increasing its use of natural gas but remains reliant on global oil markets and oil for automobile transportation. China has developed domestic sources of coal and oil, and is actively pursuing bilateral deals with oil producers such as Sudan and Angola. The United States relies on the global oil market, importing large quantities from Canada, Venezuela, Mexico, and Nigeria to augment its domestic production. It has also been willing to use military force to protect its oil interests. A RAND study clarified this position in its description of U.S. energy policy:

> Both for the health of the U.S. economy and for broader national security reasons, the United States has given ensured access to oil a high priority among its foreign and defense policy imperatives. The United States has demonstrated its willingness to go to war to prevent the domination of Persian Gulf oil-producing regions by powers hostile to the United States and its allies. (Crane et al. 2009, 60)

This policy has been in place at least since the Carter administration and is reflected in the country's official National Defense Strategy (DOD [U.S. Department of Defense] 2008). The costs of this policy are significant. Protecting oil supplies, primarily in the Persian Gulf, and their transit has been estimated to cost between $13 and $143 billion per year in 2009 dollars, depending on the particular forces and activities included in the estimate (Crane et al. 2009). That does not include the cost of either the 1991 Desert Storm mission against Iraq, or the 2003 invasion of Iraq, which was motivated at least in part by a desire to control the country's oil.

Is oil thus *strategic*? Is there a link between energy and national security? U.S. policy assumes that it is, and our past military actions reflect it. One of the few analysts to argue otherwise is Ivan Eland of the Independent Institute, who makes the straightforward market case for thinking of oil like any other commodity (2008). Reducing the U.S. military presence in the Middle East could improve our image and diplomatic posture while saving huge sums. But changing the mindset of national security analysts, political leaders, and the public will not be easy.

2003 to 2007: A Comprehensive Energy Policy Finally Emerges

Policymakers devoted more attention to the energy sector after 2003. This was due to increasing prices for oil and natural gas, concern about greater speculation in oil markets and the rise of energy demand in China and India. Partly as a result, two comprehensive energy policy acts were finally passed during President George W. Bush's second term of office, one in 2005 and another in 2007.

The Energy Policy Act of 2005

The Energy Policy Act of 2005 (EPAct 2005), finally passed by Congress on July 29 and signed by President Bush on August 8, was the first comprehensive energy policy update in 13 years; two earlier versions failed in one or the other house over the three years leading up to its final adoption. As a result, it offered a little for everyone, but did little to get the nation on the path toward energy independence. This omnibus bill signaled a significant shift in U.S. energy policy by including $14.5 billion in tax incentives, subsidies, and loan guarantees for many different types of energy sources. Most of the money was earmarked for programs to increase oil, gas, nuclear power, and coal production, but renewable and alternative energy resources were also promoted (CESA [Clean Energy States Alliance] 2005; *Petroleum Economist* 2005; Weinberger 2005).

Some of the key provisions of the bill included a new kind of federal risk insurance to promote construction of new nuclear power plants; funds for the development of hydrogen-fuel-cell technology and hydrogen-fueled vehicles; tax credits for purchasers of hybrid cars; tax incentives for continued work in clean coal technology, geothermal energy, and residential solar energy systems; increases in the amount of ethanol in motor fuel; and increasing the reliability of the electric power grid. EPAct 2005 did not include a provision for drilling for oil in the Arctic National Wildlife Refuge or increasing fuel efficiency in cars and trucks, both of which were programs desired by the president. However, it did include a provision for increasing oil and gas exploration in federal lands, as well as relaxing barriers for building liquefied natural gas (LNG) terminals and refineries. Key sections of the bill that pertained to programs in traditional sources of energy are presented in Box 4.1.

The two energy sources given most attention in EPAct 2005 were hydrogen and electricity. The G. W. Bush administration was particularly interested in funding grants, loans, and joint government–private sector research and development to create what was described as a *hydrogen society.* This involved reducing the cost of producing hydrogen as an energy carrier and hydrogen-based fuel cells, and for designing the infrastructure necessary for making it readily available and safe to produce, store, transport, and distribute. Provisions relating to renewable and alternative energy sources are displayed in Box 4.2.

The Bush energy policy was also designed to rectify the fundamental problems of the nation's electricity transmission grid. Two sets of events were behind this renewed attention to electrical energy. The first was the eastern United States and Canadian blackouts of 2002 followed by widespread summer brownouts. The second was the West Coast—particularly California—electrical energy crisis that triggered extremely high power prices and ended with the collapse of Enron Corporation.

Practically every known renewable energy source was targeted for development funding in the EPAct 2005 section on developing economically viable sources renewable energy. Wind, solar, wave, several types of biomass, geothermal, and biodiesel were among the sources targeted for funding. Programs included tax incentives, funding for research and development, joint public-

BOX 4.1 TRADITIONAL ENERGY-SPECIFIC PROVISIONS OF THE ENERGY POLICY ACT OF 2005

Oil and Gas:

Alternate Energy–Related Uses on the Outer Continental Shelf (Grants leases and distributes 27 percent of the revenues to coastal states)

Coal:

Integrated Coal/Renewable Energy System
Coal Gasification
Clean Coal Demonstration Plan Loans

Nuclear Energy:

Demonstration Hydrogen Production at Existing Nuclear Power Plants

Vehicles and Fuels:

Federal Procurement of Stationary, Portable and Micro Fuel Cells

Hydrogen:

Coordinated Plan for Programs
Research and Development Programs
Hydrogen and Fuel Cell Technical Task Force and Technical Advisor Committee
Funding of Demonstration Projects
Grants for Development of Codes, Standards
Hydrogen from Solar and Wind Technologies
Cost Sharing

Electricity:

Study on the Benefits of Economic Dispatch (To improve availability of nonutility generation resources and the benefits to consumers of such revisions)
Funding New Interconnection and Transmission Upgrades
Net Metering and Additional Standards
Cogeneration and Small Power Production Purchase and Sale Requirements
Interconnection

(*Sources:* CESA 2005; EPA 2005.)

private financing of demonstration facilities, research programs, government purchasing for fuels and building heating and air conditioning.

Increasing production, lowering production costs, and requiring an increase in the amount of ethanol added to gasoline in motor vehicles was given a great deal of attention in the policy act. A key element of the plan for promoting use of this alternative fuel was federal support for ways of producing ethanol from sources other than corn and other human food products. Funds in the form of loans, grants, and joint research projects were provided for different types of biomass, municipal solid waste, sludge, and oils skimmed from wastewater treatment plants.

BOX 4.2 EFFICIENCY, ALTERNATIVE ENERGY, AND CLIMATE SECTIONS OF THE ENERGY POLICY ACT OF 2005

Energy Efficiency:

Low Income Community Energy Efficiency Pilot Program
State Technologies Advancement Collaborative (federal and state research)
Energy Efficient Electric and Natural Gas Utilities Study

Renewable Energy:

Assessment of Renewable Energy Resources
Renewable Energy Production Incentive (REPI)
Use of Photovoltaic Energy in Public Buildings
Biobase Products (Government procurement)
Rural and Remote Community Electrification Grants
Grants to Improve Commercial Value of Forest Biomass for Electric Energy, Useful Heat, Transportation Fuels and Other Commercial Purposes (in preferred communities)
Sense of Congress Regarding Generation Capacity of Electricity from (Non-hydropower) Renewable Energy Resources on Public Lands

Ethanol and Motor Fuels:

Renewable Content of Gasoline (raises ethanol content requirements)
Commercial Byproducts from Municipal Solid Waste and Cellulosic Biomass
Renewable Fuel (Amends the Clean Air Act and funds demonstration projects)
Conversion Assistance for Cellulosic Biomass, Waste-derived Ethanol, Approved Fuels
Advanced Biofuel Technologies Program
Waste-derived Ethanol and Biodiesel (from municipal solid waste, sludge, and oils)
Sugar Ethanol Loan Guarantee Program (demonstration projects)

Incentives for Innovative Technologies:

Eligible Innovative Technologies Projects (for reducing pollution and greenhouse gas emissions and other projects)

Climate Change Initiatives:

Greenhouse Gas Intensity Reducing Technology Strategies
Climate Change Technology Deployment and Commercialization
Climate Change Technology Demonstration Program
Greenhouse Gas Intensity Reducing Technology Export Initiative

(*Sources:* CESA 2005; EPA 2005.)

BOX 4.3 ENERGY POLICY ACT OF 2005: ENERGY POLICY TAX INCENTIVES

Renewable Production Tax Credit
Application of Tax Credit to Agricultural Cooperatives
Clean Renewable Energy Bonds (municipal utilities and governments)
Energy Efficient Buildings Deduction Credit for Residential Energy Efficient Property
Credit for Business Installation of Qualified Fuel Cells and Stationary Microturbine Power Plants (investment tax credit)
Business Solar Investment Tax Credit
National Academy of Sciences Study and Report (on costs and benefits of production and consumption of energy that may not be fully incorporated into the market price of such energy)

(*Sources:* CESA 2005; EPA 2005.)

Another new direction in energy policy included in EPAct 2005 was a focus on tax incentives for private citizens, businesses, and nonprofit organizations for investments in improving energy efficiency and conservation. Box 4.3 displays some of the tax incentive elements in the policy act.

A way into the federal government's thinking about future energy policy is to review some of the projects targeted for scientific research and development funding. Box 4.4 displays some of the key research projects and studies specified in the 2005 energy policy act.

High on the list of research funding were programs relating to improving the safety, security, and efficiency of the national electricity grid. Biomass research projects, bioenergy, and other bioenergy research programs were also to be funded. Congress also required a wide variety of one-time and ongoing studies and progress reports.

Energy Independence and Security Act of 2007

This second energy policy act was signed into law by President Bush on December 19, 2007. This bill appeared to signal a shift in federal energy policy from the emphasis on hydrogen fuels in the EPAct of 2005 to include greater emphasis on (1) improving fuel efficiency, (2) expanded production and use of biofuels, (3) developing improved electrically powered vehicles, and (4) greater energy efficiency in public buildings, appliances, and lighting. It also continued to emphasize needed improvements, repairs, and upgrades to the regional electrical transmission grid. Grants and loan guarantees for alternative fuels research and development were also expanded.

The final version of the bill did not include House-version provisions cutting subsidies to the petroleum industry to promote petroleum independence and alternative energy, originally introduced in the House as the Ending Subsidies for Big Oil Act of 2007. House proposals to amend or end tax relief for oil producers in the Gulf of Mexico were also deleted from the final version of the bill. The bill was renamed the Clean Energy Act of 2007 when introduced and passed by the House.

The Energy Independence and Security Act of 2007 (EISA 2007) was proposed as a means of rectifying blanks in the 2005 energy policy act by altering royalties and tax breaks previously afforded to oil and gas companies. The Senate later amended the bill, and passed a much more comprehensive version, retaining the tax changes proposed by the House. The combined bill

BOX 4.4 ENERGY POLICY ACT OF 2005 RESEARCH AND DEVELOPMENT AND STUDY PROVISIONS

Research and Development:

R&D for Distributed Energy and Electric Energy Systems
High Power Density Industry Program (data centers, server farms, and telecommunications centers and related technology)
Micro-Cogeneration Energy Technology
Distributed Energy Technology Demonstration Programs
Renewable Energy R&D
Bioenergy Program
Concentrating Solar Power Research Program
Renewable Energy in Public Buildings
Production Incentives for Cellulosic Biofuels
Procurement of Biobased Products (for buildings in the Capitol complex)
Regional Bioeconomy Development Grants
Preprocessing and Harvesting Demonstration Grants (to agricultural producers)

Energy Policy Studies:

Study of Rapid Electrical Grid Restoration
Study of Distributed Generation
Natural Gas Supply Shortage Report (2004–2015)
Hydrogen Participation Study
Overall Employment in a Hydrogen Economy Study
Alternative Fuels Report
Fuel Cell and Hydrogen Technology Study
Passive Solar Technologies Study
Renewable Energy on Federal Land Study (potential of)

(*Sources:* CESA 2005; EPA 2005.)

included additional focus on transportation fuel economy, expanded development of biofuels, and greater energy efficiency in public buildings and all lighting.

Achieving greater energy efficiency in government buildings can result in significant reductions in our energy use. The DOD uses something like 80 percent of all the energy used by the federal government each year and roughly 1 percent of all the energy used in the nation. The DOD manages more than 500,000 buildings. To keep those facilities running in FY 2007 cost $3.8 billion, taking up 26 percent of the DOD budget. The DOD is following a three-arm strategy to improve its overall energy use: (1) develop and use new primary energy resources, including renewable resources; (2) reduce energy use through conservation; and (3) improve the efficiency of its energy use. Emphasizing conservation and focusing on improving energy efficiency not only saves the department money, it makes it possible for the DOD to direct more of its resources toward mission accomplishment.

New CAFE Standards

A key energy efficiency component of the legislation was a required increase in combined city and highway miles per gallon standard for all new automobiles and light trucks by the year 2020, when the fleet average was to 35 miles per gallon. The first increment of the new standard begins with the 2011 model year. Fuel mileage improvements for work truck and commercial medium and heavy-duty trucks were to begin in 2010. Related to mileage improvements was a requirement for the US. Department of Transportation to come up with a new rating system and labeling that makes it easier for consumers to compare fuel economy and greenhouse gas emissions of vehicles.

Presidential administrative rule makings in 2003 and 2006 sought to increase the light-truck standard to 24 mph in 2011, but those attempts were overturned by the appeals court. The first real increase in fuel efficiency standards in more than 20 years had to wait until December 19, 2007, when President George W. Bush signed the Energy Independence and Security Act. That law required new cars and light trucks to reach a combined level of 35 miles per gallon by the year 2020. Standards in 2007 were 27.5 mpg for domestic and imported cars, and 22.2 mpg for light trucks. There is widespread agreement that, while far from perfect, corporate average fuel economy (CAFE) standards have helped the country save oil. Estimates for 2002 suggest that CAFE standards for passenger vehicles contributed to saving 2.8 million barrels a day, or 14 percent of consumption that year. Similar claims extend these savings to additional years (GAO [U.S. Government Accountability Office] 2007d).

Another energy efficiency measure of the bill begins with a requirement for a 25 percent increase in the efficiency of light bulbs by phasing out some incandescent light bulbs in 2012, increasing this to a 200 percent increase in efficiency by 2020. Revised energy efficiency standards for electric and gas appliances were also included. All new and renovated federal buildings were required to decrease the use of fossil fuels by 55 percent from 2003 levels by 2010, and 80 percent by 2020. All new federal buildings must be "carbon neutral" by 2030. Some additional key elements included in the 2007 energy policy act are (AFDC [Alternative Fuels and Advanced Vehicles Data Center] 2008):

- Extension of the fuel economy tax credits for flexible fuel and dual-fuel alternative fuel vehicles through 2019.
- A grant program to fund encouraged use of plug-in electric drive vehicles and other emerging technology electric vehicles.
- Loan guarantees for plants making more efficient vehicles and parts, including advanced batteries.
- Expands renewable fuel standards (primarily ethanol) to 9 billion gallons in 2008, increasing to 36 billion gallons by 2022.
- Prohibits any new fuel retail marketing franchise agreement from restricting any provisions forbidding sale of stations to sell biofuels or renewable diesel.
- Requires each federal agency to install at least one renewable fuel pump at each federal fleet refueling center by January 1, 2010.

2008 to 2009: A Renewed Call for Energy Independence

As the second decade of the twenty-first century neared, Americans found themselves once again asking: Does America have a viable energy policy? If so, what is it? If not, why not? What can we do to make energy more affordable? Answers to those questions depended upon who was asked. The huge run-up in gasoline and other fuel prices again brought home the belief that U.S. consumers were being held hostage by greedy suppliers of imported oil; many Americans blamed leaders of the oil-rich Middle East, Russia, and Venezuela for the high prices. Moreover, Russia's use of oil and gas exports to influence Europe and several former republics quickly reminded American policy makers of the tenuous character of depending upon foreign countries for energy supplies (Aalto 2008; Romanova 2008; Tkachenko 2008; Westphal 2008).

When the price of oil dropped to less than $50 per barrel in 2008 and was projected to stay around that price through 2010, the pain at the pump was relieved but not forgotten. Politicians of both parties continued to demand policy changes that would guarantee a return to oil independence. A study of this and energy policies of the past are not reassuring, however; when prices drop, apparently so does the concern of consumers and elected leaders and a return to business as usual occurs—but not this time.

Past energy policies had focused on a variety of goals that ranged from protecting the energy industry to conserving domestic energy supplies by intentionally importing whatever supplies could be absorbed into the U.S. market. The energy policy directions of the federal government can be gleaned from the distribution of funds included in the Department of Energy's 2008–2010 budget shown in Table 4.3.

This time, energy independence was only one of the reasons for the call for a new energy policy. Problems of global warming took center stage with the call for energy independence. For the first time policy analysts were forced to recognize that energy policy had become integrally linked with environmental policy. The Obama/Biden early energy policy draft made this clear.

2010 and Beyond: Energy Efficiency, Conservation, and the Environment

The second decade of the twenty-first century began with two important energy-related developments. On February 17, 2009, President Obama kicked off his administration's efforts to stimulate the nation's economy by signing into law the American Recovery and Reinvestment Act (ARRA). ARRA includes a number of spending and tax incentive provisions designed to promote greater investment and economic activity in energy efficiency, alternative energy, and other energy-related initiatives. The second energy-related development was the president's announcement on September 15, 2009, of mandated increases in motor vehicle fuel efficiency standards that include limitations on greenhouse gas emissions. The new provisions go into effect with the 2012 model year. Responsibility for developing and monitoring the standards are, for the first time, going to be shared by the Environmental Protection Agency (EPA) and the Department of Transportation's National Highway Traffic Safety Administration (NHTSA).

Table 4.3 U.S. DOE Discretionary Fund Appropriation Summaries, 2008–2010

(US$ thousands)					
Organization	*FY 2008*	*FY 2009*	*FY 2010*[a]	*$*	*%*
National Security[b]	8,814,111	9,29,594	9,945,027	+815,433	8.9%
Energy and Environment Energy Efficiency & Renewables	1,704,112	2,178,540	2,318,602	+140,062	6.4%
Electricity Delivery	136,170	137,000	208,008	+71,008	51.8%
Fossil Energy	885,545	1,110,219	881,565	–228,654	–20.6%
Nuclear Energy	1,033,161	1,357,819	884,632	–513,187	–37.8
Energy Total	3,761,988	4,783,819	4,252,807	–530,771	–11.1%
Environment Total	6,332,142	6,465,943	6,016,327	–449,616	–7.0%
Science (R&D)	4,082,883	4,757,636	4,941,682	+184,046	+3.9%
DOE Management	301,600	7,832,046[c]	383,202	–7,448,844	–95.1%
Misc. Admin.	739,614	764,519	844,937	+818	+3.00%
Discretionary Funding Total	**24,032,338**	**33,748,319**	**38,725,000**	**–7,354,337**	**–21.8%**

Source: DOE.2009d. *FY 2010 Budget by Organization.* Washington, DC: Department of Energy. http://www.cfo.doe.gov/budget/10budget/content/ApprSum.pdf (accessed June 29, 2010.)

[a] Congressional budget request
[b] National Nuclear Security Administration
[c] Includes $7.51 billion for manufacturing advanced technology vehicles loan program

The American Recovery and Reinvestment Act (ARRA)

Energy-related provisions of the Obama economic stimulus bill total nearly $100 billion in government spending on research and development and other direct federal spending, grants to states and municipalities for energy infrastructure development, tax credits for promoting renewable energy projects and programs, loan guarantees, and other programs. The bulk of the energy-related provisions in the bill are to be distributed in the following program methods (McKinsey & Company 2009):

- Energy-related tax incentives and related programs — $22 billion
- Direct funding to the Department of Energy — $39 billion
- Energy-related funding for other agencies — $29 billion
- Energy-specific funding to other agencies — $7 billion

Of the nearly $100 billion total, $67 billion is designed to stimulate activity in four major categories: (1) renewable and clean energy, (2) energy efficiency and building retrofits, (3) smart

Table 4.4 ARRA Funding for Clean, Renewable, and Energy-Efficiency Provisions

Category	*Tax Credits*	*Loan Guarantees*	*Grants and Direct Spending*	*Total*
Renewable and Clean Energy	$8 billion	$6 billion	$17 billion	$31 billion
Energy Efficiency and Government Building Retrofits	$17 billion		$3 billion	$20 billion
Smart Grid and Infrastructure Upgrades	$5 billion	$7 billion		$12 billion
Fuel Efficiency and Electric Vehicles	$3 billion		$2 billion	$5 billion
Totals	$33 billion	$13 billion	$21 billion	$67 billion

Source: McKinsey & Co. 2009. *American Recovery and Reinvestment Act (ARRA).* March 13, 2009 Briefing presentation at the Hawaii state capitol (March 13, 2009), http://www.ARRABriefing_energy_provisions_hawaii.pdf (accessed March 11, 2010).

grid and infrastructure projects, and (4) fuel efficiency and electric vehicles. How the funds are to be distributed for these four programs is displayed in Table 4.4.

The Evolution of Fuel Efficiency Standards

U.S. peace-time efforts to legislate greater fuel efficiency for motor vehicles began in the 1970s. Congress passed the Energy Policy Conservation Act (EPCA) in 1975, two years after the first energy crisis, which followed the Arab Oil Embargo of 1973 and 1974. To reduce domestic oil consumption by the transportation sector, the EPCA established mandatory fuel economy standards for all new cars and light trucks manufactured in the United States. These minimum average fuel-efficiency standards, called the corporate average fuel economy (CAFE) standards, were to begin with the 1978 model year. By 1985, all makers of cars sold in the United States were to have achieved 27.5 miles per gallon (mpg) or higher average for their cars manufactured in the United States (Crandall and Graham 1989; Portney, Parry, Gruenspecht, and Harrington 2003).

Penalties of $50 for every one mile over the standard for each car sold were established for firms that did not meet those standards. Car makers had a significant incentive for achieving the target; it was estimated that for one mile over the average would cost General Motors upward of $200 million per year, for example. The Energy Act set standards for automobiles, but not for light trucks; the light-truck category includes most pickup trucks, minivans, and sport utility vehicles (SUVs). Instead, permission was given to NHTSA to set different rates for the various light-truck models.

The price of oil increased again after the 1978 Iranian revolution—another incentive for improving car fuel efficiency. As a result, the major U.S. automakers exceeded annual CAFE standards each year until 1981, when the price of oil began to decline. With the drop in oil prices, Ford and General Motors began making larger cars and SUVs. Both Ford and GM sought relief as they found it harder and harder to meet the CAFE standard, although Chrysler continued making smaller cars that allowed it meet the standards. As a result, in 1985 standards were eased for the 1986 to 1989 model years. The automobile average standard was lifted back to the 27.5 mpg average for the 1990 model year, where it remained until 2003 (Crandall 1992). The CAFE standard for automobiles hardly changed at all for the next 20 years, although in 2003, the light truck average of 17.2 mpg standard in 1979 was raised to 22.2 mpg in 2007 (GAO 2007c). From 1985 to the Energy Act of 2007, the fuel standard received little attention from Congress and the auto industry, with overall efficiency of new cars made in the United States actually declining for several years after 1988 (Luger 1995; Nivola 2009).

The Obama administration announced a set of new CAFE standards for fuel economy, and for the first time ever, greenhouse gas emissions for cars and trucks beginning with the 2012 model year: 35.5 mpg for cars and light trucks by 2016—four years earlier than the 2020 date set by the EISA of 2007. By the same year, CO_2 emissions for passenger cars, light-duty trucks, and medium-duty (10 or fewer passengers) vehicles would be limited to an average of 250 grams per mile per vehicle.

President Obama's proposal also included the announcement of a joint effort by the EPA and the Department of Transportation's NHTSA to work together on fuel economy and emissions control (Voorhees 2009). The two agencies were charged with the task of writing new efficiency and emissions rules by the end of March 2020.

The combined energy and environment developments seen in 2009 and 2010 of the Obama administration may have set the tone for a U.S. energy policy that will extend through the remaining years of the 2010–2019 decade. Emphasis is clearly on improving fuel efficiency with reduced emissions of greenhouse gases, strengthening the nation's electricity grid, and implementing cleaner means of generating electricity, along with continuing to research cleaner and renewable fuels sources.

Conclusion: A History of Crisis and Change

Analysis of the energy policy acts of 2005 and 2007 suggest that the focus of U.S. policy makers in the early decades of the twenty-first century is constructed around three basic goals: The first is to reduce oil use in the transportation sector and for continued economic growth. The second is to expand the production of clean electrical energy while also improving the reliability of its transmission and distribution. This appears to include a resumption of investment in nuclear-powered generating plants and renewed research into ways to produce "clean coal." The third, which is closely related to the first two, is to greatly reduce, if not eliminate, the discharge of greenhouse gases into the atmosphere and halt global warming.

A number of different solutions have been promoted as the best ways to achieve these goals. A secure, stable, reliable, and safe supply of oil and natural gas to enable Americans to continue to enjoy a high standard of living while science and industry is coming up with the solution to our dependence upon foreign supplies appears to be a nearly universal political objective. To achieve oil and natural gas independence, a number of consumer, agricultural, and industry groups are pushing for more use of ethanol, whether it is produced from corn or other biological sources.

Other groups want the nation to produce more domestic oil from Alaska and offshore drilling off states abutting oceans. And others are pushing for research to make hydrogen-based fuel cells, gasoline and diesel from coal gasification and liquidization, and a variety of sources to produce more biofuels. The way government is going about achieving this goal is to spend money on research to develop one or more alternative and renewable supplies of energy before the world's supply of oil and natural gas is gone or becomes so expensive that it might as well be no longer available.

Achieving the goal of a safe, secure, and reliable supply of electrical energy has included building more wind farms, nuclear power plants, and producing electricity from solar (photovoltaic), geothermal, and waves and tide sources, and by greatly expanding the use of nuclear power. Coal producers are supporting the use "clean coal" to power the nation's electricity generating plants. Along with generation, the current energy policy includes expanding, rebuilding, and upgrading the national electricity generation and transmission grid. A key section of this policy has been the expanded production and required use of ethanol to supplement gasoline in automobiles. This includes producing oil and gas from alternative sources, including oil shale, oil sand, and biofuels.

Each of the first two goals of energy policy are shaped and constrained by the third goal: reduce and eventually eliminate the damage to the environment caused by the burning of fossil fuels for transportation and to generate electricity. That involves finding substitutes for oil and gasoline, natural gas, and or far more carbon-clean forms of coal. A theme of the government's energy policy has been to provide extensive support in the form of tax subsidies, grants, and loans for developing a number of substitutes for fossil fuels. This includes shifting from gasoline and diesel engines to hybrid electric motor/gasoline engine vehicles.

The energy decisions being made every day by men and women in the executive and legislative branches at the federal, state, and local government levels are today shaped and constrained by policy made in industries and governments around the globe. As a result, energy policy decisions are exceptionally complex.

Chapter 5

Difficulties in Achieving a Balanced Energy Policy

> Public policy ... can be seen as the formal or stated decisions of government bodies. However, policy is better understood as the linkage between intentions, actions and results. At the level of intentions, policy is ... what government says it will do. At the level of actions, policy is ... what government actually does. At the level of results, policy is ... the impact of government upon ... society.
>
> —**Andrew Heywood (2000)**

What lessons do the various energy eras explored in Chapter 4 teach us about the challenges inherent in crafting energy policy in the United States today? An analysis of policy making in the energy arena suggests that there are six factors that make efforts to change our energy policies particularly difficult:

1. Multiple stakeholders
2. Politics I: Influence
3. Ambiguous and conflicting policy goals
4. The nature of energy policy interventions
5. The innovators: States, regions, compacts
6. Environmental policy, energy policy, and Politics II

This chapter will survey these factors and examine the general case for intervening in the energy sector. As we consider the economic, political, structural, and other forces complicating change in the energy sector, it is useful to remember Heywood's warning. We need to pay attention to goals, action "on the ground," and the actual results of policy, since citizens indeed perceive the gap between goals, expectations, and reality.

Why Intervene in Energy?

Why do government entities fund and deliver energy programs? The public finance literature, with its foundation in microeconomics, focuses on factors that cause various types of energy markets to fail. The primary reason is that energy prices may not reflect the true social costs of their consumption due to externalities. Externalities are "non-market costs or benefits not accounted for by the producers, importers or consumers (and therefore not measured in the marketplace) that spill over to people who are not a party to the transaction" (CRS [Congressional Research Service] 2005, 8). In the case of a negative externality, energy producer (or consumer) A, makes B (an individual, the environment, society) worse off, but shoulders none of the costs.

The most obvious examples are pollution related, as inefficient processes transforming energy sources into useful products and services leave nasty residuals and by-products in common-pool resources such as the air and water. These by-products are not priced into the good, and are thus "external" costs. Rationales for government action in the energy sector may also be based on one or more of the following:

- Split incentives; principal agent problems
- Lack of information
- Prospect theory—behavioral failures
- Normative preferences
- Average cost pricing
- Inequity of energy consumption
- Energy security

With concern about rising CO_2 concentrations and climate change now the primary drivers of energy policy, these energy producing and consuming activities are perhaps the most important generators of environmental externalities in our society. A variety of approaches are used to restrict the quantities of these by-products released or to force producers and consumers to reflect the costs in their production and consumption choices. Yet these programs are the province of the U.S. Environmental Protection Agency and its various state government counterparts. What about the many programs supported by state governments, and those of the U.S. Department of Energy?

One of the major challenges in economic development is the reality that firms face disincentives to invest in research and development. This is an example of a positive externality that has significant and damaging ramifications, especially in the energy sector. Innovations reflect new knowledge resulting from basic research into energy phenomena or applications of research findings that create new products and services. But the benefits of innovation accrue not only to the investing firm, but also to its competitors and consumers, since much of the knowledge is, or becomes, a public, not private good. As a consequence, they invest much less than what is socially optimal.

Estimates of the social rate of return of research and development are thus 2 to 4 times the rate of return to private entities (Gillingham, Newell, and Palmer 2009). Support for research is a major direct form of energy-related expenditure for the network of U.S. Department of Energy (DOE) laboratories, plus the new Advanced Research Projects Agency-Energy (ARPA-E) program while lowering the cost of demonstrating, deploying, and helping to commercialize new research findings through subsidies of various kinds is equally important. Many specific types of research, especially relating to renewable energy sources such as wind and solar, are also supported out of a belief that market forces systematically underinvest in these alternatives.

Perhaps the most challenging set of market failures and barriers lie in the realm of energy efficiency. Individuals, households, and firms tend not to take advantage of opportunities to "invest" in options that would save large amounts of energy. The size and causes of this "energy-efficiency gap" are a matter of great debate (Charles 2009; Gillingham, Newell, and Palmer 2010).

The evidence is clear that consumers tend not to select efficient alternatives that generate energy savings—from appliances to insulation to light bulbs—but at a somewhat higher total cost, they pay for existing solutions. Major challenges are the lack of recognition of the importance of the problem of household energy efficiency (since the cost is relatively low), lack of information needed to make an informed decision, and the short time period within which they expect the investment to "pay back" through energy savings—often two years, implying an astronomical 50 percent rate of return. These challenges of consumer and firm behavior lie outside of the usual economic framework for thinking about government intervention. They seem to be best described by the application of Kahneman and Tversky's prospect theory, which identifies systematic problems with the ways humans tend to make decisions (Thaler and Sunstein 2009).

The Principal-Agent Problem

Several additional rationales for energy interventions are worth mentioning. One in the energy efficiency arena is the landlord-tenant or *split incentive* problem. Landlords have little incentive to purchase energy-efficient appliances or invest in insulation because in most cases their tenants pay their own energy costs. This is an instance of the classic *principal-agent problem*, in which the interests of the decision maker or principal and the user or agent often diverge. Cable TV companies, for example, now require a set-top box that offers enhanced services or converts a digital signal to analog televisions. Since cable companies don't pay the power bill, they pay little attention to the power cost—now as much as 40 watts—which is rarely turned off (Charles 2009). At an even larger scale, this problem plagues the construction sector, since architects and construction firms are rarely the ultimate owners of the homes and buildings they construct.

A second are a series of interrelated reasons that energy prices may be incorrect, and thus lead to over- (or at times under-) consumption. One factor is the ubiquity of "postage stamp" or average cost pricing of electricity under state regulation. The marginal cost of produced electricity changes throughout the day, peaking in the morning and evening as households ramp up their energy use and forcing utilities to bring higher-cost sources online. But households pay the same for kilowatts consumed at 8 a.m., 5 p.m., and midnight. Another we will return to in Chapter 8 is the "fairness" of gasoline prices and whether they reflect all relevant costs to consumers and society.

An often overlooked rationale for intervening in energy markets is to ensure equity. As we explained in Chapter 2, energy consumption in the United States and throughout the world is highly inequitable, with households enjoying higher incomes and wealth able to consume much more energy. Government action is needed to ensure that people have heat and access to reasonably priced energy sources.

These mainly economic justifications are helpful but do not cover all of the country's important energy policy goals. Regulation of electricity and natural gas provision at the state level assumes that the high capital costs required to install poles, wires, and gas connections to each household make competition wasteful. As a result, most states regulate these utilities as monopolies within a defined service area.

Also, energy sources are not just any other commodity. Important sources are depletable and their discovery and development thus entails a considerable amount of risk and the challenge of managing their extraction over time. Critical sources, particularly petroleum and natural gas, are important factors of production in their own right, and changes in their availability and price influence the economy's overall rate of growth. These concerns are tied to the notion of energy security—lack of secure supplies leaves a country economically, politically, and militarily vulnerable. Consequently, supply-side policies have been crafted to ensure the availability of steady supplies at reasonable prices, plus demand-side policies that aim to limit our reliance on imports as well as the harmful emissions from our transportation fleet.

Multiple Stakeholders

One way to conceptualize the energy policy arena is to emphasize the pluralist model of the U.S. political system and the participants or stakeholders in the energy policy development process. During the decade from 2001 to 2010, the goal of those—mostly in Congress—who carry out the task of energy policy development was to forge a public policy in which all parties with a stake in the energy pool win some victories and no group loses all. The 2005 and 2007 Energy Acts were ecumenical in providing benefits for virtually the entire menu of U.S. energy sources, with the overriding theme of these bills being the updating of U.S. energy infrastructure. This model emphasizes the real world of negotiation, collaboration, competition, and cooperation that exists among the myriad stakeholders and public servants so that our wheels keep rolling and the economy keeps growing.

Art of the Compromise

How energy policy that is formed is often, although not always, an example of the *art of the compromise*—the complex give and take that government and the governed go through to form the policies and strategies for solving a nation's big problems. One glaring exception to this approach was the Cheney Energy Task Force, which produced a National Energy Policy proposal in 2001 emphasizing supply expansion, with input provided almost solely by oil and other energy producing corporations. Another is the long-running conflict over the opening of the Arctic National Wildlife Refuge to oil and gas exploration.

Stakeholders in Forming Energy Policy

Stakeholders are individuals or groups that have an interest in or may be affected by a decision or action. A list of the special interest groups with a stake in the energy policy-making process and that lobby governments for various energy programs include (1) the petroleum industry, which includes the world's oil producing countries, oil producing companies (both private and owned by foreign governments), oil and gas refiners and shippers, the petrochemical industry, and distributors of all oil and gas products; (2) coal mine owners, coal users and shippers, and mine labor unions; (3) electricity producers and distributors, both public and private; (4) U.S. and foreign-owned motor vehicle producers and marketers; and (5) fossil fuel–consuming commercial and consumer user groups.

Other groups with a stake in the energy policy debate include conservationists and climate "greens"; the nuclear lobby; the national security community; the farm lobby, both for producers of biofuels products and consumers of fossil fuels; research laboratories and organizations working on one or more of each of the renewable energy products, programs, and sources; public transport member organizations; participant groups for all modes of transportation; consumer protection, health, and safety groups; taxing bodies and environmental protection agencies at all levels of government; and most likely twice as many as these or more. Given the diversity of these groups it is a wonder that any energy policy legislation ever gets passed. These energy stakeholders may be placed into four categories:

***Prime Movers*:** These are the energy industry producers and distributors. They include the major oil companies; electricity, oil, and gas producing companies; and the pipeline, storing and marketing groups, and organizations that ensure energy is where it is needed when it is needed. To these groups must be added the energy-rich countries, and their national oil companies.

***Industry Shapers*:** These are the facilitating organizations that keep the flow of energy working efficiently and effectively. They include banks, insurance companies, commodity exchanges, transportation groups, industry professional and management organizations, and energy research organizations, among others. It is important to note that this group also includes a large and growing group of organizations working *against* policies endorsed by prime mover organizations and industry sectors.

***Energy Users*:** The three fundamental classes of energy users are (1) consumers, (2) industrial and commercial users, and (3) governments. The groups working on behalf of consumers include organizations that lobby in behalf of special consumer groups, independent consumer protection organizations, and other bodies with a stake in protecting consumers' quality of life. Groups standing up for commercial users of energy are largely comprised of industry and management associations, their hired lobbyists and public relations counselors, and similar bodies organized to protect the interests of business and industry. Governments are very large consumers of all types of energy. State and large municipal and county governments lobby on their own behalf during policy-making sessions. They also join smaller government bodies through groups and associations.

***Energy Industry Regulators*:** The fourth group of stakeholders includes the federal, state, and local governments charged with regulating the three prior groups to ensure a steady, reliable, and affordable supply of energy. A paradox of governments as industry regulators is that they are also large consumers of all types of energy. They therefore often find themselves in conflicting positions as they develop regulatory policy upon the very organizations upon whom they depend for uninterrupted supplies of energy at stable prices—while also being pressured by interest groups with opposing and/or negative goals.

The federal government regulates nuclear energy and is the primary regulator of domestic production of oil and gas. State and local governments regulate most electrical energy distribution within their borders, while federal and regional bodies regulate production and transmission of electricity. Since the pluralist perspective emphasizes staking and advocating for particular positions, and aiming to win in negotiations, each of these groups at one time or another must receive answers to one or more of the questions that concern their stakeholder group. Typical examples of such questions include:

1. What's in it for us?
2. What will be required of us as energy supplies continue to dwindle?
3. How much is it going to cost?
4. Who is going to pay for it?
5. How much will the policy that emerges disrupt the status quo?
6. How long will it be before government adopts a new policy direction?

BOX 5.1 POLICY ANALYSIS FOR ENERGY DECISIONS

Decision makers in the energy sector and the political system need the best possible data to inform their perspectives on the issues, lay out the range of options, and select the best alternatives. Energy policy analysts from a variety of backgrounds seek to provide them with a range of products that will help them make better decisions. These products include data on energy sources and demand, forecasts of energy use, descriptions of the energy system and available energy technologies, and models that help to capture the complexity of energy systems, evaluate ongoing programs and predict the effects of new or revised policies.

Energy is grounded in the action of physical systems, and is also fundamental to our society and economy, making its analysis complex and inherently interdisciplinary. This creates a fascinating challenge for policy analysts at two levels. First, the research itself is difficult. Models of energy policy impact may start at a simple and descriptive level, but quickly become more complicated. They must accommodate the physical properties of fuels, economic assumptions about cost, the reactions of the overall economy to prices and availability of energy sources, taxes, consumer and firm behavior, and the reaction of regulatory institutions. The more complex the proposal or policy, the more powerful the model must be to reflect that complexity.

The second challenge is how to communicate the results to policy makers; managing the interface among scientists, economists, and policy makers thus becomes important in its own right (Engels 2005). Findings must be presented in formats that will be helpful for decision makers with a variety of preferences in terms of complexity and detail.

Perhaps the most ambitious energy models yet created are the integrated assessment models for climate change policy. These combine the scientific, economic, and social components of the climate problem to analyze relevant policy options. A well-known example is the DICE model created by Nordhaus (2008). The large number of variables involved and the particular strengths of the analyst lead such models to generate somewhat different results. Yet this is a healthy outcome since users may examine the sensitivity of the results to varying assumptions and datasets.

Basic data on energy in the United States and worldwide are collected and distributed by the DOE's Energy Information Administration (EIA). The EIA and other DOE divisions, notably the National Renewable Energy Laboratory, plus the main congressional agencies, the CRS, Congressional Budget Office (CBO), Government Accountability Office (GAO), and congressional staff members perform extensive analysis of energy issues and policy problems. At a global level, the International Energy Agency (IEA) is a key data and analysis resource. Most governmental analysts attempt to be thorough and objective and the quality of analysis provided by the congressional agencies is generally very high. The CBO has the particularly awkward task of providing estimates of the costs of legislation. In a PAYGO world where costly new programs passed by Congress must be linked to new revenues, this is a setup for inevitable friction.

The EIA and IEA also regularly publish comprehensive energy analyses that include forecasts for future energy demand and supply, and energy scenarios. The various assessment reports of the Intergovernmental Panel on Climate Change (IPCC) both summarize the data on climate change and provide detailed analysis of emissions scenarios under various assumptions.

The world of policy analysis also includes decidedly nonobjective analysis. Think tanks, policy institutes, and trade associations regularly put out analyses and proposals that support or attack particular policies. Carbon taxes and cap and trade are popular energy topics for such analysis. And the release of even an outline of a new energy bill on Capitol Hill is met eagerly by lobbyists and energy activists, whose first task is to hand the proposal off to policy analysts who parse the language, try to figure out what is being proposed, and estimate its costs and effects. Dueling analysts work from a set of ideologically driven assumptions to arrive at such estimates and generate conclusions relatively supportive or dismissive of the proposal; these analyses are now rapidly posted to the Internet. The challenge for the consumer of all such studies is to carefully consider the likely preferences of those analysts, their assumptions and thoroughness of their approach, and how much credibility to give their results.

7. How can we participate in the many research and development programs being funded by government?
8. How can we be sure our position is heard and considered as government policy makers hammer out a policy that hurts the fewest while it helps the most?
9. How long will it really be before we run out of fuel for our vehicles?
10. How can I prepare for that day when we can no longer afford the high price of fossil fuels?
11. If we can't turn environmental warming around in time, how can government help us prepare for its adverse effects?

Stakeholders also seek to influence the analysis of energy issues, as examined by Box 5.1

The Energy Scope Challenge

One of the biggest challenges facing energy policy makers is the great scope of the forms or sources of energy—each with its several special interest groups lobbying for government precedence. Indicative of this complexity, the U.S. Department of Energy official Web page lists fourteen energy topics under its administrative oversight: bioenergy, coal, electric power, fossil fuels, fusion, geothermal, hydrogen, hydropower, natural gas, nuclear, oil, renewables, solar, and wind.

Government agencies and individual states are also lobbyists for the energy policies they believe will best serve their missions and responsibilities. This issue was manifested more than 60 years ago in the Interior Department's acquiescence to pressure from the petroleum industry to cancel President Truman's proposed synthetic fuels research and demonstration plant program. The National Petroleum Council, representing oil producers, completed a cost analysis using Bureau of Mines data, and conveniently concluded that oil made from coal was uneconomic. Interior officials meekly agreed and passed the report on to Congress—despite the president's support for the program (See Box 5.2).

BOX 5.2 EARLY MISMANAGEMENT IN ENERGY POLICY

Perhaps the Interior Department was at fault for not defining appropriate premises or articulating a rationale for the program against which cons and benefits could be measured. Perhaps it erred by not interpreting the Council's report when it passed it on to the Congress. If a public bureaucracy acts as a mere pipeline, rather than a filter, for advice from the private sector, then it is not fulfilling its own institutional responsibilities that are necessary to make the advisory-council mechanism work. ... In the broadest sense, energy-market conditions framed the public policy issue. ... A mix of intragovernmental and interfuel politics characterized the political process (and) bureaucratic mismanagement was certainly in evidence ... the Bureau unnecessarily minimized its cost estimates and the Interior Department provided inadequate direction.

(From Vietor, R. H. K. 1984. *Energy policy in America since 1945.*

Energy Stakeholder Policy Preferences

The "stages" model of public policy suggests that crafting public policy occurs through a four-stage process: initiation, formulation, implementation, and evaluation (Heywood 2000). During the initiation stage, policy makers and various stakeholders come together to define social, economic, or political events as problems needing concerted attention and resolution, and then begin to formulate appropriate responses. Each group of stakeholders in the energy arena, however, tends to define the policy "problem" in ways that lead to policy options that will benefit them in the long run. There are four problem and solution sets analyzed in detail below. These focus on supply, demand, national security, and the environment.

Focus on Supply

Stakeholder groups pushing for public policy that focuses on supply issues can be grouped into just two different classes according to what they think is the best way to solve the "energy crisis." Because of their focus on supply, stakeholders in this class may consist largely of producers and distributors of energy supplies.

A very large and powerful group of stakeholders comprise primary energy producers and distributors who have a vested interest in maintaining the continued emphasis on oil as our primary energy source for as long as possible. Hence, they promote boosting supply by increasing exploration and drilling for more oil and gas. They can further be grouped in one of the following categories based on their beliefs regarding the best way to secure additional oil or gas: (1) drilling in the Arctic National Wildlife Refuge, (2) drilling for more domestic oil and gas offshore of all U.S. shoreline states and on more public lands, (3) placing their faith in improvements in science and technology will make it possible to get more oil and gas out of existing fields, or (4) developing more efficient extraction of oil and gas from the United States and Canada's extensive oil sands and oil shale reserves.

Another group believes that the U.S. energy supply issue is best addressed by a policy that would focus on developing efficient and economic electrical power from U.S. fuel sources. One solution promoted is the production of liquid and gas fuels from the very large deposits of coal in the United States. Maintaining a steady supply of reliable, affordable energy can best be achieved if this source were more rigorously exploited. Another group—now becoming more vocal than in the last decades of the twentieth century—promotes production of electrical energy by nuclear power. Not surprisingly, little mention of how to permanently and safely dispose of nuclear waste is heard from this group since much of this material will remain hazardous for 250,000 years (Sovacool 2008a).

Focus on Demand

Stakeholder groups that are pushing for public policies that focus on demand issues can also be grouped into two classes based on what they see is the best solution for the energy crisis. The first group belongs to the extreme conservation class, arguing that reductions in per capita electricity and energy use are needed.

The second group emphasizes greater efficiency in energy consumption as the most attainable and economically feasible solution. Because of the focus on demand, the groups in this class are often affiliated with or represent consumer or industrial users of energy products. Groups promoting conservation believe that conservation and energy efficiency is the best short-term solution to the energy crisis, and most of the available data on the cost-effectiveness of energy interventions supports this. They have also been very effective at the state level in promoting alternative means of transportation, building insulation, and lower lighting levels, for example. Stakeholder groups promoting greater fuel efficiency continue to push for fuel efficiency in motor vehicles, appliances, lighting, public buildings, and boosting the energy content of some alternative fuels.

Focus on National Security

The third energy policy focus is driven by stakeholder groups who hang their solutions to the energy problem on national security. These groups are both highly vocal and ubiquitous. They belong to two closely related goal groups and tend to promote any policy that moves the country toward the ethereal, but politically powerful, goal of *total energy independence.*

The stakeholders in this class are the professionals directly involved in some way with national security. As a result, many groups in this class have public sector connections or political affiliations. They also often have the ear of senior policy makers.

The second group in this class uses the security issue as a shield behind which they promote substituting domestically produced alternative fuels for imported oil and gas for their own benefit, however meritorious. Some of these groups have issues with economic globalization; they see dependence upon foreign energy as another manifestation of the evils of foreign trade.

Focus on the Environment

The fourth class of stakeholders consists of the many groups and organizations that associate a large portion of environmental degradation with the burning of fossil fuels. They see coal-fired production of electrical energy as one of the most insidious contributors to global warming and acid rain. Two classes of interest groups emphasize an environmental solution to the energy problem. One group focuses on the supply side of the question; another addresses demand issues.

Supply-side environmentalists (often citing Pacala and Socolow 2004) push for such diverse solutions as "clean coal" and carbon sequestration, development of a hydrogen-based fuel infrastructure, and greater investment in wind and solar power, among others. This body of stakeholders accepts the idea that the nation's energy needs are going to increase despite what damage to the environment results. Their solution to this problem is mitigation of the damage by substituting cleaner-burning fuels. Recognizing that an economically viable means of producing enough nonpolluting energy is still not available, they place a great deal of confidence in the world's scientific and technology community to be able to solve some if not all supply problems in time to fend off any energy shortage. Some identify 2020 as the year that economically viable clean energy will be available; others name 2030 or 2050 as the year for sufficient supplies of clean energy to be available. No one really knows.

Demand-side environmentalists place their faith in the power of conservation and more efficient use of renewable sources such as wind power, concentrated solar, and photovoltaics as solutions to the energy problem. This body of stakeholders promotes such conservation and energy efficiency programs as more energy-efficient appliances and buildings, more insulation, lower winter and higher summer thermostats, banning of incandescent light bulbs, higher mileage standards for motor vehicles, and similar programs.

Widely Different Perspectives

The four stakeholder perspectives examined above are not completely mutually exclusive, but they are based on very different perspectives, values, and goals for the United States as a whole and its energy systems in particular. As discussed earlier, the battle in Congress, state legislatures, and in the court of public opinion is often over the most basic issue, the framing of the problem.

But whose frame will "win" and why? Who will have the ear of important members of Congress, their staff members, or their equivalents at the state level? How can public opinion be swayed? One long-run strategy is to support the development of ideas and data that support your preferred frame and, not incidentally, aim to undermine the arguments made by your opponents.

The think tank sector has expanded to support this strategy. In recent years a number of progressive think tanks, including the Center for American Progress, have emerged in an attempt to counter the decisive advantage in funding and size held by conservative and libertarian organizations such as the Heritage Foundation and the Cato Institute.

New organizations are created regularly. These groups produce research reports, position papers, and blogs that usually make no pretense of academic balance or objectivity. But they are highly effective at distilling complex data and issues into clear conclusions and policy proposals that happen to favor particular ideological points of view, and are usually tailored to be chopped up into sound bites for a willing and eager mass media. This strategy allows stakeholders to avoid the inevitable charges of self-interest that result when they produce such reports on their own.

By funding multiple organizations, interest groups can create what Burton (2007) termed an "echo chamber" effect, as the same policy message and prescription is repeated by different groups. Government agencies are cautious participants in this game, as they produce position papers that support administration policy, or more rarely, grapple with a problem in depth. In the energy sector, battles over energy supplies, regulation, auto fuel economy standards, the cost and efficacy of renewables, and climate change are staples of the think tank sector.

But there are two other time-honored means to influence both the bureaucracy and policy makers in Congress and statehouses: contribute to political campaigns and directly lobby individuals with clout. Campaign contributions influence which individuals and parties will be elected and thus have power. Electing members sympathetic to your industry and ideological preferences increases the likelihood that policies supportive of your industry and company will remain in place. Lobbying means monitoring the ongoing process of policy development and intervening every time changes are pondered that may hurt your group or company, or when policy windows open that imply that the system is receptive to change that could benefit you.

The Center for Responsive Politics (CRP) analyzes electoral contribution data compiled by the Federal Elections Commission, and lobbying data collected by the U.S. Senate, to show which sectors, industries, corporations, and lobbying organizations spend the most on lobbying and elections contributions. Overall, during the period 2000 to 2008, individuals and organizations in the Energy and Natural Resources sector contributed $304 million to candidates in federal elections, with 72 percent of those funds going to Republicans.

Slightly less than half of those contributions ($141 million) came from individuals and organizations in the oil and gas sector. Typically, these contributions go to candidates from the states where they have headquarters and refineries, produce oil and gas, and to individuals in leadership positions in both houses of Congress and on important committees—as well as to presidential candidates from each party (Juhasz 2008). During the same period, individuals and companies in the alternative energy sector also contributed. But their contributions to federal candidates totaled $3.8 million, about 1.5 percent of the donations in this sector.

Although these are impressive figures, they are dwarfed by the amounts spent on lobbying. Our analysis of the data shows that during the same period (2000 to 2008), firms and lobbying organizations in only the top 13 sectors tracked by the CRP spent a total of $20.2 billion on lobbying activities. The top two sectors, unsurprisingly, were Finance, Insurance, and Real Estate ($3.1 billion) and Health ($2.9 billion), with the Energy and Natural Resources sector ranking fifth, with $2.0 billion spent on lobbying over that period. The amount on lobbying spent per year by these sectors doubled over that period, from $1.5 billion to $3.2 billion.

What do they accomplish through spending such vast sums? At times, direct support for their positions. One analysis of campaign contributions and its impact on energy legislation found that members of Congress voting against the bill (HR 2776, The Renewable Energy and Energy Conservation Act of 2007) received four times the campaign contributions from the oil and gas industry compared to those who voted for the bill (Weiss and Wingate 2007). Yet most analyses of the influence of campaign contributions on legislative behavior are highly nuanced.

Difficulty Showing Links

Showing direct links between campaign contributions and voting patterns is surprisingly difficult. The influence of corporate contributions, however, may be enhanced in the future by the recent Supreme Court decision *Citizens United v. Federal Election Commission*, the effect of which will be to allow corporations to purchase advertisements in favor of particular candidates.

The real results come from lobbying. Here, the policy network metaphor is instructive. Well-financed firms will attempt to overload the policy network to limit the risk that policy changes will not be in their favor. The implications for energy policy are clear: proposals that threaten the status quo will be fiercely resisted unless they accommodate the threatened parties.

Vietor (1987) came to the fantastical conclusion that "despite its superior resources, business exercises limited influence on public policy." That is no longer the case, if it ever was. Although the policy changes in this arena since 2005 show that the blocking power of firms and lobbyists in the energy sector is not absolute, it is still substantial. Most recently, the Obama administration's attempts to phase out subsidies to oil and natural gas production starting with the 2011 budget were dead on arrival in Congress. The collision of an activist government with a powerful and well-financed corporate sector makes it highly likely that lobbying activities will continue to grow. As the debate over climate change policy becomes more heated, the amounts spent in the energy arena will increase.

Ambiguous and Conflicting Policy Goals

What, exactly, are the goals of U.S. energy policy? And is there clear alignment and consistency between the goals and methods of the particular policy approaches enacted and implemented in behalf of those goals? In an ideal world, the country would focus on clear goals and craft strategies and policies that would help us to attain them at the lowest cost, in the process avoiding policies that work at cross-purposes. Other countries, notably Japan and France, have generally succeeded at implementing such comprehensive strategies.

Alas, the crafting of policy goals in the United States is a highly political process, the results of which tend to enmesh public administrators in ambiguity. Scholars of public administration argue that public agencies are caught in a bind (Wilson 1989). When crafting legislation, members of Congress must satisfy the interests of multiple committees, bill sponsors, and administration officials, and focus on creating the coalitions needed to pass a bill.

This means that broad goals are emphasized, politically powerful constituencies are appeased, and inconsistencies are ignored. The exceptions occur when legislators don't trust the agency and instead use legislation to micromanage the implementation of a policy. The result is that legislatures mostly give agencies broad and ambiguous legislative mandates, expecting them to sort out the meanings and intent of the legislation. Agency personnel must then puzzle through how to implement their new mandate.

Setting National Goals

One way to avoid some of this ambiguity is to set clear national goals. Formal statements of national goals and intentions appear most consistently in each presidential administration's January budget proposal to Congress. Occasionally they appear in policy documents. In recent years, they are also stated in documents on the White House website. The considerable shift in approach to energy policy between the Bush and Obama administrations is captured in two documents. The report of the National Energy Policy group chaired by Vice President Cheney made clear the group's preferences: "[W]e urge action to meet five specific national goals. America must modernize conservation, modernize our energy infrastructure, increase energy supplies, accelerate the protection and improvement of the environment, and increase our nation's energy security" (Cheney 2001).

The Obama administration's change in policy is made clear in this statement from the White House website: "To take this country in a new direction, the President is working with Congress to pass comprehensive legislation to protect our nation from the serious economic and strategic risks associated with our reliance on foreign oil and the destabilizing effects of a changing climate. Policies to advance energy and climate security should promote economic recovery efforts, accelerate job creation, and drive clean energy manufacturing." (White House 2010).

Both of these statements acknowledge the dominant U.S. energy policy goal, which may be captured in one word: *Enough*. Americans expect to be able to fill their gas tanks when they're empty, and to have light when they flick the switch. The overwhelming emphasis of our energy system at all levels is on ensuring sufficient supplies are available to meet the needs of consumers, businesses, and government. Further, we have chosen to do that through reliance on markets, and corporations that provide almost all of our energy supplies.

There are other important goals. One is to provide our energy supplies at the lowest possible cost. The design of our transport system, in particular, is predicated on relatively cheap gasoline. Another is the Holy Grail of energy independence. This mantra is repeated by politicians of all ideologies, but their favored paths to reaching a state of energy nirvana differ widely. A third is to place some limits on the market power of corporations supplying energy.

Strategic Limitations

A strategy based on "enough, at low cost" has serious limitations. Since we only have about 2 percent of world oil reserves, it means we must import vast quantities of oil and petroleum products (now about 57 percent of our consumption) to run our transport system. In the electricity sector, we continue to rely on public utilities that are guaranteed a rate of return in exchange for meeting demand within their service area. Most utilities have incentives to sell more, not less, electricity, and to generate it using the lowest-cost source. That is still coal, if the external costs are not factored in. In both of these cases, we satisfy "needs" for mobility and electricity, but at a very high environmental and social cost, in part due to the staggering levels of technical inefficiency discussed in Chapters 1 and 2.

Thus, the second-order energy policy goals: to meet needs while using fewer energy inputs, or inputs that are perceived to be less wasteful and more environmentally friendly. In addition, the process of meeting these goals is usually intended to have important spillover effects, in the form of generating economic growth and jobs for U.S. workers.

The challenge for energy policy is that the main goals espoused by leaders from both parties—energy independence, job creation, low cost, and increasingly, minimizing carbon—are mutually exclusive, at least in the short run. There are plausible scenarios in which the transition to a lower-carbon economy may be brought about with only a slight increase in cost, as we'll discuss in the next chapter. But higher energy costs are likely. And it is difficult to design policy strategies that cause energy-related investments to encourage growth in U.S. employment, without leaking into demand for foreign products.

The Buy American provisions of the 2009 American Recovery and Reinvestment Act (ARRA) or the Stimulus Act sought to encourage demand for U.S. iron, steel, and manufactured goods. But this caused considerable complications for those who had to implement the provisions, and ran counter to the main objective of ARRA, which was to spend money on infrastructure projects quickly to stimulate the economy. And they also caused consternation with U.S. trading partners, who rightly criticized this as a protectionist measure inconsistent with our obligations under the world trading system.

The Nature of Energy Policy Interventions

Chapters 7 and 8 will examine in detail many of the energy policy interventions that are likely to be relied on as we transform our energy system away from its reliance on carbon-based sources. Here we will focus on a broader topic: the general features of energy policy interventions that make them imperfect means to accomplish important energy goals. The available approaches to meeting those goals each have significant limitations. Yet once enacted in law and regulation, changes in the resulting constellation of programs and benefits is often difficult to revise in response to changing conditions. The very difficulty of changing the behavior of complex systems means that it can be challenging to calibrate the size of intervention needed, with the result that double-barreled approaches to change may contain inherent contradictions.

The traditional way in which most of us conceive of *government* means government agencies that implement programs designed to accomplish public goals. This includes provision of direct services by government employees, or regulations crafted and implemented by those employees. That perception, always an oversimplification, has been blown away by the transformation of the federal government into a vast mechanism for intervening more indirectly in the various sectors of our society and economy, using what Lester Salamon (2002) terms "tools."

These tools are systemic and encompass the intervention, a way of delivering that intervention, organizations in that delivery system, and rules for defining the goals of the activity/intervention and the interrelationships between the various entities involved. The activities include social and economic regulation, contracting, grants, loans and bonds, loan guarantees, subsidies/tax expenditures, liability law, government corporations and ventures, and vouchers. The delivery system may include public and nonprofit organizations, universities, the banking system, the tax system, or the legal system.

U.S. government programs in the energy arena may be pictured as a toolkit, mostly filled with levers of varying size, each designed to make use of a fulcrum to move the behavior of some element of the energy system in a desired direction. Tax credits, for example, are a favored means to encourage households and businesses to invest in energy efficiency or new forms of energy such as solar or wind. A portion of eligible expenditures on more efficient appliances or high-mileage vehicles may be directly deducted from taxes.

There are, metaphorically, also a few hammers used to apply brute force and some sketchpads and pencils for doodling out solutions to problems. One such hammer is an implicit policy, reliance on high market prices, to curb demand. And a "doodle" is the extensive set of programs designed to encourage research and development into new technologies.

Table 5.1 summarizes the federal government's spending on energy-related programs in fiscal year 2010. These expenditures reflect programs in several different agencies, including the Departments of Energy and Agriculture, and regulatory agencies including the Federal Energy Regulatory Commission and Nuclear Regulatory Commission. Also included are an estimate of energy tax expenditures, the budget for EPA programs with a direct link to energy, offsetting receipts, such as various revenues generated by the programs through user fees, and the cost of the Climate Change Research Program. This is a sizable figure, and given our earlier discussion about the federal government's fiscal stability, one that is cause for concern.

Table 5.1 US Government Budget Allocations for Energy Programs FY 2010

Executive Branch programs	$ 24,235,392,187
Offsetting receipts	$ (2,500,000,000)
Energy Tax Expenditures	$ 18,600,000,000
EPA and Climate Change Research	$ 2,497,877,000
Total	$ 42,833,269,187

Sources: Agency budget requests; U.S. Congress, Joint Committee on Taxation 2010; Office of Science and Technology Policy, 2010; CBO, 2009.

The "offsetting receipts" row of Table 5.1 reflects income earned from wholesale electricity sales, sales of uranium, oil from the Strategic Petroleum Reserve, and user fees for a variety of regulatory programs. The "tax expenditures" estimate includes tax credits and a variety of other programs that aim to encourage desired energy activities through reducing tax. Estimates of the costs of those programs vary considerably. Congress's Joint Committee on Taxation estimated $14.7 billion in energy-related tax expenditures in 2009; the Office of Management and Budget (OMB) estimate was $5.33 billion.

The breadth and scope of U.S. government energy programs is impressive. The U.S. Department of Energy official Web page, for example, lists fourteen energy subarenas under its administrative oversight: bioenergy, coal, electric power, fossil fuels, fusion, geothermal, hydrogen, hydropower, natural gas, nuclear, oil, renewables, solar, and wind. Within each of these there are a variety of programs, using many of the tools mentioned previously, from direct grants to loans, loan guarantees, tax credits, and insurance.

The 2010 DOE budget contains seventeen different line items supporting its energy programs, including substantial support for research and development through the department's twenty major laboratories and the new Advanced Research Projects-Energy program, plus the power marketing administrations. The specific programs financed by the Recovery Act provide a sense of the high-priority initiatives, including $16.8 billion for investments in energy conservation and renewable energy sources, $6 billion for environmental management, $6 billion in loan guarantees for renewable energy and electric power transmission projects, $4.5 billion for power grid modernization, $3.4 billion for carbon capture and sequestration, and $1.6 billion for basic science research. And that does not include tax expenditures, programs in other departments, or the regulatory activities of the Federal Energy Regulatory Commission and Nuclear Regulatory Commission.

Complex and Unwieldy Interventions

How did we end up with such a complex and unwieldy set of interventions? Since a primary energy policy goal has been to guarantee low-cost supplies, policy makers must consider ways to influence supply, either directly or indirectly, by reducing demand. Bamburger (2006) suggested that many energy interventions seek to influence either supply or demand for particular energy sources, with a focus on a particular time horizon, as shown in Table 5.2.

Why these particular interventions? First, we rely on markets or arenas for transactions that bring together buyers and sellers. More direct interventions, such as state ownership of the means of production, are not culturally acceptable. So the primary lever available to policy makers is to influence the cost and terms of market exchanges. Also, Americans are allergic to taxes that directly raise prices for goods. Gasoline taxes are grudgingly accepted as a user fee that pays for highway construction and repair. That leaves the use of loans, various other subsidies, and the tax system as an arena for lowering costs and encouraging desired investments and expenditures.

Another interpretation is that individual policies were added over time, ad hoc, with no attempt at coordination of the many policies designed for different energy sources and uses (Switzer 2001). This is supported by an analysis of energy law, which is very fuel specific (Tomain and Cudahy 2004).

The positive spin on the breadth of these programs is that the U.S. political system has acknowledged and accommodated the need to invest in energy efficiency, renewables, improving the efficiency of the U.S. electricity system, and for basic research and diffusion of the results of

Table 5.2 An Energy Policy Framework

	Focus	
Time Horizon	*Supply*	*Demand*
Short-term	Strategic Petroleum Reserve High prices	High prices Conservation and efficiency programs
Long term	Promote production—tax incentives Promote transition to cleaner sources Open federal land to exploration or leasing (ANWR, Continental Shelf, etc.) Research and development—new sources of energy Rely on market prices	Auto fuel economy standards Use other forms of rationing Conservation and efficiency programs Research and development in efficiency and conservation High prices through tax or markets

Source: Adapted from Bamberger, R. 2006. *Energy policy framework and continuing issues.* (May 11). Washington, DC: Congressional Research Service (CRS), Library of Congress.

that research into the economy. The less flattering interpretation is that the system is incapable of setting clear priorities and thus continues to fund everything. And the influence of the logrolling, pluralist and money-laden political system is evident. New subsidies and programs are added, but old ones rarely go away. A new presidential administration takes over, and new priorities and initiatives are emphasized, and old ones, if authorized and funded by Congress, continue.

The FutureGen Program

A recent example is the FutureGen program. The goal of the program, begun in 2003, was to show the feasibility of carbon capture and storage (CCS) of emissions from coal power plants, through the use of advanced technology integrating coal gasification with combined cycle electricity generation. In January 2008, the Bush administration decided to restructure the complex $1 billion program from a DOE R&D project to a commercial demonstration project, citing cost increases.

This shift would have killed development of a planned facility, likely to be located in Illinois, supported smaller carbon sequestration projects at coal power plants around the country, and reduced the program's cost. This left Illinois politicians very unhappy, particularly Senator Dick Durban (the Senate Majority Whip and thus its second-ranking Democratic member). Sen. Durban convinced his former colleague, now President Barack Obama, to revive the program in mid-2009. Sen. Durban's case was enhanced by a GAO report criticizing the DOE for making the decision to restructure the project without a comprehensive analysis of the two options (GAO 2009b). The question, however, is really how the Congress and DOE make strategic decisions about research. We need to know, quickly, whether CCS is at all viable as a means of reducing CO_2 emissions from coal, and it is hardly clear that FutureGen is the best way to meet that objective.

Need to Rethink Energy Subsidies

Another challenge in the energy arena is whether, and how, to rethink the subsidies provided to energy producers in the United States. The data on federal subsidies for energy are contested, and estimates can vary dramatically based on assumptions about which programs to include and how to value them. An EIA study of 2007 energy subsidies, shown in Table 5.3, estimated total subsidies that year at about $16.6 billion, a more than 100 percent increase from the $8.2 billion in 2004 (both figures in 2007 dollars). But this estimate omits some significant subsidies for oil companies, including the Foreign Tax Credit and the intentional reduced "take" or royalty rate for oil and gas taken from federal lands and on the Continental Shelf (Adeyeye et al. 2009).

Koplow's 2006 analysis concluded that fossil fuels received about 66 percent of the estimated $74 billion in U.S. energy subsidies that year; nuclear received $9 billion, with ethanol and renewables receiving about $6 billion each (Koplow 2006b). More recent studies have found a trend away from fossil fuel subsidies and toward renewables, including solar, wind, and ethanol (Bezdek and Wendling 2004; 2007). And with the Recovery Act, spending on energy subsidies for renewables has increased dramatically.

Although Chapter 7 will examine specific subsidy programs in detail, there are two problematic aspects of the subsidy issue that are serious challenges to U.S. energy policy. The first is how to design subsidies and other energy programs to optimize their impact and avoid costly policy errors. How do we accomplish the most abatement of greenhouse gases with the minimum expenditure of tax dollars, while causing the least distortion to markets and production processes? We will discuss this further in Chapter 7, but one answer is to avoid favoring particular energy source solutions, and to instead encourage the lowest cost abatement solutions—often, investing in energy efficiency. But it is difficult for the political system to resist favored solutions.

Table 5.3 U.S. Energy Subsidies, 2007 (US$ millions)

Beneficiary	*Direct Expenditures*	*Tax Expenditures*	*R&D*	*Electricity Subsidies*	*Total*	*Percent*
Coal	0	$2,660	$574	$69	$3,302	20%
Oil and Gas	0	$2,090	$39	$20	$2,149	13%
Nuclear	0	$199	$922	$146	$1,267	8%
Renewables	$5	$3,970	$727	$173	$4,875	29%
Electricity	0	$735	$140	$360	$1,235	7%
End Use	$2,290	$120	$418	0	$2,828	17%
Conservation	$256	$670	0	0	$926	6%
Total	$2,551	$10,444	$2,820	$768	$16,582	100%

Source: EIA. 2008b. *Federal Financial Interventions and Subsidies in Energy Markets 2007.* Washington, DC: US Department of Energy Office of Coal, Nuclear, Electric and Alternate Fuels.

Problems with Biofuels

For example, U.S. biofuels policies are particularly problematic. U.S. federal biofuels policy now has three main components. The first is a renewable fuel standard, or RFS, put in place by the Energy Policy Act of 2005, and expanded in the Energy Independence and Security Act (EISA), which mandates a gradual increase in biofuels production, up to 36 billion gallons per year by 2022. This legislation includes additional requirements for production of advanced biofuels, which are not made from corn.

Second is the Volumetric Ethanol Excise Tax Credit (VEETC), enacted in 2004, a tax credit of 45 cents paid to oil refiners, blenders, and marketers of gasoline on each gallon of ethanol blended with gasoline. This credit is due to expire at the end of 2010. Third are the tariffs now applied on ethanol imports. As many observers have pointed out, it makes little sense to both require use of a fuel, through a mandate, and to subsidize its use through a tax credit (Koplow 2009; de Gorter and Just 2010). Moreover, the advanced biofuels mandated under EISA are not currently economically feasible, meaning that additional levels of required production will be met with corn ethanol. By one estimate, U.S. government biofuels subsidies will total $92 billion between 2006 and 2012 (Koplow 2006a) and this does not even reflect the many biofuels subsidies at the state level.

Can We Afford All Planned Energy Programs?

This leads to the second issue. Can we afford to continue all the federal energy programs now in place? The deteriorating fiscal status of the federal government means that there will be increasing pressure on the Obama administration to produce a viable plan to reduce future deficits, now $1 trillion per year for the foreseeable future. Some of the nation's energy programs will be cut, and the critical question will be, how will these cuts be decided? The department needs to engage in planning now to identify programs that will meet the country's energy goals while minimizing cost and economic distortions. This could be a backward means to accomplish the Holy Grail of energy analysts in the United States since the 1970s, the creation of a sensible and comprehensive U.S. energy policy. Changing directions in energy policy is also difficult, as described in Box 5.3.

But is it either possible or desirable to integrate the many strands of U.S. energy policy, organized around the great variety of forms and sources of energy, with specific laws and regulations by fuel and goal, into one comprehensive national energy policy? Some observers have concluded that it is an impossible task. Each subarena now has its several special interest groups lobbying for government precedence, supported by the offices and divisions of the DOE and other federal agencies. We will return to this issue in the next chapter.

The Innovators: States, Regions, Compacts

Much of the action in the U.S. energy policy arena over the past two decades has occurred at the state government level. Sawyer (1985b; 1985c) concluded in the mid-1980s that states were moving into the vacuum left by budget cuts at the federal level, as the Reagan administration slashed funding for conservation and renewable energy programs. Since then, states have acted for a variety of reasons, including concern over climate change and federal inaction on that problem, air quality

BOX 5.3 PROBLEMS IN CHANGING DIRECTIONS IN ENERGY POLICY

An essential transition to inexhaustible and renewable sources of energy has yet to begin. Because of the importance of energy in the structure of national economies, we face the possibility of major political and social upheavals if energy need and expectations are not met. But very little seems to be happening to deal effectively with any of these situations, for reasons that are social, economic and institutional…

Most fundamentally, our [energy policy] crisis is one of irresolution. We find it agonizingly difficult to form a broad consensus in American society about our objectives and how to reach them. We are undecided about institutional roles and relationships, uncertain about impacts and risks, and unclear about the connections among social, economic, environmental, and energy policy goals.

Meanwhile, adversarial decision-making structures convert uncertainties into disagreements or, worse, into antagonisms, and prospects for action fade away time after time.

(From Gentemann, K. M., ed. 1981. *Social and political perspectives on energy policy*).

issues, and a desire to have greater control over their own energy futures. Another contributing factor was a belief that state action would encourage growth in green energy jobs.

There has been wide variation in state involvement in energy issues. Chandler (2009) concluded in his review of state renewable portfolio standards that states with "high renewable potential, educated and affluent citizenry, poor air quality, low industrial economic dependence, and more liberal government ideology" were more likely to support such programs. The result has been either a profusion of experimentation, or a hodgepodge of inconsistent and conflicting standards, depending on one's perspective.

The breadth of action has been considerable: 29 states plus the District of Columbia have renewable portfolio standards, and another six states have renewable portfolio goals; 43 have net metering laws; many have energy efficiency programs including 25 offering rebates and 38 loans to support various efficiency programs. Incentives for investments in renewables are also very common: 27 states offer sales tax rebates for renewable energy investments, 31 have property tax rebates, and another 37 states provide loans for renewables projects. Other kinds of production incentives and rebates are offered by over half the states. In addition, as of August 2009, 34 states have completed climate action plans that describe actions planned to reduce their impact on climate change; four additional states are in the process of creating such plans. At the local level, many cities and municipalities have also taken action to reduce their energy use and greenhouse gas emissions.

Regional GHG Initiatives Underway

There are also three major regional greenhouse gas emission initiatives underway, as shown in Table 5.3. The Northeastern states' Regional Greenhouse Gas Initiative (RGGI) is an ongoing cap and trade system for power plants in the region, while the Western and Midwestern regional programs are still under development. Notable nonparticipants in these initiatives are major oil and coal producing states, with the exceptions of California and Wyoming (Selin and VanDeveer 2009).

Table 5.4 Status of Regional Climate Initiatives

Initiative	*States*	*Program and Status*
Regional Greenhouse Gas Initiative (RGGI)	Connecticut, Delaware, Maine, Maryland, Massachusetts, New Hampshire, New Jersey, New York, Rhode Island, and Vermont.	Regional cap and trade system underway. Goal is 10 percent reduction in power sector emissions by 2018.
Midwestern Greenhouse Gas Reduction Accord (MGGRA)	Members: Iowa, Illinois, Kansas, Manitoba, Michigan, Minnesota, Wisconsin. Observers: Indiana, Ohio, Ontario, South Dakota.	Design of regional cap and trade system underway.
Western Climate Initiative	Arizona, California, Montana, New Mexico, Oregon, Utah, and Washington, and the Canadian provinces of British Columbia, Manitoba, Ontario, and Quebec. US observers are Alaska, Colorado, Idaho, Kansas, Nevada, and Wyoming.	Emissions targets; planned cap and trade program to be implemented in 2012.
	Florida	Individual cap and trade system. Planning underway.

Source: EPA. 2009b. *2009 Greenhouse Gas Inventory Report: Energy*. Washington, DC: US Environmental Protection Agency. http://www.epa.gov/climatechange/emissions/downloads09/GHG2007-03-508.pdf (accessed January 5, 2010).

Although the extent of state action in the energy sector has been impressive, it is difficult to assess its overall efficacy. California offers a useful example. The state's wide array of programs supporting energy efficiency and conservation has kept per capita electricity consumption within California level for 30 years. However, as of 2008, California's three investor-owned utilities were supplying about 13 percent of their total sales from eligible renewable resources. This is below the renewable portfolio standard (RPS) of 20 percent by 2010. Based on planned additions to renewable capacity, the 20 percent requirement is not likely to be met until 2013 or 2014 (California Public Utilities Commission 2009). A significant portion of the renewable power sold in California is generated in wind energy farms in Oregon and Washington—states that also have RPS standards to meet.

Many states initially set goals that were easy to reach, but recently they set more ambitious targets. As the dates of reckoning for renewables begin to arrive, how many will meet their goals, and will there be a scramble by states and utilities to claim renewable generating capacity, including renewable energy credits? Engel (2006, 1024) concluded after her review of state climate change programs, "The activity at the state and local level seems to be driven by the prospect of local economic benefits, political opportunism, and genuine concern that some government response to climate change should be forthcoming in the absence of strong federal leadership."

Another issue is the fiscal sustainability of state-based programs supporting investments in energy efficiency, conservation, and renewables. The recent Great Recession has mauled state government budgets, and energy programs have not been spared the ax. In 2010, for example, New Jersey Governor Chris Christie froze $286 million from two of the state's energy funds, and proposed for 2011 $68 million in cuts in programs funded by the Regional Greenhouse Gas Initiative, and $52 million in cuts to the Clean Energy Fund.

Oregon recently cut its Business Energy Tax Credit program by an estimated $140 million over three years. The program, which primarily subsidized construction of wind energy projects, became a target for cuts thanks to both its unexpectedly high cost and allegations of poor program design and implementation. But not all the news is bad; California approved a $3.1 billion budget for energy efficiency programs for its three public utilities for 2010–2012.

Contentious State–Federal Relationships

The relationship between the states and federal government over energy issues is occasionally contentious, but overall features clear and divided responsibilities. Since the arrival of deregulation, the federal government has focused on establishing market and regulatory frameworks, within which states have considerable scope for independent action. The respective roles of state public utility commissions and the Federal Energy Regulatory Commission (FERC) in the regulation of the electricity and natural gas sectors are relatively uncontroversial.

In the "not in my backyard" (NIMBY) era, the new potential battlegrounds are over siting of energy projects, including power plants and transmission lines. Some powerful energy stakeholders—particularly the prime movers described above—believe they may benefit from a move away from local control and toward federal preemption and the use of eminent domain to push through transmission lines needed to support a more comprehensive national grid.

With some form of climate change legislation likely to emerge from the U.S. Congress over the next few years, there are significant questions about the role and scope of state action, and what limits, if any, should be set on state energy subsidy and regulatory programs. The argument for federal action is clear: national-level action would provide a comprehensive and consistent commitment to a policy goal and support a clear national strategy. States and regions, however, can make the case that they share regional climate change threats, such as wildfires in the West and hurricanes in the Southeast. A significant risk of state or regional action is the threat of competition weakening standards, as some states seek a cost advantage in attracting industry and employment. For federal policy makers, a related question is whether new regulations should provide a floor on state action (Andreen 2009). Should states be allowed to aim for more stringent greenhouse gas (GHG) emissions targets?

Sharing Responsibility

Given a likely upcoming era of tight budgets, should the states and federal government seek a grand agreement on how to divvy up responsibility for energy programs? Two open questions after two-plus decades of state support for energy programs are, what the most effective interventions are and what an "optimal" overall level of intervention might be for energy investments. The states continue to experiment, but the track record appears uneven. The data suggest that investments in conservation and efficiency have a high payoff, and that renewable portfolio standards have encouraged investments in alternative energy. But the renewable fuels market is in trouble, and there are few data on a wide range of other state-level interventions. One policy option likely to be included in federal legislation is a national RPS, in which case one strategy would be for the federal government to support a national feed-in-tariff, and leave other direct energy subsidies to the states. This would allow state governments to design programs to best fit their particular energy profile.

Environmental Policy, Energy Policy, and Politics II

Energy policies are environmental policies. Our energy production and consumption choices, and strategies to influence those choices through public policy, inevitably have environmental consequences. Our current framework of energy policies does not reflect this reality, as described above in the section on interventions. There are good reasons for this. Using Kingdon's terminology, the policy windows for each policy arena tend to open for different reasons.

Energy policy has been driven primarily by reliability of supply and environmental policy by concern about pollution and a desire to reduce its health impacts. And the types of policy interventions that resulted traditionally differed. As Lowry (2008, 1196) suggests, "energy and environmental politics have been traditionally different in that energy policies had a stronger distributive component than environmental policies. ... the politics associated with energy policy were relatively consensual and focused on resource allocation, not the behavior-altering regulations of environmental policy."

It is true that energy policy has been relatively bipartisan. Major bills passed over the last 30 years have enjoyed support from members of both major parties. It's relatively easy to gather support for bills that are adorned like Christmas trees with shiny new programs and benefits for a broad array of energy stakeholders. Environmental programs, conversely, usually require new government regulations that aim to reduce or eliminate the actions that result in the release of harmful substances. Costs externalized to the environment are made explicit, and internalized, with the threat of government compulsion for noncompliance always in the background. The regulated—usually businesses—must bear higher costs and the challenge of implementation. Such programs are costly for business and strengthen the state (i.e., government) thus making them a difficult sell for political conservatives.

The exceptions confirm these generalizations. Fuel economy standards are a quintessential regulatory program, and have been the single most controversial energy policy issue (other than the Arctic National Wildlife Refuge [ANWR]) over the last 30 years. The major bureaucracies that are responsible for these two policy arenas at the federal level, the U.S. Department of Energy and the Environmental Protection Agency (EPA), are also very different. As suggested above, the civilian energy programs of the DOE, a relatively new agency, use virtually the entire toolkit of new public management interventions.

Managers are aware of the extent to which energy subarenas (e.g., solar, wind, efficiency, etc.) are ultimately reliant on leveraging state and private sector funds, and there is, arguably, an understanding of the need for quicker decisions and less suspicion of the private sector. The EPA, however, is a more cautious and bureaucratic agency. It is responsible for implementing some of the most difficult and costly legislation ever created by Congress, including the Clean Air Act and Clean Water Act. It is often forced to take an adversarial posture with its regulated entities, since taking the agency to court is often the preferred response to new regulation. The market-based cap and trade approach used for the acid rain program was thus a substantial change for the agency.

The Climate Change Challenge

Yet the primary challenge faced by the energy sector is that of climate change. What are the likely implications of making the primary focus of energy policy, environmental? Lowry suggests that energy policy will begin to look like environmental policy, with a more regulatory approach, making the arena increasingly partisan and divisive. Given the highly partisan tone of the 111th Congress, that is hard to bet against. Michael E. Kraft (2010) described how the two policy issues became so closely related during the Clinton and George W. Bush administrations.

Table 5.5 Primary Fuel for Generating Electricity by Region and State, 2008

Fuel	*Regions, States*	*Number of States*	*Average Retail Cost (cents/ kWh)*
Coal	**East:** Delaware, Maryland, Pennsylvania **South:** Alabama, Arkansas, Georgia, Kentucky, North Carolina, Oklahoma, Tennessee, Virginia, West Virginia **Midwest:** Illinois, Indiana, Iowa, Kansas, Michigan, Minnesota, Nebraska, North Dakota, Missouri, Ohio, South Dakota, Wisconsin **West:** Arizona, Colorado, Montana, New Mexico, Utah, Wyoming	30	8.05
Gas	**East:** Maine, Massachusetts, New York **South:** Florida, Louisiana, Mississippi, Rhode Island, Texas **West:** Alaska, California, Nevada, Texas	12	12.17
Nuclear	**East:** Connecticut, New Hampshire, New Jersey, Vermont **South:** South Carolina	5	13.41
Hydro	**West:** Idaho, Oregon, Washington	3	6.49
Oil	**East:** Washington, D.C. West: Hawaii	2	28.15

Source: EIA. 2009h. *Electric Power Industry 2009: Year in Review*. Washington, DC: U.S. Department of Energy. http://www.eia.doe.gov/cneaf/electricity/epa/epa_sum.html (accessed January 12, 2011).

And, the politics of energy and the environment are becoming increasingly complex. This is already evident in the bickering over the EPA's initial steps toward regulating CO_2 emissions, as it is allowed to do following the 2007 U.S. Supreme Court decision *Massachusetts vs. EPA*. Many members of Congress are against the action, including both Democrats and Republicans. Why? The data in Table 5.5 suggests some reasons. Politics in the energy sector is tied to many different factors. One is resource endowment. Members of Congress from energy producing states generally work to protect their energy producers. Texas and Louisiana are strong in oil and gas, while U.S. coal supplies are concentrated in the Midwest and Intermountain West, particularly Wyoming and Montana.

Not surprisingly, these coal producing states are also heavily reliant on coal for generating electricity. Many of the coal-fired plants in the region are old and relatively high polluting. And the average cost of retail power in these states (in 2008) was much lower than in states reliant on more price-volatile natural gas. Power in natural gas–reliant states was over 50 percent more costly. Although there is now a glut of natural gas and prices have dropped sharply, the residents of these states have reason to be concerned that regulation of CO_2 will impact their electricity prices. Given the fragile state of the economies of many of these same Midwestern states, higher electricity prices, which raise the cost of doing business, could be damaging. Employment is also an issue:

coal mining employed over 22,000 people in West Virginia in 2008, nearly 19,000 in Kentucky, and 8,000 in Pennsylvania (EIA 2009f).

The political appeal of cap and trade–type pollution control policies is based in part on the belief that they can mitigate CO_2 emissions much more cost-effectively than the source-point regulation likely to be used by the EPA, with fewer impacts on growth and employment. We will further explore the political challenges of climate change in the next chapter.

Conclusion: A Complex, Interrelated Energy Policy Result

Several broad themes in relating public policy theory to energy policy are addressed in this chapter. First, the six challenges to energy policy development and implementation we emphasize reflect the complexity of policy development and implementation in the energy arena. Second is the concept that public policy requires negotiation and compromise among various stakeholder groups in what ideally should not be a zero-sum game for any one faction. In public policy making, not every group gets all that it wants and no group should go away from the negotiating table empty handed. Third, the energy policy process does not exist in a vacuum; forging energy policy takes place in a complex, shifting institutional context. Third, the attempt to impose a purely scientific or technological approach to achieve a national consensus on the "best" way to solve the energy question has been a failure for more than 50 years, and will likely be for the recognizable future.

Scientists have known how to make gasoline from coal or oil sands for decades (Vietor 1984). It has just been too costly to do so. The same can be said for eliminating the carbon emissions from burning coal to generate electricity. It is cost effective to remove some of these emissions today, possibly even most; but is not easy to remove all. So we continue to pollute the air and poison the land and water with acid rain.

In 2008, the European division of the Ford Motor Company produced a car in Germany that achieved something like 60 miles to a gallon of gasoline. That car was not available in the United States because of transportation safety concerns. Evidently, we can have safe highways but must pay for them in inefficient but safe automobiles. Again, this is an example of the trade-offs that underlie policy making. Our next chapter will shift to a critical and related issue: the appropriate decision rules for energy policy development in the current era.

Chapter 6

What's on the Current Energy Policy Agenda?

> Policies first come into being through being put on an agenda—a notational list of topics that people involved in policy making are interested in, and which they seek to address through developing, or exploring the possibility of developing, policies.
>
> **—Edward C. Page (2006)**

Following his historic 2009 inauguration, President Barack Obama and his incoming administration faced the most daunting policy agenda since Franklin Roosevelt in 1932. A long list of matters demanded attention, including the stalled economy, a failing financial and banking system, massive job losses, wars in Afghanistan and Iraq, the threat of a nuclear Iran, and the stalled Middle East peace process. During his campaign, candidate Obama had also pledged to take on health care reform, a challenge that had frustrated Democratic presidents since Harry Truman. But despite this overload of foreign and domestic priorities—and criticism that his agenda was far too ambitious—action on energy remained high on his list of priorities.

As a candidate for president, Barack Obama's energy platform combined a set of populist proposals aimed to salve the pain at the pump from 2008's spike in gasoline prices, with longer-term proposals to take action on climate change and the task of lessening U.S. dependence on foreign oil. Specifically, the Obama/Biden 2008 energy policy statement—New Energy for America—spelled out the mid- to long-term programs they would implement if elected. These were grouped into six broad categories:

1. Tackle climate change.
2. Invest in a secure energy future and create five million "green" jobs.
3. Make transportation more fuel efficient.
4. Increase production and use of domestic energy sources to cut imports.
5. Diversify energy sources.
6. Implement programs to improve energy efficiency in all applications.

Following the election, President-elect Obama reiterated his commitment to action on energy, stating in a *60 Minutes* interview that the United States was stuck in a "shock to trance" pattern of energy use and policy. The shock of higher prices leads to a whirlwind of activity, which peters out as increasing supplies lower prices and "we start, you know, filling up our SUV's again." He committed his incoming administration to taking action on the nation's long-term energy problems, which one commentator likened to an "energy quest" (Revkin 2010).

As we describe in Chapters 4 and 5, the 2009 Stimulus Act provided a significant start toward the accomplishment of the administration's energy goals, with $80 billion for research and development, weatherization, incentives for investments in renewable energy, a variety of energy-related tax cuts, and many other provisions. Other actions by the administration have also supported its energy agenda, including the bailout of the auto industry (particularly the support for General Motors, with the launch of the Chevrolet Volt, with its innovative series-hybrid drive train), opening of offshore areas on the Gulf Coast and Atlantic seaboard to oil and gas exploration, and implementation of higher fuel efficiency standards, aiming to increase average fuel economy to 36.5 mpg by 2015. These actions, while significant, hardly amount to an energy "quest"; there is a significant amount of unfinished business.

Assessing what is on the agenda at any given moment is an interpretive process. We observed what is being discussed at energy conferences, reviewed reports from congressional committee hearings, talked to a small number of energy professionals and politicians about their concerns, and followed discussions online and in print about topics central to the process of crafting climate change legislation. We found, not surprisingly, that many energy issues are hardy perennials that remain in play year after year. These include the reliability of oil and electricity supplies, enhancing U.S. energy infrastructure, "energy independence," access to public lands, and many other source-specific issues. Figure 6.1 illustrates the interdependence of primary and secondary energy sources.

What is relatively new in energy policy setting is the *urgency* of the climate change agenda. Although the first congressional hearings on the topic were held back in the 1970s (thanks to then Congressman Al Gore, Jr.) the issue was largely ignored throughout the G. W. Bush administration. House Speaker Nancy Pelosi emphasized its importance with the creation of a special committee in January 2007, the Select Committee on Energy Independence and Global Warming.

After two and a half years of hard legislative work, the House narrowly passed H.R. 2454, the American Clean Energy and Security Act (known as Waxman-Markey after its two main sponsors) in June 2009. The main features of this complex bill are a cap and trade system for CO_2 emissions that includes provisions granting a high proportion of permits to industry in early years, and a wide array of provisions designed to encourage clean energy investments and enable increased regulation of the energy sector.

In the spring of 2010, the action was in the Senate, where Senators Kerry, Graham, and Lieberman crafted a comprehensive climate change bill. The breadth of these bills and need to reach out to Republican members of the Senate means that few energy options are off the table—making the "agenda" virtually unlimited. Areas of emphasis include a mechanism for pricing carbon and reducing CO_2 emissions, how to expand investment in nuclear energy, the role of the Environmental Protection Agency (EPA) in regulating emissions, how to encourage continued investment in renewable energy and conservation, access to the continental shelf—the list is extensive. It is lengthy in part because carbon emissions are deeply woven into the U.S. economy and the day-to-day lives of Americans.

A climate change policy analysis by McKinsey identified over 70 greenhouse gas abatement opportunities, with no single opportunity representing more than 10 percent of the total (McKinsey 2007). The Pacala and Socolow analysis discussed in Chapter 2 focuses on 15 options for their stabilization wedges. Arguably, all of these options are somewhere on the agenda. Yet the country's ongoing economic crisis links all of these to the need for energy policy to make a contribution to the country's recovery and generate new jobs. This chapter will examine these and other elements of the U.S. energy agenda. Because it is now driving the energy policy process, we will begin with the complex debate over how the United States should respond to evidence of climate change.

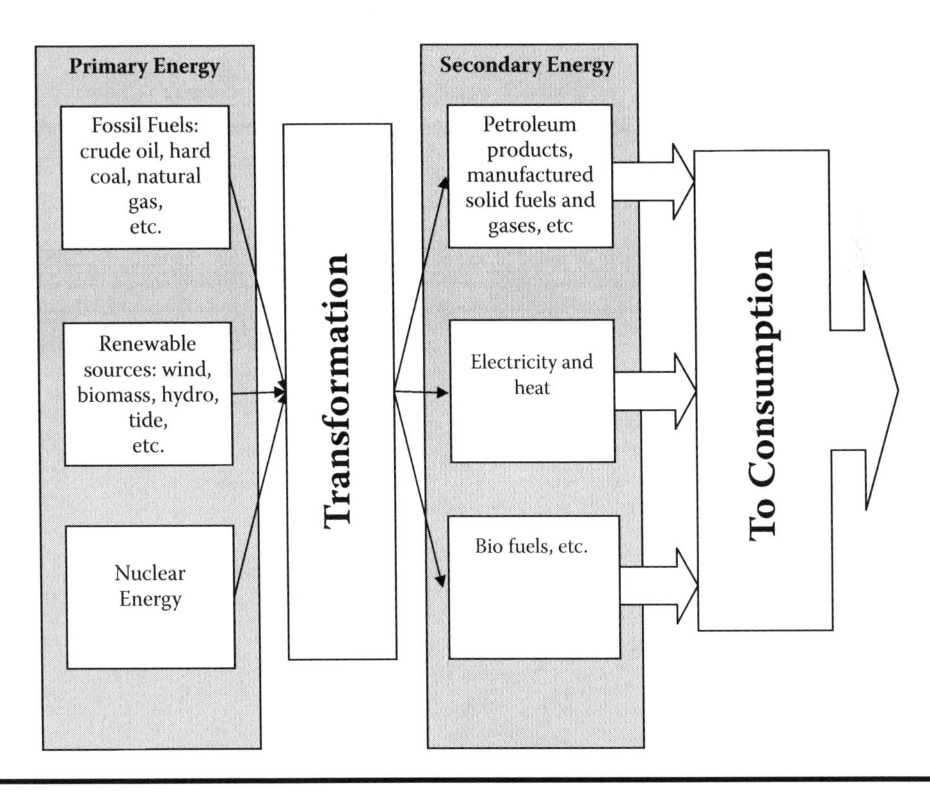

Figure 6.1 Primary and secondary energy relationships. (Adapted from UN [United Nations]. 2008. *Issue paper: Definition of primary and secondary energy.* United Nations Statistics Division: Energy Statistics, http://unstats.un.org/UNSD/envaccounting/londongroup/meeting13/LG13_12a.pdf, accessed December 28, 2009.)

Table 6.1 Numbers of Nuclear Power Plants in Operation and under Construction, 2010

Country	*No. Plants in Operation*	*Elec. Net Output (MW)*	*No. under Construction*	*Elec. Net Output (MW)*
United States	104	100,683	1	1,165
France	58	63,130	1	1,600
Japan	54	46,823	1	1,325
Russian Federation	32	22,263	8	5,944
Korea Republic	20	17,705	6	6,520
United Kingdom	19	10,137	—	—
Canada	18	12,569	—	—
India	18	3,984	5	2,708
Germany	17	20,470	—	—
Ukraine	15	13,107	2	1,900
China	11	8,438	21	20,920
Sweden	10	9,043	—	—
Spain	8	7,450	—	—
Belgium	7	5,902	—	—
Czech Republic	6	3,678	—	—
Taiwan	6	4,980	2	2,600
Switzerland	5	3,238	—	—
Finland	4	2,696	1	1,600
Hungary	4	1,889	—	—
Slovakian Rep.	4	1,762	2	810
Argentina	2	935	1	692
Brazil	2	1,884	—	—
Bulgaria	2	1,906	2	1,906
Mexico	2	1,300	—	—
Pakistan	2	425	1	300
Romania	2	1,300	—	—
South Africa	2	1,800	—	—
Armenia	1	376	—	—

(continued)

Table 6.1 Numbers of Nuclear Power Plants in Operation and under Construction, 2010 (Continued)

Country	*No. Plants in Operation*	*Elec. Net Output (MW)*	*No. under Construction*	*Elec. Net Output (MW)*
Netherlands	1	482	—	—
Slovenia	1	666	—	—
Iran	—	—	1	915
Totals	437	371,450	55	50,905

Source: ENS (European Nuclear Society). Nuclear Power Plants, World-wide. Brussels, Belgium: European Nuclear Society. http://www.euronuclear.org/info/encyclopedia/n/nuclear-power-plant-world-wide.htm (accessed January 15, 2011).

Climate Change and the U.S. Economy

From the start of the Obama Administration, legislation aimed at reducing U.S. greenhouse gas emissions was firmly on the agenda. The June 2009 passage of the Waxman-Markey bill by the House raised the hopes of climate change activists. But Democratic leaders could not find the 60 votes in 2010 to avoid a filibuster on similar legislation in the Senate. With Republican control of the House during the 112th Congress, action on climate change is off the national agenda.

Why was a climate bill on the energy agenda? The administration and many members of Congress were convinced, as summarized in Chapter 2, that the risks associated with our use of energy and unchecked climate change are compelling enough to justify action. Action on climate change is still actively being sought by many domestic constituencies including environmental groups, many unions, and most important, an array of business leaders in the energy sector. Many of them have carefully examined the evidence and concluded, first, that climate change is a real problem, and second, that with a reasonable set of policies in place, including a price on carbon, they can adapt their business models and still make money.

In fact, the lack of a clear carbon price is at this moment a major impediment to energy sector investments. By one calculation, between $1.5 and $2 trillion in investments are needed in the power sector alone between 2010 and 2030 (Chupka et al. 2008). The lack of clear price signals on carbon creates uncertainty that makes committing to such investments more challenging. In effect, many participants in energy markets are waiting for the government to provide a policy framework for the industry that will structure future opportunities.

A final factor is that the rest of the world is watching us. A global climate change agreement will be impossible without U.S. action. But the complexity of the agenda is reflected in a presentation given by Energy Secretary Steven Chu at a U.S. Energy Information Administration conference in April 2010. The second slide in his PowerPoint presentation stated (Chu 2010): "America has the opportunity to lead the world in a new industrial revolution: To ensure American competitiveness, decrease dependency on foreign oil, and mitigate climate change."

As we have noted above, other countries, notably Germany and Denmark, aggressively used energy policy to change their energy systems, in the process creating opportunities for private companies to become successful in the alternative energy market. These companies, including Vestas and Enercon in wind turbines and Q-Cells, and Sharp Solar and Suntech in photovoltaic production, initially had a significant competitive advantage that is now eroding as Chinese competitors are ramping up production. The Holy Grail for U.S. energy pundits (yes, it is a quest) is to implement a similar set of policies in this country.

The U.S. economy in the early months of 2010 was emerging from a deep recession with high unemployment that decimated many industries (including automobiles, construction, and the retail sector); the pressure was on in Congress to find new sources of growth that would kick-start the economy and generate significant numbers of jobs. A main argument used in behalf of this strategy was that investments in alternative energy sources generate more jobs than investments in fossil fuel production (Pollin et al. 2008). A side benefit of advancing efficiency technologies in the transportation energy sector will reduce our need for imported oil, which will lower our $450 billion annual bill.

For several years, conventional wisdom held that there were limited policy options available to tackle this problem. Regulatory approaches were dismissed with little consideration, while carbon taxes were believed to be political suicide, given the American anathema to increasing taxes of any kind. The market-based cap and trade approach, whose viability was proved with the EPA's acid rain program, was perceived to be a better alternative.

The broad-based cap and trade approach was declared dead in late February 2010 by Senator Lindsey Graham of South Carolina, the only Republican in the Senate willing to help craft a comprehensive bill on climate change. Increased skepticism toward markets after the 2007 and 2008 collapse of the financial sector and opposition from farm state and coal state senators, made it unlikely that a bill featuring such an approach could pass the Senate.

Although as mentioned above, the Kerry/Graham/Lieberman energy proposal reflects a broad series of agenda items that are swirling around the Congress as of the spring of 2010, several are more critical than others and have long-term staying power as energy issues. These include:

- The future role of nuclear energy
- Continuing concern about our reliance on imported oil
- Renewables and conservation policy issues
- Is carbon capture and storage viable?
- Access to federal lands
- EPA regulation of greenhouse gases
- Federal primacy vs. state freedom
- Concern about a global climate treaty

One issue omitted from the list is what to do about the nation's oldest coal power plants (discussed at length in Chapter 4). In the following text we provide brief summaries of the key issues within each of these agenda topics.

The Future Role of Nuclear Energy

What to do about nuclear generation remains one of the most problematic issues on the U.S. energy agenda. As of early 2010, there were 65 nuclear power plants with 104 reactors operating in the United States. Located in 31 different states, together they generate about 19 percent of U.S. electricity output. No new plants have come online in the United States since 1979, although as of April 2010, twenty-six companies plus the Tennessee Valley Authority have announced their intent to submit applications to the Nuclear Regulatory Commission for new nuclear power plant licenses.

Nuclear power received a boost in the Energy Policy Act of 2005, which included a combination of regulatory initiatives to shorten the licensing process, a process for creating standardized designs, and financial incentives in the form of tax credits for additional capacity and federal loan guarantees for project costs. The U.S. Department of Energy (DOE) received 19 applications for such guarantees for a total amount of $122 billion, with only $18.5 billion authorized to be spent on the program. Early in 2010, President Obama proposed tripling that amount, to $55 billion.

There is a case to be made for nuclear as a noncarbon source of base load power. We will discuss the pros and cons of this case in more detail in Chapter 10. Here we will mention briefly that despite concerns about the lack of a long-term repository for nuclear waste and the high cost of nuclear power despite significant federal subsidies, the industry retains significant support. Passage of an energy bill will likely be impossible in 2011 without additional funds for nuclear energy, because it will be a minimum price of Republican Party cooperation.

Concern about Our Reliance on Imported Oil

The world is running out of oil. That is a fact. It is not going to happen tomorrow, or the day after, or a decade later, or even by 2050. Reasonably affordable supplies of oil are likely to be around for decades to come. The problem is that the world is using more of it and the rate of use is increasing. In 2009, fossil fuels—predominantly petroleum—had an 81 percent share of world primary energy, renewable energy 13 percent, and nuclear energy 6 percent. Studies by the U.S. Energy Information Administration, the European Union, and three other energy information sources agree that at least as far ahead as 2050, little change will occur in this distribution: fossil carbon energy sources will continue their roughly 80 percent share. More important, the sources also agree that availability of fossil fuels will not be a problem in the foreseeable future (Aalto 2008b; Moriarty and Honnery 2009).

Moreover, the fossil fuel industry is continually expanding capacity and bringing new supplies online to meet growing global demand for liquid fuels. It is also expected to participate in developing alternative liquid fuel supplies to make up for the expected decline of traditional resources. In one form or another, or from one source or another, petroleum products are likely to be important for at least the next 50 years and probably longer.

Issues to Address Immediately

Among the immediate issues that policy makers must address in their deliberations on energy policy are: (1) supplying growing world demand for petroleum products to fuel the growing demand for fuel for transportation, as the recognized decline in oil reserves is threatening to pit nation against nation, society against society, as competition for fewer supplies increases; (2) the country's dependence upon foreign oil, now 57 percent of our national oil consumption, has resulted in a significant shift of wealth from oil-consuming to oil-producing countries (Wirth, Gray, and Podesta 2003).

Who are these foreign energy suppliers? The United States imports more than 70 percent of its foreign oil from just five countries, only two of which are problematic. The biggest sources of foreign supply are our neighbors, Canada and Mexico—both of which are among our largest trading partners. Together, Canada and Mexico provide more than 30 percent, Saudi Arabia 11 percent, Venezuela 10 percent, and Nigeria about 9 percent (Figure 6.2). The rest comes from whoever has it to sell at a price refiners are willing to pay.

Although current projections by both the DOE and major oil companies, including Exxon-Mobil and BP, suggest that U.S. oil use may already have peaked, the most likely scenario is for

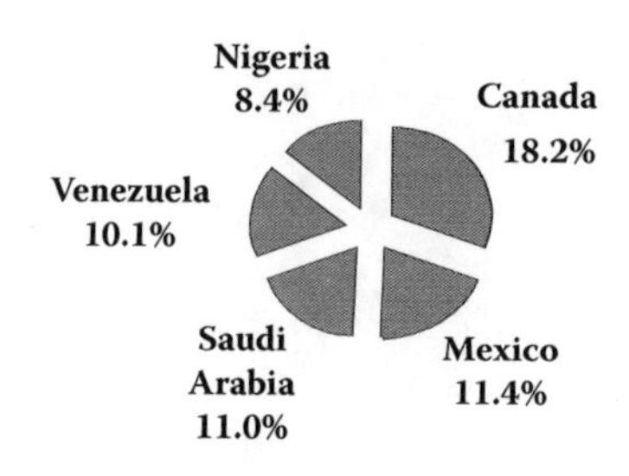

Figure 6.2 Five largest suppliers of oil imported by the United States, 2009. (From *EBR [Energy Business Review]*. 2009. How dependent are we on foreign oil? Energy Business Daily (September), http://energybusinessdaily.com/oil/how-dependent-are-we-on-foreign-oil, accessed June 18, 2010.)

further increases in price. Given the limits of the ability of developed countries to increase their own production, this will result in continuing the transfer of wealth from consuming nations to producing nations, with the potential to further impact the economic well-being and quality of life of all Americans. As a result, developing affordable substitutes for transportation remains a key consideration in ongoing deliberations on energy policy.

One additional important item on the energy agenda is the continuing effort to make transportation more fuel efficient. In December 2009, the Energy Information Administration reported that the transportation sector was the second largest contributor of CO_2 emissions to the atmosphere (EIA 2009e). Cutting carbon and other greenhouse emissions by this sector was one of the early energy policy initiatives by President Obama. He announced stringent new standards on May 18, 2009, less than six months after taking office.

Because the Obama energy policy platform was most likely created during 2007 when consumers were paying upward of $4 a gallon at the pump and the price of crude oil was at the all-time high of nearly $150 a barrel, it is not surprising that the first section spelled out a set of short-term actions collected under the label: "Immediate Relief from Pain at the Pump." Speculators were blamed for causing the run-up in the price of crude oil. The promised actions included providing an immediate emergency rebate to consumers; a crackdown on excessive energy speculation and forced cuts in market prices by releasing oil stored in the nation's strategic petroleum reserve and allowing refineries and electricity generators to swap less-expensive heavy crude for the higher-priced light crude stored in the reserve (see Box 6.1).

Increasing fuel economy standards, converting government fleets to hybrid and electricity-powered vehicles, and mandating flexible fuel capability in new vehicles are key elements of this policy initiative. In December 2009, the Energy Information Administration (EIA) reported that the transportation sector was the second largest contributor of CO_2 emissions to the atmosphere (EIA 2009e). Emissions by sector are displayed in Figure 6.3.

Also included was a commitment to invest in development of the next generation of sustainable biofuels and the infrastructure needed to collect, produce, and distribute these sustainable fuels effectively and efficiently. And, for the first time, corporate average fuel economy (CAFE) standards included a national low-carbon fuel standard. Average fleet fuel efficiency standards were raised to 39 miles per gallon (mpg) for cars and 30 mpg for light trucks by 2016. The new standards cover model years 2012 through 2016.

BOX 6.1 CHANGES IN PRICE DIFFERENTIALS FOR HEAVY CRUDE OIL

Two important characteristics of crude oil influence the mix of final petroleum products produced from the crude and, hence, its price. One is its viscosity or density; the other is its sulfur content. Oils with low density (light crude) yield a higher proportion of light petroleum products like gasoline. Heavy crude oils have lower amounts of hydrocarbons making them more difficult and costly to refine; they produce more heating oil and asphalt. Crude oils with heavy concentrations of sulfur are referred to as "sour"; low sulfur oils are "sweet." Heavy investments are required by refiners to remove undesirable sulfur; just prior to the collapse of the U.S. economy, the U.S refiner Valero had invested billions to process more heavy crude.

Because refined light/sweet crude oils produce more desired products than heavy/sour crudes, refiners are willing to pay a premium (a price differential) for them. Spot prices are compared against West Texas Intermediate (WTI), the industry standard for light/sweet crude. The price differential for Saudi Arabia's Arab Heavy crude was $11.80 a barrel below the spot WTI price in August of 2008, with five year average difference of $13.15 a barrel. Reflecting the damage by Hurricane Katrina to the few U.S. refineries able to refine heavy crude, the discount for heavy crude reached as high as $14.50 a barrel in November of 2005. A year later, in August of 2009, the global economic slowdown had driven the discount for Mexico's Maya heavy crude dropped to just $4.63 a barrel below the WTI price. According to a Reuters 2010 financial report, some industry experts do not expect big discounts for heavy crude to ever return.

(From Fattouh, B. 2006; Schneyer, J. 2009.)

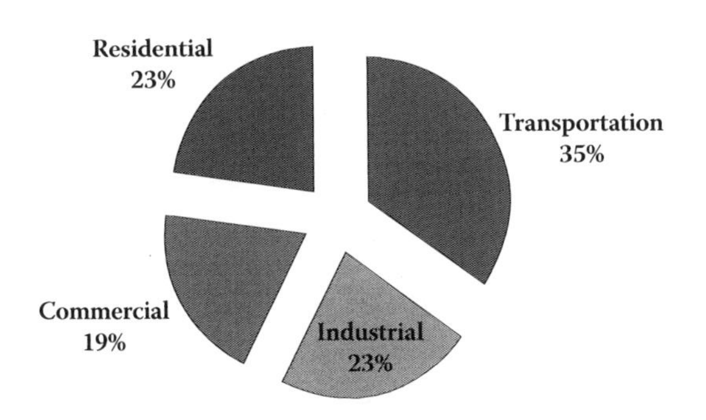

Figure 6.3 Transportation sector CO_2 emissions in 2007. (From EIA. 2009e. *Annual Energy Outlook 2009 with Projections to 2030: Electricity Demand.* Paris: Organization for Economic Cooperation and Development. http://www.eia.doe.gov/oiaf/ieo/pdf/0484(2009).pdf, accessed February 2, 2010).

Renewables and Conservation Policy Issues

The last three major pieces of U.S. energy legislation, the 2005 and 2007 energy acts plus the 2009 stimulus bill, included many provisions relevant to renewables and conservation. These federal provisions, plus the torrent of actions at the state level described in Chapters 4 and 5, have led to considerable increases in wind and solar energy capacity. However, as of 2010 these and other forms of non-hydro renewable energy total only about 3.5 percent of overall U.S. electricity generation. Still, if we are going to be successful in reversing greenhouse gas emissions, we must increase this proportion significantly. Figure 6.1 displays proportions of energy contributed by the major renewable sources; Table 6.2 lists recent government-supported research efforts in renewable energy.

Tackling the Major Agenda Issues

Continuing to decarbonize the U.S. power industry will require tackling an interlinked series of issues, the most important of which is establishing a mechanism for creating a carbon price. This will allow renewable sources to be more competitive against both coal and natural gas. Other issues include (1) how to encourage the transition to renewable sources while reducing the cost of the power they produce, (2) how to encourage research and more effective development of the new technologies that emerge, and (3) how to expand U.S. production of alternative energy products such as wind turbines and photovoltaics without violating trade agreements. The vital need for improved technologies for storage of electricity is also a critical focus for research and development since the intermittency of these sources makes their management as part of the electricity grid more challenging than traditional sources.

The enormous scale and cost of this task—reshaping the entire U.S. energy system over several decades, with a price tag in the trillions—means that policy makers must carefully consider the impact of each policy proposal on market structures, incentives, private investment, and costs to consumers.

The way this so-called ecosystem of energy investments works today is that the DOE provides financial support for research and development through its own research centers, working with universities, and other entities; then, private sector companies take the resulting technologies and create products capable of commercialization. Although significant progress in lowering carbon emissions in the United States and worldwide is possible now using current products, further advances in photovoltaics, batteries, and storage are needed. Important examples include the payoff from DOE support for research in shale gas exploration techniques that helped lead to recent increases in U.S. gas reserves; and the support for battery research and loans for development helped put battery electric vehicles (BEVs) on the road much sooner than would have been the case without them.

Changing Energy Investment Strategies

An important issue relating to the scale and cost of this task is that managers of utility and energy companies should be encouraged to invest in efficiency and renewable energy sources, and discouraged from investing in the worst sources, especially coal. Yet consumers must be protected from sudden price spikes. This is the scenario that has led many states to implement renewable portfolio standards (RPS) as a way of gently encouraging a transition toward low-carbon power sources. Additional programs outlined in this portion of the Obama energy policy announcement included requiring 10 percent of electricity to be generated from renewable resources by 2012 and

Table 6.2 Research Funding Summary by Program (US$ thousands)

	FY 2007 Current Appropriations	*FY 2008 Current Appropriations*	*FY 2009 Request to Congress*	*FY 2009 vs. FY 2008 Appropriations*
Energy Efficiency and Renewable Energy Programs:				
Biomass and Bio–refinery Systems R&D	$196,277	198,180	225,000	+26,820
Building Technologies	102,983	108,999	123,765	+14,766
Federal Energy Management Program	19,480	19,818	22,000	+2,182
Geothermal Technology	5,000	19,818	30,000	+10,182
Hydrogen Technology	189,511	211,062	146,213	–64,849
Industrial Technologies	55,763	64,408	62,119	–2,289
Solar Energy	157,028	168,453	156,120	–12,333
Vehicle Technologies	183,580	213,043	221,086	+8,043
Water Power	0	9,909	3,000	–6,909
Wind Energy	48,659	49,545	52,500	+2,955
Subtotal, Programs	958,281	1,063,235	1,041,803	–21,432
State and Other Supporting Activities:				
Weatherization and Intergovernmental Activities	281,731	282,217	58,500	–223,717
Facilities and Infrastructure	107,035	76,176	13,982	–62,194
Program Direction	99,264	104,057	121,846	+17,789
Program Support	10,930	10,801	20,000	+9,199
Congressionally Directed Activities	0	186,664	0	–186,664
Adjustments	0	–743	–738	+5

(continued)

Table 6.2 Research Funding Summary by Program (US$ thousands) (Continued)

	FY 2007 Current Appropriations	*FY 2008 Current Appropriations*	*FY 2009 Request to Congress*	*FY 2009 vs. FY 2008 Appropriations*
Subtotal, State and Other Supporting Activities	498,960	659,172	213,590	–445,582
Total, Energy Efficiency and Renewable Energy	$1,457,241	1,722,407	1,255,393	–467,014

Source: EERE [Energy Efficiency and Renewable Energy, U.S. Department of Energy]. 2009. *Research funding summary by program ($US thousands).* Washington, DC: U.S. Department of Energy, http://www1.eere.energy.gov/ba/pba/pdfs.FY2009-budget-brief.pdf, accessed June 8, 2010.)

expanded efforts to develop and deploy clean coal technology. This is not as optimistic a target as it may seem; the U.S. energy information administration reported in 2010 that renewables were responsible for a little more than 9 percent of fuels used to generate electricity in 2008. Of this total, hydroelectric, biomass, and wind accounted for about 1 percent each; and both solar and geothermal sources accounted for less than 1 percent each (Figure 6.4).

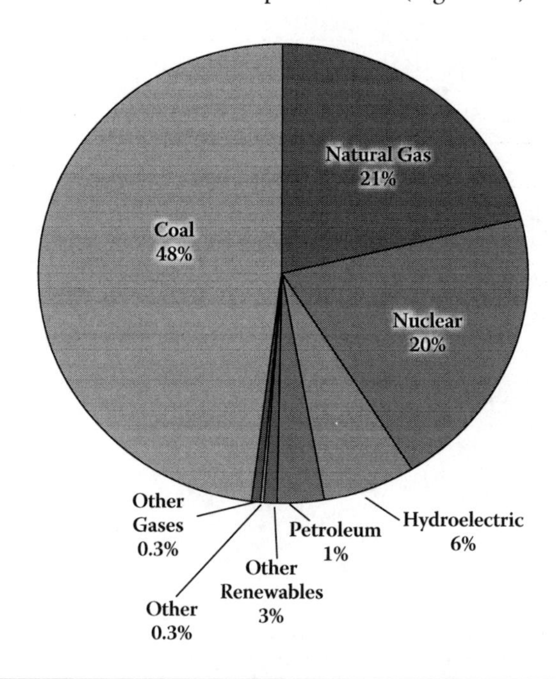

Figure 6.4 Electricity fuels in the United States. (From EIA [U.S. Department of Energy, Energy Information Administration]. 2010a. *Electricity explained: Your guide to understanding energy.* Washington, DC: U.S. Department of Energy, Energy Information Administration, http://www.eia.doe.gov/energyexplained/print.cfm?page=electricity_in_the_united-states.pdf, accessed March 31, 2010.)

The RPS standards require that a minimum amount of renewable energy is included in each energy supplier's portfolio of energy resources. These numeric targets generally increase over time. However, each state is allowed to set its own minimum standards. A national RPS would be a sensible approach, but would reflect yet another federal intrusion into an arena traditionally controlled by the states—utility regulation.

Another issue is how to continue to encourage investments in conservation by households and businesses, so that reducing energy use in homes, residential construction, commercial buildings, and production processes becomes second nature. The challenge here is that households demand very quick payback periods for their energy efficiency investments, between 2 and 3 years. And the menu of possible efficiency improvements is lengthy, creating a policy dilemma—is it best to craft interventions on a program-by-program basis, such as a focus on weatherization, efficient appliances, and the like. Or is finding a market-based means, such as higher prices, to encourage reductions in use a better path?

A major, long-run problem is that key policy instruments at the federal level have suffered from a start-stop-start pattern that has in the past been highly detrimental to investment flows. For example, the production tax credit was approved by Congress in 1992, but lapsed in 1999, 2001, and 2003. Between 2005 and 2008, the credit remained in place and encouraged a huge surge in wind energy investments. To be effective, subsidies must receive a serious long-term commitment. But under PAYGO budget rules, such subsidies will need a corresponding revenue source. This suggests that a means to generate revenues, either through an auction of cap and trade allowances, or various taxes or fees, will be essential.

Major Investments in Smart Grid Technologies

A final piece of the renewables puzzle on the agenda at both the state and national levels is the question of how to nurture the so-called smart grid into existence. The technology is available to give households and businesses significant control over their decisions about energy services and the timing and cost of energy use. Households willing to agree to dynamic pricing—with lower rates for electricity use at off-peak hours, and higher (possibly *much* higher) rates at high demand periods—will have high-tech meters, thermostats, appliances, and electronic devices that will generate a vast stream of data about their energy use and its costs. These data will flow back to utilities to enable more efficient generation, transmission, and will also be available for analysis by a variety of energy services companies that can help users increase the efficiency of their energy use. The DOE, thanks to projects authorized in the 2009 American Recovery and Reinvestment Act (ARRA)/stimulus bill, has already funded 100 smart grid grant projects nationwide worth $3.4 billion, which are matched by $4.8 billion of applicant funds, plus another $100 million for smart grid training for energy professionals.

Without the smart grid, increasing U.S. reliance on solar or wind distributed generation or household charging of electric cars will be a much more challenging task. The latest generation of electric cars will require about 8 kWh to recharge, which is roughly one-quarter of the daily electricity consumption of an average U.S. household. It's essential that as much as possible of this additional electricity use occur at off-peak hours, to avoid adding to overall power demand. In sufficient quantities, electric autos could also become a substantial reserve store of energy at times of peak demand.

This vision of the U.S. electricity sector has major implications for its organization, structure, and regulatory regime. Recent experience in Europe suggests that it leads over time to a disaggregated business model with firms specializing in generation, transmission, distribution, and a wide variety of intermediary services. What now-extinct firm would have been best placed to succeed in such a marketplace? Enron!

The vision of the smart grid will require a corresponding reconceptualization of the competitive framework and state-based regulatory structures that have ruled in this sector for over a century. How do we protect consumers and businesses while "freeing" the electricity marketplace? So far there are no clear answers to this question. The smart grid vision also assumes that businesses and households will be able and willing to take an active role in the hour-by-hour management of their energy use, but that is a questionable assumption. A combination of incentives and penalties will be needed to change the entrenched habits around household energy use.

Access to Federal Lands

The pushmi-pullyu at the heart of U.S. energy policy is not Dr. Doolittle's mythical two-headed beast, but the tension between advocates of conservation and efficiency, and those who emphasize increasing energy supplies. By 2010 this two-headed beast had pulled the nation's electrical grid a considerable distance toward increased energy efficiency, although the capacity for vast improvement remains. Yet the reflexive reaction in the oil and gas sector has been to pull toward a supply response that includes gaining greater access to federal lands for oil companies searching for new supplies. The tension over this issue makes it a prime source of political controversy and keeps it constantly on the national agenda.

Access is an issue in part because of the sheer magnitude of federal land holdings in the western United States. Overall, the U.S. government controls about 30 percent of the country's land area, about 650 million acres. Much of this land is in the West, particularly the intermountain west where many residents resent the extent of the federal government's control over land management. These lands are managed by different agencies, including the U.S. Forest Service of the Department of Agriculture, and the Department of the Interior's Bureau of Land Management (BLM), National Park Service (NPS), and Fish and Wildlife Service (FWS). Much of these lands are blessed with abundant resources. By one estimate, over 30 billion barrels of oil and 231 trillion cubic feet (tcf) of natural gas lie under federal lands, although only 38 percent of the oil and 59 percent of the gas is believed to be accessible with existing technologies (BLM 2008).

Several related subissues confound this agenda item. One is the extent to which these lands are available for resource exploitation. Over 40 percent of these lands are set aside for conservation. These include Park Service lands, substantial areas within national forests, and increasingly within BLM lands. Conservationists would like to continue to create additional wilderness areas, including on BLM lands where large reserves of oil and gas are believed to be located. Generally, the energy industry opposes such set-asides.

Additional agenda problems include the process for determining what lands will be opened for energy development, and then who will establish and manage the requirements for obtaining the permits required to engage in resource exploration and development. The George W. Bush administration's National Energy Policy Development Group report (2001, 3–13) recommended steps to "rationalize permitting for energy production in an environmentally sound manner by directing federal agencies to expedite permits and other federal actions necessary for energy related project approvals on a national basis." This was followed by an Executive Order 13212, which created an interagency task force to implement the recommendations. The result, in part, was a vast increase in the number of approved permits, about 41,700 from fiscal years 2001 to 2009, an increase of 250 percent over the previous eight-year period (Kenworthy 2010).

Not surprisingly, this increase made it difficult for the BLM to properly manage the environmental impacts of this expanded development. Under the Obama administration, Interior Secretary Ken Salazar has tightened up the leasing process and cancelled some leases in Utah that are near national parks and monuments. Although public lands are still being opened to development, many companies have been unwilling or unable to explore lands for which they have permits. This is likely related to the recession and to the very large supplies of oil shale gas that has depressed prices.

A final important land issue on the energy policy agenda is the development of lands on the country's outer continental shelf (OCS), generally between U.S. shores and the country's 200-mile limit. A framework for the resource development of this vast area was established in the Outer Continental Shelf Lands Act of 1953, as amended. Oil and gas exploration in this zone, with the exception of large areas of the western Gulf of Mexico, has been generally off limits since the 1980s due to congressional action—primarily due to concerns about potential pollution from oil spills. That policy has now changed.

In March 2010, President Obama announced a plan to allow leasing in a zone of the OCS stretching from the northern border of Delaware down to central Florida, a total of 167 million acres. Areas in the Northeast and on the Pacific coast would remain off limits. A 2006 Department of the Interior analysis estimated reserves of 8.5 billion barrels of oil, plus another 86 billion barrels as undiscovered resources, plus trillions of cubic feet of natural gas.

Predictably, prosupply advocates criticized the action as insufficient, while conservation groups, as well as some governors in the region, expressed concern about the implications for the health of the oceans and wildlife and potential degradation of beaches and the tourist economies reliant on them. The move was widely interpreted as a gambit by the president to win conservative votes for a proposed climate change energy bill. But the Challenger Horizon oil spill in April 2010 led to a moratorium on U.S. offshore drilling. This was lifted in October 2010 following implementation of more stringent regulations. With the risks of offshore oil drilling clear, the politics of this issue are now vastly more complex.

One item not now on the agenda is the opening of the Arctic National Wildlife Refuge (ANWR) to development. Although this was a hot topic throughout the Bush administration, opposition by President Obama makes it unlikely that Congress would take action on this issue. But passage of a bill opening ANWR to exploration and development could be an early, symbolic action by the 2011 Republican-controlled House.

Although this issue will remain part of the national conversation on energy, several realities make it likely that the prosupply and proconservation forces will continue to talk past each other. Oil prices are set on a world market, so new production from federal lands will do nothing about gasoline prices. At best, new domestic production could replace declining fields elsewhere in the United States and modestly decrease U.S. imports. And the recent expansion of shale gas reserves changes the equation on natural gas exploration on public lands, making it much more of a long-term proposition.

Is Carbon Capture and Storage Viable?

America's coal supplies are one of the few energy sources in the country that are not in danger of near-term exhaustion; the United States has enough coal reserves that it could be an important energy resource for hundreds of years. The problem with coal is it is the dirtiest of all the fossil fuels. As such, burning coal also does the most damage to the environment. Still, recent shortages and spikes in the price of natural gas forced some utilities to begin designing and/or building new coal-fired generators. However, if coal is to retain its dominant role in generation, a permanent resolution of environmental concerns associated with this fuel must be resolved.

The three key technology challenges barring the path to more effective and cleaner burning of coal are (1) developing advanced coal gasification technologies to reduce harmful emissions; (2) reducing the high capital costs of building electric power generating facilities that are capable of producing and consuming the new coal gas fuel (GAO 2006); and (3) developing a viable, long-term solution for storing underground the CO_2 generated by burning coal to generate electricity. This process, called *sequestering*, is one part of the proposed "clean coal" concept.

While there are still large deposits of coal around the globe—for example, the United States is said to have enough accessible coal to last another 250 years—the world's supply of oil may only last for another 50 years or so; therefore, an alternative source is absolutely necessary for economic security. Increasing the use of nuclear energy is one proposed solution.

EPA Regulatory Authority

The Supreme Court's 5–4 decision in the 2007 case *Massachusetts vs. EPA* concluded that carbon dioxide met the Clean Air Act's definition of a pollutant and that the agency thus had the responsibility to regulate it. The possible ramifications of this decision are profound. As the EPA stated in its 2008 Advance Notice of Proposed Rulemaking, "the potential regulation of greenhouse gases under any portion of the Clean Air Act could result in an unprecedented expansion of EPA authority that would have a profound effect on virtually every sector of the economy and touch every household in the land" (EPA 2008, 5).

Under the Obama administration, the EPA was given the go-ahead to proceed to implement this new authority, which it did initially with its December 2009 endangerment finding: "The Administrator finds that six greenhouse gases taken in combination endanger both the public health and the public welfare of current and future generations" (EPA 2009a). This clarified the agency's intentions to proceed with related regulations that would initially affect emissions requirements for passenger cars.

The long-run implications of this authority include questioning the desirability of taking a regulatory approach to greenhouse gas regulation, the idea of which is anathema to most business leaders in the U.S. energy sector. Potentially, the agency's expanded regulatory scope could be applied to large emitters, including power plants, chemical plants, and oil refineries. A proposed rule issued in September 2009 would apply to new sources or major modifications at major emitters—defined as 25,000 tons per year CO_2 equivalent. The core underlying issue, of course, is how the country will go about reducing its carbon emissions, and which industries will have to bear the costs of this transformation. Generally, the business sector wants to prevent the EPA from having regulatory authority in this area, as well as to limit states from having the right to set higher standards. As mentioned in Chapter 4, several senators have already introduced legislation that would prevent the EPA from implementing this mandate.

Failure to Produce a Global Climate Treaty

Climate change is a global issue. The process of crafting a global climate change agreement is on the U.S. energy agenda since it is intimately tied to whether the United States passes comprehensive climate change legislation, and the terms of that legislation.

Prior to the UN Climate Change Conference held in Copenhagen in December 2009, there were high expectations that the meeting could conclude with a significant and binding agreement limiting greenhouse gas emissions and providing funds for climate change in developing countries. However, given that the United States had never passed comprehensive climate change legislation and that several large developing countries, including China and India, were opposed to such a binding accord, expectations should have been more modest.

Thanks to last-minute intervention from President Obama, working with representatives from the United States, China, India, Brazil, and South Africa, the conference did generate an agreement: the Copenhagen Accord, which allowed the 193 signatory countries to make voluntary pledges to reduce emissions by 2020. A system for monitoring those pledges was also created, along with a fund that would enable wealthy countries to provide money for developing countries to tackle climate change mitigation and impacts. The relative weakness of the Accord is reflected in its UN status; countries were only expected to "take note" of the agreement, meaning they could decide at a later date whether to accept it.

The weak results of the Copenhagen Conference were a powerful reminder of the extreme difficulty of reaching global agreements in an arena where there are diverging national interests. The likelihood of crafting a global agreement will be strongly influenced by whether the United States is able to pass comprehensive climate change legislation; the terms of the U.S. legislation will also influence what can be credibly agreed to. A weak law (or no law) will make it very difficult to meet the goal suggested by many developing countries of limiting the peak rise in average global temperature to 1.5°C. From a domestic political perspective, President Obama also has to consider the likelihood of Senate approval of any global climate change treaty, which would require 67 votes for passage. We will return to a discussion and analysis of the global challenge of climate change in Chapter 9.

What's on the Periphery of the Energy Agenda?

A number of important topics in the long run are receiving attention but are not as central to the immediate agenda. These include U.S. biofuels policies, natural gas issues, and myriad environmental concerns, including the complex challenges of increasing energy development in a world increasingly short of fresh water. The latter has become a major topic at energy conferences and we will give it further attention in Chapter 9.

A Need for More Cost-Effective Biofuels

The Government Accountability Office (GAO), after surveying the U.S. Department of Energy and six sovereign countries committed to expanding use of renewable energy, identified three key challenges slowing full achievement of the promise held out for renewable energy. The most pressing of these was the need to develop cost-effective technologies for producing ethanol using agricultural by-products and other biomass materials instead of relying of corn and other food products. Associated with this is the need to develop the infrastructure required for effective and efficient distribution of ethanol. Other challenges include developing new wind technologies that are effective in low-wind and offshore sites, and improving solar technology to make the technology more cost effective (GAO 2006).

The Department of Energy's 2030 goal for biomass production in the United States is to produce enough biofuels to meet 30 percent of the energy demand in 2006. This is the equivalent of about 60 billion gallons of biofuels a year. Corn-based ethanol cannot meet this goal. To come even close to meeting this goal, developing ethanol from such sources as farm and forest wastes and energy crops such as grasses is also needed (GAO 2008b).

Should policy makers continue to promote expanding the use of corn and soybeans to the making of ethanol for replacing petroleum-based fuels for transportation? The question is an important one, and will become increasingly important as their use continues to drive up their prices. With more than two billion people in the world going hungry, using food crops for fuel takes on a moral question as well.

Natural Gas Back on the Agenda

One issue that has sidled its way onto the energy policy agenda is how to structure energy legislation so as to best use newly discovered reserves of natural gas. Throughout the 1990s and early 2000s, clean-burning natural gas became an important fuel for electric power generators and industrial processors. As a result, natural gas is now the fuel of choice for peak-power, high-efficiency gas turbine generators in both of these applications. This fuel is desirable in part because when it is burned it releases more than 50 percent less carbon dioxide than what is released by existing coal-fired generating plants. It is also ash free, which further increases its efficiency as a fuel for power generation (Bartis 2009). Although the move to gas power generation was slowed by the rising cost of oil in 2007 and 2008 that also resulted in an increase in the price of natural gas and increased its price volatility, new supplies have brought the price back in line and increased its attractiveness.

The move to natural gas is aided by increasing global acceptance that something must be done now to halt global climate change associated with the burning of coal for power generation and petroleum products for transportation and other uses. Natural gas has been touted as an interim solution to these problems. The large increases in U.S. natural gas reserves thanks to advances in shale gas and coal bed methane drilling technology have changed the dynamics of this discussion. Natural gas production is now forecast to be in excess of demand. One potential new market is to expand use of natural gas as a transportation fuel.

The conversation about expanded gas supplies is also on the agenda since it presents a potential risk to the timing and cost of the long-term transition to renewable sources. The policy dilemma will be how to structure carbon prices, renewable portfolio standards, and other means for supporting renewables, so as to ensure that natural gas replaces coal production and not additional increments of wind and solar electricity production.

Conclusion: Sorting out the Energy Agenda

Crafting a sensible energy strategy and set of policies out of this complex and contradictory set of national concerns and interests is a very challenging task. Previous attempts to create a national energy strategy focused on relatively discrete pieces of the agenda such as enhancing energy infrastructure (2005) and expanding incentives for alternative energy (2007). Even so, the need to create a bipartisan coalition required a sacrifice of strategic clarity for political feasibility, and as discussed in Chapter 5, created bills with benefits for every major energy interest. With action on climate change no longer dominating the energy agenda, the focus will move to more focused legislation. Continuing ongoing federal investments in research and development, support for nuclear power, and a national renewable portfolio standard are possible agenda items. The States will continue to be the most creative energy policy innovators, although costly state programs will collide with recession-tightened budgets.

This chapter has examined some of the issues that are shaping the energy policy agenda, and briefly introduced some agenda issues for the major sources of energy. The energy industry is facing one of the most trying periods of its more than a century of existence. Five urgent challenges to the energy industry and government policy makers and regulators over the next several decades are: (1) how and where to acquire enough reasonably priced money to maintain all the key sectors in good shape technically, operationally, managerially, economically, and financially; (2) forging a balance between the interests and goals of legislative bodies, local authorities, regulators, utility industry segments, consumers, and the general public; (3) finding acceptable solutions to fix the inefficiency and volatility of energy markets; (4) continuing to fund research and development in each of the existing sectors, especially coal, gas, oil, wind, and other renewable producers, while also finding research in new and untried energy sources; and (5) resolving the debate that still makes construction of electricity generating plants fueled by nuclear power problematic.

With the pollution costs of current coal-fired facilities and dwindling supplies and resultant price increases for natural gas, policy makers must look toward other means of maintaining existing and adding new electricity generating capacity. Despite the controversial nature of nuclear generators, the remaining facilities continue to supply more than 20 percent of the nation's electricity; they cannot be eliminated. The heavy environmental costs associated with coal-fired plants are causing many utilities to take another look at employing new ways of using coal as a source of energy for generating electricity. Many new coal-fired plants are now planned or in early stages of construction; emissions controls have improved, but are not yet capable of removing all pollutants.

Among the big issues in the electricity industry are: moving ahead with stalled restructuring; replacing older generating, transmitting, and distributing facilities; the rising costs of natural gas and whether to turn to coal or nuclear power for the new generating capacity; and adding new technology.

The next three chapters will examine the most important interventions available to policy makers. Chapter 7 will focus on the many programs designed to influence supply, demand, and improve energy efficiency in the short and long run, including taxes, subsidies, and many other programs. Chapter 8 will focus on the many interventions designed to create new energy markets, including cap and trade systems, feed-in tariffs, and many technological subsidies. Chapter 9 will wrap up this section with an analysis of the global and international aspects of energy policy. As a global problem, climate change requires a global solution, but substantial barriers to such action have yet to be surmounted.

2 POLICIES FOR ENERGY TRANSITION

Chapter 7

Crafting Policy with Subsidies and Regulations

> Decisions about energy and energy policy are inextricably linked to economic, environmental, and national security considerations, and have significant consequences for all three areas.
>
> **—Sandia National Laboratories (2009)**

> Research in energy economics and policy has not kept pace with marketplace developments and policymaking needs—not has it been well disseminated to policy audiences.
>
> **—Phil Sharp, president of Resources for the Future (2009)**

No one book on energy policy can possibly provide comprehensive coverage of the universe of U.S. energy interventions. However, in this chapter we attempt to describe the major categories of interventions and discuss their strengths and weaknesses. In addition, we will provide short case studies of several specific policies and policy situations, and the interaction of particular interventions. National energy goals include increasing U.S. production of renewable energy, encouraging investment in new nuclear generating capacity, enhancing energy security, and increasing energy efficiency. For these energy policy goals, a mix of instruments is being used. As a result, considering any one particular program in isolation is misleading. Policy makers must use economic, financial, and regulatory analysis to consider the discrete impact of particular policies, plus the same tools to assess the likely impact of a mix of policies on the likelihood of meeting these goals. Individual investors and stakeholders use similar tools to consider the impact of these policies on the viability of particular projects (Gillingham, Newell, and Palmer 2009).

The Nature of Government Interventions

How does government in the United States intervene in the economics of the energy sector? A better question may be: How *doesn't* it intervene? There are literally hundreds of interventions through which federal, state, and local governments use their powers to tax, spend, and regulate in order to influence energy outcomes. Chapters 7 and 8 will examine the arguments justifying government intervention in this sector, and consider in detail how an important set of these interventions aim to accomplish their results. This chapter will consider the role of subsidies in the energy arena, and will examine the general case for tax expenditures, regulations, and a variety of other expenditures. We will restrict our discussion to nonmarket policies; approaches such as carbon taxes, cap and trade, and renewable portfolio standards will be examined in Chapter 8.

Even a cursory analysis of this set of interventions leads to many difficult questions. Do we have data on the effectiveness of these interventions? Are we receiving our money's worth from these programs? After all, even small, local, nonprofit organizations are now expected to provide evaluation results to support their programs; shouldn't those benefiting from much larger sums be able to show concrete benefits? Collectively, do these interventions make sense? Are they internally consistent, and is there any grand design in evidence? How do these various programs interact? And finally—should we attempt to sustain all of these programs, given the upcoming fiscal tsunami?

The Importance of Subsidies

Perhaps the single most contested issue in the energy policy arena has to do with the nature, extent, and impact of energy subsidies. There is no commonly accepted definition of a *subsidy* broad enough to encompass the many activities to which the term is applied. The core of the concept is of a transfer of resources from government to a public or private entity. This may be done with the goal of lowering the cost of some good or service and thereby increasing its production or consumption, or to enable a firm to increase or decrease its costs or revenues.

Subsidy Categories

A 2006 analysis by the World Trade Organization suggested three main categories of subsidies: (a) direct government expenditures or transfers to producers or consumers, (b) provision of goods and services at no cost or below market price, and (c) government regulations that create transfers between one group and another.

This is helpful, but does not go far enough. From an economic perspective, policies that do not treat all members of a group equally (such as fuels or types of automobile) also are arguably subsidies since they favor one member of the category over others and lower its relative cost. For example, diesel fuel is more lightly taxed than gasoline in many European countries to encourage the purchase of diesel cars. And the aggregate effect of tax policy is to produce overall tax rates that differ substantially by energy source and change the relative incentives for investing. In the past, oil and gas benefited from relatively low overall tax rates; nuclear and renewables have benefited from more recent federal legislation.

One approach is to distinguish between *what* is being subsidized, and the *method of action* or policy instrument used to accomplish the subsidy, as shown in Table 7.1. Examine each specific piece of legislation or policy closely enough and it is usually (but not always) possible to discern a program theory that uses a mechanism of action to lower the cost of some important energy-related good (a primary energy source, information, R&D, etc.), thereby accomplishing a public purpose. For example, the Department of Energy (DOE)'s Generation IV program is engaging in research designed to design and test new nuclear reactor designs; nuclear power producers will

Table 7.1 Subsidy Analysis: What Is Subsidized, Policy Instruments, and Examples

	Policy Instrument				
Type of Subsidy	*Primary Energy Sources*	*Information*	*Research and Development*	*Risk*	*Capital*
Tax rates	Differential effective tax rates				
Tax expenditures/ credits	Production tax credit Consumer tax credits				Investment Tax Credit
Direct spending	LIHEAP	State Energy Program	Nuclear reactor research, DOE labs		
Loans, Loan Guarantees				Title 17, EPACT of 2005	Advanced Tech Vehicles
Access to federal lands; royalty provisions	Deep Water Royalty Relief Act of 1995			Various royalty provisions	
Tariffs, trade restrictions	Ethanol import tariff				
Insurance/ Indemnity				Price-Anderson Act	
Public ownership	Federal dams; Power Market Administrations				
Reserves				Strategic Petroleum Reserve	

benefit substantially from this research. Another example is the substantial difference in effective tax rates paid by producers of various kinds of energy (Metcalf 2009).

Slavin (2009) suggests a simpler approach, focusing on tax credits, targeted disbursements (including direct spending, grants, and loan guarantees), and regulatory subsidies. A study of federal subsidies between 1950 and 2003 concluded that these categories amounted to 45 percent, 20 percent, and 35 percent of U.S. energy subsidies during that period (Bezdek and Wendling 2007). This reinforces the substantial extent to which the tax system is relied on as an instrument for influencing energy development and decision making.

Few of the resulting cells in this table receive unanimous support in the energy policy community, and most are hotly contested. The broad debate is over whether the United States should provide *any* subsidies to energy (and thus move toward a completely market-based system), or whether a "rational" system would support some subsidies (e.g., for renewables) and eliminate others (e.g., oil and gas). Several of these policies are especially controversial. One of these is the power (and irrigation) produced by the system of federal dams in the western United States, and marketed by several Power Marketing Administrations in the DOE. Whether their power sales are subsidized is the subject of a long running debate.

Another is the Strategic Petroleum Reserve, which now holds 727 million barrels of oil in Louisiana and Texas salt caverns. Established following the 1973 oil embargo, the reserve was created to provide a supply of oil that could be auctioned in the event of a severe energy supply interruption. The definition of such an interruption is nebulous. The reserve has been drawn down on four occasions, including the aftermath of Hurricane Katrina. It is costly to maintain and fill, but does provide some cushion to the oil market in the event of a short-term disruption.

Nuclear Power Subsidies

Finally, several subsidies to the nuclear power industry deserve some discussion. First is the Price-Anderson Act, which provides indemnity insurance and a liability framework for the nuclear power industry. The act was included as part of the Nuclear Power Act of 1957, due to concerns that investors would not provide capital for nuclear power development if they faced unlimited liability. The 1957 act was most recently renewed to 2025 as part of the Energy Policy Act (EPAct) of 2005.

Under this act, licensees must obtain the maximum private insurance available, now $300 million per plant. These liability policies are with American Nuclear Insurers (ANI), a joint underwriting association of 21 major U.S. insurers. In the event of an accident, reactor companies would first pay out from this insurance. Claims greater than that amount, but under $11.6 billion, are covered by the Price-Anderson Fund, to which every reactor licensee company must contribute a total of $111.9 million in payments limited to $17.5 million per year; both of these amounts are adjusted for inflation every 5 years. Claims over the $11.6 billion fund maximum would be covered by the U.S. government. The act also places the legal jurisdiction for nuclear accidents in federal court, and prohibits punitive damage settlements. So far, about $200 million in claims under Price-Anderson have been paid out, including damages from the 1979 Three Mile Island accident.

Exactly how much of a subsidy Price-Anderson provides the nuclear power industry is a matter of debate, subject to assumptions about the risk of a nuclear plant accident, how much damage would result, and about how much insurance a plant licensee would need to obtain in the absence of the law (CBO [Congressional Budget Office] 2008a). The CBO estimated that the overall benefit of the act was minimal—1 percent of the levelized cost. Other analyses have come up with much higher estimates (*Public Citizen* 2004).

Several other subsidies included in EPAct 2005 influence decisions to invest in nuclear power, including a production tax credit, loan guarantees, delay insurance (covering delays in starting operation), and subsidized decommissioning costs (CBO 2008a). Of these, the federal loan guarantee program is the most important. EPAct 2005 provided the Department of Energy with the authority to issue loan guarantees "for projects that avoid, reduce, or sequester air pollutants or anthropogenic emissions of greenhouse gases and employ new or significantly-improved technologies as compared to technologies in service in the United States at the time the guarantee is issued" (White House 2010).

Under such programs, the federal government agrees, in effect, to cosign project loans and reimburse lenders if the borrowers default. Project loans may cover 80 percent of the project costs, and guarantees may cover 100 percent of this amount. This can significantly ease the process of obtaining credit and lower its cost. The law requires the DOE to either receive an appropriation to cover the cost of the projected subsidy reflected in the guarantee, or receive fees from the borrower to cover the cost. The DOE was instructed by Congress in recent appropriations bills to follow the latter option (GAO [Government Accountability Office] 2008a). This makes the formula for estimating that subsidy a highly political calculation—too low and it is an even larger subsidy; too high and the cost may deter potential borrowers from using the program.

A total of $42.5 billion in guarantees is available, of which $18.5 billion is authorized for nuclear power under the program. The proposed Obama budget for FY 2011 would expand that amount by $36 billion, to $54.5 billion. As of May 2010, $8.33 billion in loan guarantees for nuclear construction and operation have been approved, for two new nuclear reactors at the Vogtle Electric Generating Plant in Burke, Georgia. Four other, nonnuclear projects have also received loan guarantees.

Risky Business

Loan guarantee programs are risky to the government. They are subject to an array of classic principal-agent problems, since borrowers have—or should have—a better understanding of the potential risks of a project, and have incentives to hide that risk from the government. Underestimated risks result in a lower estimate of subsidy cost, and thus in fees due from borrowers.

There are additional risks from this particular program. For example, it is aimed at encouraging the application of new reactor technologies, and the Nuclear Regulatory Commission (NRC) has a related program to expedite the licensing of these new technologies. Also, nuclear projects have a consistent track record of underestimating total costs, often by billions of dollars. In an earlier study, the CBO estimated the risk of default of such projects at 50 percent (CBO 2003). The GAO (2008) has also criticized the DOE's earlier management of the program. The longer-term weakness of this approach to energy development is that it obscures the costs of the new technologies being developed (UCS [Union of Concerned Scientists] 2009). We will consider the broader case for nuclear power in Chapter 10.

Tax Expenditure Interventions

Why pay attention to so-called *tax expenditures*? The main reason is that they now reflect a high proportion of the federal government's yearly support to the energy sector—60 percent in a 2007 analysis. These are "provisions in the Federal tax code that reduce the tax liability of firms or individuals who engage in specific economic activities that affect energy production, consumption, or conservation in ways deemed to be in the public interest" (EIA 2008b, 1). Our perceptual bias is to think of governmental action in terms of the spending side of the ledger: the funded "programs" implemented by government employees, or grants provided directly to firms or nonprofits. The receipts forgone on the revenue side of the government's accounts generally receive much less attention. Details about these tax breaks (including estimates of their yearly cost) are buried in the fine print of the Appendix of the federal budget and in a cryptic document produced by the Joint Committee on Taxation of the U.S. Congress.

There are several types of tax expenditure of importance to the energy sector. The most important are (1) tax credits (reductions in the taxes owed), (2) deductions (reduction in the total income subject to tax), (3) exclusions (a source of income that is allowed not be counted), and (4) preferential changes in depreciation so that preferred investments may be written off more quickly (Caperton and Gandhi 2010; Koplow 2008). Of these, tax credits are the most advantageous and costly, since a credit directly reduces tax liability on a dollar-for-dollar basis. The best available data, from 2007 for example, shows that the top two energy tax expenditures are the volumetric ethanol excise tax credit and the alternative fuel production credit. In 2007 these programs had costs of $2.99 billion and $2.37 billion, respectively.

Expensing of exploration and development costs by oil and gas companies makes up the third most costly energy sector benefit, with an estimated cost of $860 million in 2007. Oil and gas companies are allowed to deduct in the current tax year most "intangible drilling costs" (labor and other expenses involved in drilling for oil and gas) rather than amortizing them over time.

Another well-known tax expenditure program for the energy sector is the production tax credit (PTC); this program provides companies a renewable energy credit of 2.1 cents per kilowatt hour as an incentive to invest in alternative energy projects. Several studies have shown a strong correlation between the availability of the credit and the level of investments in wind energy (Shaffer, Rode, and Dean 2010).

An Incomplete Portrait

But considering tax credits for one industry sector in isolation gives an incomplete portrait. From the perspective of a potential developer of wind energy, for example, the federal government has created a menu of options that include not just the PTC, but also in the American Recovery and Reinvestment Act (ARRA) of 2009 a renewed investment tax credit of 30 percent; it also allows for significant accelerated depreciation of the required equipment.

Complex considerations of total construction cost and expected energy production (the PTC is granted on a production basis) influence the choice of subsidy approach taken by an energy sector or an individual company (Shaffer, Rode, and Dean 2010). Moreover, developers may join with tax equity investors (typically large investment banks) to harvest the tax credit benefits that might otherwise be left on the table for large projects. Since the developers themselves don't usually generate enough tax liabilities to use all of their credits, they may find firms that have significant tax liabilities and have them provide capital for the project. This enables the firms to both use a portion of the available tax credits and earn a return on their investment.

The ARRA also included a provision for $6 billion in loan guarantees for renewable energy technologies. Developers also generally generate project revenue by selling both the electricity from the projects and renewable energy certificates. This is covered in greater detail in Chapter 8.

Problems Financing Renewable Energy Projects

Since the 2008–2009 financial collapse slowed down the tax equity market, the process of financing renewables projects with debt became more difficult and caused a shakeout in the renewables industry (Schwabe, Cory, and Newcomb 2009). One response from Congress was to include Section 1603 in the Recovery Act, allowing developers to receive a grant equal to 30 percent of a project's cost in lieu of either the then-available tax credits.

According to a DOE study, 64 percent of wind energy developers, and 100 percent of geothermal developers chose to accept the grant. Those projects accounted for 92 percent of the $2.8 billion in grants given up to March 1, 2010 (DOE 2010). The good news is that this supported the development of 2.4 MW of wind development and thousands of short-term and long-term jobs. The less good news is that a large proportion of the grant funds were paid to foreign companies, including an estimated 80 percent of the first $1 billion distributed (Choma 2010).

Policy Challenges of Tax Subsidies

From a policy perspective, tax expenditures present a number of challenges to policy makers as well as recipients. One, subsidizing investment rather than production is risky because it encourages companies to pay less attention to project costs. Subsidies complicate both the tax code and the process of decision making around the types of projects being incentivized. As a result, they present a very difficult-to-determine "dosage" problem. That is: what is the best level of rate, write-off, or exemption to accomplish the desired effect?

Evaluating the impact of these dosage decisions may be difficult given the lack of data about the use of particular tax expenditures. However, the stop-and-start nature of the PTC has provided a natural experiment suggesting the solid impact of the credit on wind energy capacity in the country. Once enacted, these subsidies remain on the books until their expiration date. Still, PAYGO budget scoring requirements have had an impact on the length of time such tax provisions have remained in effect.

The lack of transparency of tax expenditures is also a problem. For confidentiality reasons, the Internal Revenue Service (IRS) does not disclose extensive data on use of tax credits and other expenditures to the public, making it difficult to assess their effects. For this reason, many analysts suggest revising the process for the budgeting, analysis, and evaluation of energy-related tax expenditures (Caperton and Gandhi 2010).

A close examination of the Energy Information Administration (EIA)'s 2007 study of the DOE's energy interventions also shows that different energy sources and program beneficiaries tend to receive very different proportions of the total pie of either tax expenditures, or other DOE program spending. Oil and gas, for example, received 20 percent of the $10.4 billion in energy tax expenditures that year, but only 1 percent of the $6.2 billion in other spending (Table 7.2).

Energy Research and Development

The federal government is a major funder of research and development (R&D) activities. The 2009 federal budget included $147 billion in support of R&D programs. The top four recipients of research funds were: the Department of Defense, $80.8 billion; Health and Human Services, $30.1 billion; National Aeronautics and Space Administration (NASA), $10.9 billion; and the Energy Department, $10.3 billion. The case for federal involvement in R&D is that in the absence of government support, society's overall investment in these activities will be far less than optimal.

There is also a belief among some economists that federal efforts need to be spread across the spectrum of activities in the innovation chain—basic research, applied research, and development of innovative ideas into products that will be useful in the marketplace. The last portion of the chain has been especially troublesome, thanks to the so-called Death Valley between laboratory findings, successful piloting of a concept, and its ultimate commercialization. The repeated failure

Table 7.2 Proportion of 2007 Energy Tax Expenditures and Energy Program Spending

Beneficiary	*Tax Expenditures*	*Other Programs*	*Overall*
Coal and Refined Coal	25%	10%	20%
Oil and Natural Gas	20%	1%	13%
Nuclear	2%	17%	8%
Renewables	38%	15%	29%
Electricity	7%	8%	7%
End Use	1%	44%	17%
Conservation	6%	4%	6%
Totals	100.0%	100.0%	100%

Source: EIA 2008a, Table ES1.

of potentially useful findings to meet with commercial success is evident in the federal funding pattern, as over half—$83.9 billion, or 57 percent—of federal R&D spending in 2009 went to development activities, with about $29.6 billion to fund basic research and $29 billion to fund applied research (CRS 2010a).

A Major Presidential Theme

President Obama has made support for energy research a major theme of his administration. In an address to the National Academy of Sciences in 2009, he established the goal of devoting 3 percent of gross domestic product (GDP) overall to R&D, and confirmed that "energy is our great project, our generation's great project. And that's why I've set a goal for our nation that we will reduce our carbon pollution by more than 80 percent by 2050." The Obama 2010 budget proposal included $150 billion in funding over 10 years to support renewable energy, energy efficiency, climate research, and the creation of the Advanced Research Projects Agency for Energy, or ARPA-E.

The Department of Energy's research and development programs are substantial, including 17 national laboratories and many specific programs focusing on basic science and particular technologies. The overall budget for the department's science, discovery, and innovation programs for fiscal year 2009 was $3.7 billion; this omits $2.2 billion from the Energy Efficiency and Renewable Energy division of the DOE, plus $5.5 billion in Recovery Act funds for R&D.

Basic science programs include Basic Energy Sciences, Advanced Scientific Computing, Biological and Environmental, High Energy Physics, Nuclear Physics, and Fusion Energy Sciences. Energy efficiency and electricity programs include Energy Efficiency and Renewable Energy, Electricity Delivery and Energy Reliability, The Energy Transformation Acceleration Fund, and Nuclear Energy. In addition, several other initiatives are underway. The DOE is establishing several Energy Innovation Hubs ($135 million proposed over 5 years) that will bring together researchers with science, engineering, and policy expertise to focus on critical national needs.

Energy Frontier Research Centers are being created to research the fundamental science needed to move past barriers to advance revolutionary energy technologies, such as energy storage. Finally, ARPA-E will support research that has the potential for high commercial impact but is too risky to be funded by private investors. ARPA-E initially received $400 million in ARRA funding and a further $300 million is requested for the program in FY 2011. This is an impressive amount of activity.

There are several underlying assumptions behind the large increase in government energy R&D funding. One is that a policy creating a clear carbon price will not be sufficient to transform the U.S. energy sector. It would help to change the profile of electricity generating sources in the country. But additional breakthroughs in wind power, concentrated solar power photovoltaics, vehicle batteries, and energy storage will be needed to accelerate the pace of change.

A second is that the historical pattern of government spending on energy R&D has been inadequate and erratic. Overall, R&D in the energy sector has been problematic since the inception of the Department of Energy. In the United States as a whole, the private sector provided about two-thirds of the country's total of $389 billion in R&D expenditures in 2009 (about $253 billion), with the U.S. government picking up 28 percent ($109 billion). But the energy sector has been a laggard, with 2009 investments of only about $3 billion, in an industry with about $1.8 trillion in revenues that year. This computes to an *innovation intensity* (R&D as a percent of revenues) of only 0.23 percent, compared to a national average across all industrial sectors of 2.6 percent.

Prior to recent increases in R&D funding, the federal commitment to energy research and development was as paltry as that of the private sector. Figure 7.1 shows the results of a GAO analysis of the DOE's budget authority for R&D on renewable, fossil fuel, and nuclear between 1978 and 2008. Support for renewables research was slashed in the early years of the Reagan administration and did not recover until the America COMPETES Act of 2007 (Public Law 110-69) with the support of the Bush administration, which increased support for the related basic science research.

Although a general case can be made for increases in funding for federal energy R&D, exactly how much is needed is difficult to assess. Various sources have suggested doubling the amount from its 2008 baseline or increasing it to as much as $15 billion or $30 billion per year (Reicher 2009). But without a rethinking of how the DOE spends these funds, such increases may not be used productively.

The DOE's Spotty Record of Success

The DOE's record in the R&D arena is uneven. On the one hand, there have been clear successes: DOE research led to the low-energy ballast for florescent lights, the fluidized bed combustion process for burning coal, greatly accelerated efficiency improvements in solar panels and wind turbines, and many other efficiency advances.

But on the other hand, over the past ten years a series of reports have concluded that the overall system of support for public energy innovation is in need of transformation. Criticisms include: (1) the lack of focus at the DOE national laboratories (Anadon et al. 2010); (2) the politicized nature of DOE's R&D funding, which prevented the department from cutting fossil fuel research (GAO 2008); (3) the lack of an interdisciplinary and integrated approach to innovation (Ogden, Podesta, and Deutch 2008); (4) concern about the poor performance of DOE demonstration projects; and (5) the narrow nature of legislation authorizing energy research that leads to incentive structures overly tailored to particular technologies (Bonvillian and Weiss 2009).

To be fair, the process of intervening in energy innovation is extremely complicated. Federal innovation successes such as the Manhattan Project, Apollo Program, and the Internet had money to burn, and their focus on governmental goals made product commercialization strictly an afterthought. In contrast, energy research, development, and demonstration is primarily market driven,

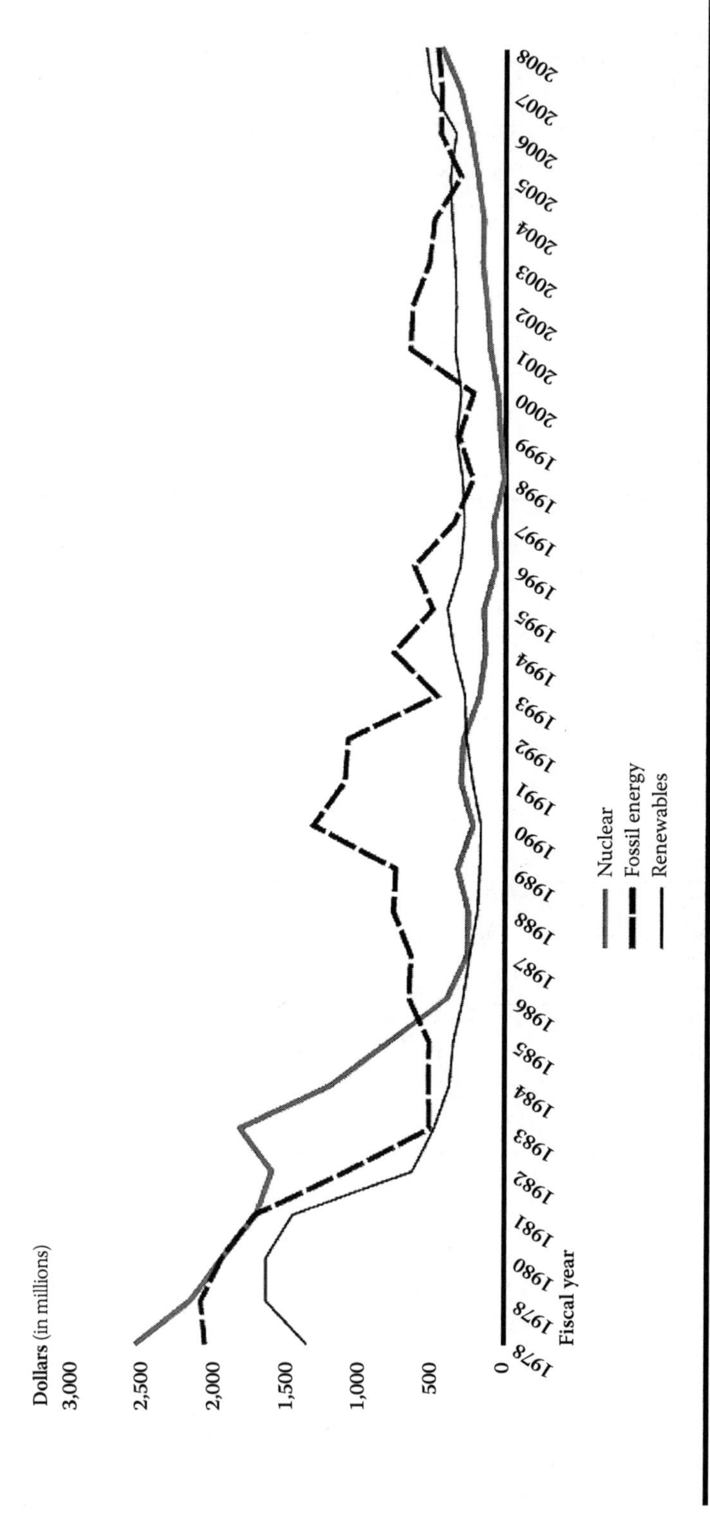

Figure 7.1 DOE's budget authority for renewable, fossil, and nuclear R&D, FY 1978 through FY 2008 (2008 dollars). (From GAO, 2008.)

nonsequential, and takes place in a startlingly complex network of public, private, university, and nonprofit actors all seeking to capitalize on subsidized research.

There is also a complex overlay of related programs (as we are examining in this chapter) that provide incentives for new technologies through tax expenditures, standards, mandates, loan guarantees, and similar programs. The pervasiveness of government intervention and regulation in the energy arena (particularly the electricity sector, with regulated rates) also makes it difficult to argue that it should begin to act like a high tech sector, with R&D around 10 or 20 percent of sales. Some of the DOE's recent actions, including its creation of the Energy Innovation Hubs and ARPA-E, have occurred in response to the criticisms noted above, suggesting a more focused and interdisciplinary approach to energy research and innovation.

Federal Regulation of the Energy Sector

Chapter 6 provided a brief introduction to the topic of regulation in the energy sector. Regulation is the use of the government's sovereign power to promulgate rules on the behavior of individuals and firms. Failure to comply with these rules may result in fines or civil penalties. The concept of energy regulation encompasses a gamut of governmental activities. These include laws and rules structuring energy markets, the particular rules governing the use, supply, and pricing of particular energy sources and commodities, and specific programs that aim to change consumer and producer behavior to increase efficiency and reduce pollution.

Much of the literature on energy regulation focuses on the challenges in structuring and regulating public utilities, particularly electricity and natural gas provision, at the state level. We will briefly consider those issues in our discussion of renewable portfolio standards in Chapter 8. Here we will broadly examine the benefits and problems associated with a regulatory approach to energy policy, and consider in some detail a sample of important issues and specific regulatory policies, including the Energy Star program.

Federal action through regulation occurs at the end of a complex and often conflicted process (Pelast, Oppenheim, and MacGregor 2003). Congress establishes a framework for the intervention through major legislation. Then regulatory agencies use procedures established in the Administrative Procedures Act (Public Law 79-404; 5 U.S.C. 500) to gain input for the drafting of the required regulations, put them out for comment in draft form, and then incorporate this feedback prior to releasing the final rule, or regulation. Under the 1969 National Environmental Policy Act (NEPA), federal activities that may have environmental impacts must complete an environmental assessment and an environmental impact statement (EIS) if the activity will impact the environment. This requires careful consideration of project alternatives and minimizing environmental impacts.

Early Requirements for Federal Agencies

President Clinton's Executive Order 12866, issued on September 30, 1993, provides detailed requirements that federal agencies must follow to establish both their overall regulatory agenda, and to promulgate particular regulations. Most federal agencies have analysis staffs that complete preliminary economic and policy analyses of proposed rules, to determine if they are "major," or likely to have an impact of $100 million or more. Such major rules are subject to detailed cost–benefit analysis, and the Office of Management and Budget's (OMB) Office of Information and Regulatory Affairs (OIRA) is given substantial power to intervene in the rule-making process if proposed rules are perceived as excessively costly, cumbersome, and not in the public interest. Finally, the Congressional Review Act (CRA) gives Congress 60 days to review new federal regulations and a legislative mechanism to reject such regulations, and requires the GAO to complete analyses of every major rule.

Over the past three decades the federal regulatory process has been subject to increasing political pressure. The critique of federal regulation as costly, intrusive, and overly burdensome for businesses and households first gained traction in the Carter administration, which began a push for deregulation that continued under subsequent administrations and brought sweeping changes to the airline, trucking, railroad, telecommunications, and energy industries. Reagan's 1981 Executive Order 12291 began the process of giving the OMB increasing authority to monitor the regulatory process and use the results of cost–benefit analysis to intervene and reject regulations believed to be too costly.

Regulatory process gained further attention in the George H. W. Bush administration, with Vice President Dan Quayle's Competitiveness Council repeatedly intervening in the regulatory process against environmental and health and safety regulations believed to be overly burdensome to business. Despite this, many strong regulations were implemented during the first Bush administration.

The use of the CRA was further entrenched under John Graham, director of the OMB's OIRA during the George W. Bush administration, who sought to implement a "smart regulation" approach focused on strengthening regulatory analysis and increasing the scrutiny of major rules. Part of his strategy included passage of the 2001 Data Quality Act, a two sentence rider in the Consolidated Appropriations Act of 2001 (Public Law 106-554). This enabled OMB to direct federal agencies on how to ensure and maximize "the quality, objectivity, utility, and integrity of information (including statistical information) disseminated by Federal agencies" (P.L. 106-554, Sec. 515a). The act was used by corporations on several occasions during the Bush administration to prevent implementation of regulations, by challenging data and information used to support regulatory decisions.

Regulation as Command and Control

Regulation as an approach to governmental action is often perceived as synonymous with a particular method, the command-and-control or prescriptive model. This approach requires the regulator to set a standard for the matter to be regulated (often around a particular pollutant such as nitrous oxide), specify how the firm will respond, and establish a monitoring system. Implementation of the Resource Conservation and Recovery Act (RCRA 42 U.S.C. §6901), which gave the EPA authority to regulate hazardous waste, followed this approach.

The method is appropriate for fixed or point source pollution featuring large emitters such as factories and power plants, but it is less effective with mobile sources such as motor vehicles. Other frameworks have largely supplemented such a strict command-and-control approach, including a performance-oriented framework that establishes a regulatory goal and gives the regulated entity flexibility in how it goes about meeting that goal. Market-based approaches have also proved effective. One approach that generally has proved to be problematic is self-regulation by firms, which often have too many incentives to cheat in the absence of strong oversight.

Regulation is ubiquitous in the energy sector due to the many externalities produced in energy transformations, plus government attempts to influence the demand and supply of energy. The complexity of energy systems and the wide reach of energy legislation in recent decades led legislators to craft many programs narrowly around particular sources and approaches. Regulators, including officials from the DOE, Federal Energy Regulatory Commission (FERC), NRC, EPA, the Department of Interior's Minerals Management Service, which regulates offshore energy, including drilling for oil and gas, and the Department of Transportation's Office of Pipeline Safety (oil and natural gas pipelines), then attempt to tailor the regulatory approach to the demands of the situation.

Government must be particularly demanding of energy activities that present risks of catastrophic failure, such as the nuclear power industry or offshore oil wells, or damage to human health. Many other activities in the energy realm present lesser, but still significant risks to the environment, workers, or society. Mining and refinery work is especially dangerous, and in all these cases, firms have incentives to reduce costs in ways that raise those risks. Over recent months in early 2010, the BP Gulf oil spill, Upper Big Branch coal mine disaster, and the Anacortes, Washington, refinery explosion are reminders that work in the energy sector is dangerous and that regulatory systems are needed.

Limitations of Regulation

But regulation has serious limitations. When intervening in complex and unpredictable systems, regulators cannot anticipate every eventuality, including creative means of evasion on the part of those regulated. This was one reason Wildavsky (1988) in the context of safety regulation, suggested a strategy of resilience and recovery, rather than anticipation, when uncertainty is high. Regulating in advance against every possibility will be very costly. From an economic perspective, we have learned about the poor record of government management of pricing in complex and volatile energy markets.

Regulatory systems tend to veer between adversarial relationships between the government and the regulated organizations, and excessively cozy relationships, which are often termed regulatory "capture." Adversarial relationships create incentives for those regulated to hide problems, diffuse responsibility, and use legal action, rather than embrace accountability and search for solutions.

Regulatory capture can damage the credibility of government generally. One factor cited in the 2010 Gulf of Mexico oil disaster was the willingness of the Department of the Interior (DOI)'s Minerals Management Service (MMS) to grant hundreds of permits for offshore drilling without requiring the completion and submission of environmental assessments (Cooper and Broder 2010). MMS officials accepted the assurances of various company officials that the drilling presented little risk. In fairness, the number of severe offshore spills from oil drilling has been minimal over the last few decades, yet the willingness of regulators to look the other way meant that specific features that may have made that particular site more risky were never explored.

The initial process of setting a regulatory standard or deciding what is "safe" and what is not may be based on science and analysis, yet it still may be perceived as arbitrary. Uniform requirements are challenged from an economic perspective, since the costs of compliance are likely to differ between firms, and thus penalize some much more than others.

Political Implications of Regulation

The application of regulation has long had significant political implications. Although the term *regulation* isn't quite as vilified as the term *liberal*, proponents of increasing government intervention in American life must battle against a decidedly skeptical population. A Gallup poll in January 2010 found 50 percent of Americans agreed with the statement "Government should become less involved in regulating business," while only 24 percent thought the government should become more involved in regulation (Newport 2010).

But that general sentiment did not apply to every sector; April 2010 polls found significant support for more stringent banking regulation in the aftermath of the financial crisis—particularly when Wall Street was the target (Saad 2010). Although there have been no recent polls on energy regulation, respondents have (as noted previously) been generally sympathetic to the concept of increasing the country's reliance on renewable energy—a process that will continue to require many complex regulations.

Appliance Efficiency Standards and Energy Star

A major analysis of U.S. energy issues published in 2009, *America's Energy Future*, identified nine major energy programs responsible for over 13 quadrillion BTUs (quads) of energy savings in the United States each year. These savings are significant given that U.S. energy use has stabilized at about 99 quads per year. The first two of these, auto fuel economy or corporate average fuel economy (CAFE) standards and appliance efficiency standards, were responsible for 55 percent of the savings, with the latter program alone estimated to save 2.58 quads per year (Committee on America's Energy Future 2009).

U.S. households are full of appliances that provide useful services, from kitchen appliances (refrigerators, microwaves, dishwashers), clothes washers, hot water heaters, lighting, heating and cooling devices—heaters and air conditioners—and increasingly, electronic devices from computers to TVs. Collectively, about 20 percent of the electricity use of U.S. households is required to power these devices, a total of about 1.14 trillion kWh in 2001. Although on a monthly basis the electricity cost of these appliances appears low, over the course of their useful life, the operating cost may exceed its initial purchase price. The steady integration of consumer electronics into daily life has also brought the potential for substantial increases in household energy consumption, as people alter their habits and personal space to incorporate larger, digital TVs and related equipment (e.g., digital video recorders, etc.) that are constantly in operation or standby mode (Crosbie 2008).

The case for regulating the electricity use by these devices is based on a series of arguments about market barriers to energy efficiency. Third-party purchases (landlords, for their tenants) are common and purchasers have few incentives to consider electricity use by these products; in the absence of government action, little information would be available about their electricity use, and many purchases of these products, especially large appliances, are made under time pressure after the current product has failed (Nadel 2010). The assumption is that in the absence of government action, the market will do little to encourage products to become energy efficient.

The appliance efficiency movement began in the early 1970s, as increasing power demand and higher energy prices led several states, including California, to set state standards for important appliances, particularly refrigerators. But state-by-state standards are bothersome to manufacturers, who must monitor state requirements, revise specifications to match, and in the process lose out on the economies of scale of the huge overall U.S. market. Congress reacted to requests to set national standards with the National Appliance Energy Conservation Act of 1987, which established minimum efficiency standards for many household appliances, including kitchen appliances and clothes washers and dryers, among others.

A series of additional laws, including the Energy Policy Act of 1992, and the Energy Policy Act of 2005, added new device standards. In each of these pieces of legislation, Congress established initial efficiency standards and gave the DOE responsibility to review and update them through the regulatory process. These standards are set based on research performed at the DOE's national laboratories, with input from manufacturers, and following a series of analyses of the life-cycle cost and payback period of the product with the new standard, and analyses of its national impact and overall impact of the regulation.

Test procedures are established so manufacturers may document that their products meet the standards. The Federal Trade Commission (FTC) is responsible for the associated rules that govern labeling of the resulting residential products; the DOE and FTC work jointly on commercial product labels. The annual budget for the appliance standards program was $20 million in FY 2009.

EPA's Energy Star Program

A second federal program, Energy Star, managed by the EPA with the cooperation of the DOE, aims to encourage consumers to purchase energy-efficient products through a voluntary labeling program. The chief goal of the program is reduction of greenhouse gas emissions. Begun in 1992 to encourage consumers to buy greener computers, the program now covers about 3,000 manufacturers, 40,000 different product models, 1,500 retail partners selling qualified products, 8,500 home builders, and 700 utilities (EPA 2010). EPA staff finds products that would benefit from the program, develops specifications in conjunction with the DOE; manufacturers sign up for the program, test their products, and may use the Energy Star label if they meet the program guidelines.

The benefit of the program is to lower transaction costs to consumers researching new products, as well as reduce the risk that products will not perform as advertised. The EPA estimates that the program prevented the release of 45 million metric tons of GHG emissions in 2009 and saved consumers $17 billion in utility costs. The annual budget for the program is about $50 million.

Criticism of Energy Star

What's not to like about these programs? They lower electricity use at a low budgetary cost and, in the process, reduce GHG emissions and lower utility bills for households and businesses. However, Energy Star has been repeatedly criticized for failure to properly monitor and protect its certification process (Wald and Kaufman 2010). Economists are also less than keen on this regulatory approach to changing consumer behavior. Generally, more efficient products are believed to incorporate more costly designs and materials than less efficient models, and thus are sold at a higher price. Mandatory standards remove the option of purchasing a less expensive model from lower-income consumers or those who don't value energy efficiency—perhaps because they won't use the product often (Gillingham, Newell, and Palmer 2009).

Increases in efficiency may also be associated with rebound effects if consumers perceive the greater efficiency and increase their use of the product. Detailed evaluations and analyses of these programs are also scarce. One implicit assumption of the program is that technological change incorporating efficiency will occur too slowly. Newell, Jaffe, and Stavins (1998) analyzed the factors influencing the energy use of room air conditioners, central air conditioners, and gas water heaters. They concluded that increases in energy prices, labeling standards, and efficiency standards contributed to documented increases in efficiency, but that "a substantial portion of the overall change in energy efficiency for all three products cannot be associated with either price changes or government regulations" (20). Technological change—encouraged by the research occurring at the DOE's laboratories—was the likely factor strongly influencing increased product efficiency.

Energy Interventions for Households: LIHEAP

Most energy interventions are aimed at corporations operating in the energy sector. But several programs are designed to encourage American households to invest in energy-efficient homes, home improvements, autos, and appliances. Household energy subsidies are common throughout the world. Energy sources vital to everyday life—gasoline, coal, fuel oil, charcoal, propane, electricity—priced at market rates often require low-income people to pay a high proportion of their income for them, squeezing out other necessities.

One major program, the LIHEAP or Low Income Home Energy Assistance Program, enables the federal government to provide energy assistance to low-income households. Under the LIHEAP program, the federal government provides energy assistance to households "that pay a high proportion of household income for home energy, primarily in meeting their immediate home energy needs" (Public Law 103-252, Sec. 2602(a) as amended). The program was created in 1981 as the rise of inflation-boosted energy prices left many low-income households and families unable to pay for heat or electricity.

The Department of Health and Human Services (HHS), Administration for Children and Families, manages the LIHEAP program, which is a block grant for which states, U.S. territories, and Native American tribes must apply. The formula for distributing funds is based on weather and the size of the state's low-income population. Each state crafts criteria for the program, but they may not go below 110 percent or above 150 percent of the poverty level (in 2009 the ceiling was 75 percent of state median income). Because funds for the program are generally not sufficient for all potentially eligible households, they are generally targeted to vulnerable households (with children, people with disabilities, or frail elders), or low-income families with particularly high energy costs. Funds are distributed to states based on a complex formula that calculates each state's proportion of home energy expenditures. For FY 2010, the total program budget was just under $5 billion.

A portion of LIHEAP funds are reserved for the national Weatherization Assistance Program, which enables low-income households to obtain funds for home repairs and investments that will reduce their energy demand. Many states also supplement their LIHEAP block grant with additional funds from state general funds, systems benefit charges included in utility bills (electricity and natural gas) donated funds, and funds established separately by utilities. About $1 billion was budgeted for this program in 2008, with funds provided by LIHEAP/HHS, the DOE, and state sources, primarily public utilities. This program has been evaluated extensively since its inception and repeatedly found to be cost effective.

Additional Programs for Households

Two other sets of programs are aimed largely at households. One is the State Energy Program, which provides grant funding to states to support household and institutional investments in energy efficiency. Each state must provide a 20 percent match, and a designated state energy office allocates the funding to eligible projects. The program, which began in 1975, received $3.1 billion in Stimulus/Recovery Act funds that did not require matching grants. The DOE estimates that each dollar spent on the program produces $7.23 in energy savings.

The Energy Policy Act of 2005 also authorized and funded several programs designed to encourage people to invest in energy efficiency for both their houses and autos. These include:

- The residential energy efficiency tax credit (for consumers that purchase and install specific products such as energy-efficient windows, insulation, doors, etc.).
- Residential renewable energy tax credit (for consumers installing solar power, wind power, and other renewable energy systems.
- Several automobile tax credits, for hybrid vehicles, alternative fuel vehicles, plug-in electric vehicles, conversion kits for plug-in hybrids, and even low-speed and two- or three-wheeled vehicles.
- Exclusion from income of subsidies from utilities for energy conservation.
- Energy-efficient mortgages.

Funding in the ARRA/Stimulus bill focused attention on programs designed to encourage families and individuals to invest in energy efficiency. One of these was the Cash for Appliances program, which provided $300 million for states to establish rebate programs to consumers purchasing efficient household appliances. The biggest of these, however, was Cash for Clunkers, the Car Allowance Rebate System (CARS), administered by the Department of Transportation's National Highway Traffic Safety Administration (NHTSA). This program was similar to a program operated successfully in several European countries.

Originally supposed to be in effect between July 1 and November 1, 2009, CARS enabled new car buyers to receive a credit of $3,500 or $4,500 toward the purchase of a high-mileage vehicle, under several restrictions; the "clunker" could not be newer than 2001, had to be drivable, and when new, obtained not more than 18 miles per gallon. The amount of the credit depended on the amount of improved fuel efficiency and was paid directly from NHTSA to the car dealer.

The program was wildly popular and the original $1 billion for the program was spent within days of the program going online. Congress appropriated another $2 billion and that, too, was snapped up, and the program was halted on August 24. Despite problems administering the rebates—many dealers complained of slow processing—the agency eventually honored 677,842 vouchers and 12,272 were rejected; a total of $2.85 billion was paid out. Fifty-nine percent of the vehicles were passenger cars and the remainder light or heavy trucks; just under half of these vehicles were made in the United States. Overall, their mileage was 24.9 miles per gallon, replacing vehicles with a collective mileage of 15.8 per gallon. The program is estimated to save 825 million gallons of gasoline over the next 25 years and reduce CO_2 emissions by 9 million metric tons (DOT 2009). It also provided a notable lift to the auto industry in the middle of the recession, as intended. And what happened to the clunkers? Their engines were disabled by the auto dealers and were to be crushed or shredded by disposal shops within 180 days.

Conclusion: The U.S. Energy Policy Blunderbuss

The blunderbuss was one of the first shotguns; eighteenth-century weapons with a fluted muzzle that when fired, produced a loud roar and flung its shot out in a wide pattern. It could be effective at close range but was not the weapon of choice to hit a long-range target. U.S. energy policy is a blunderbuss. It looks impressive from a distance. We fire many policies at our targets—energy efficiency, expanding domestic oil and natural gas production, support for renewables and nuclear energy—but collectively, they don't form a coherent pattern. In part, that is because the gun must attempt to hit multiple, nebulous targets, some of which are of secondary importance yet have displaced more legitimate concerns.

Only the logic of politics, for example, can explain the continuing support for the two major ethanol subsidies. Also, only the need to craft bipartisan support for omnibus legislation can explain linking such energy supply-expansion policies as tax expenditures, royalty policies, and public land access for fossil fuels with support for nuclear power and demand-constricting efficiency measures and support for science and renewables.

A perceived need to link energy policy to economic development and job creation also distorts decisions. Although there is a case to be made for nuclear power, it is likely that the astounding $55 billion in loan guarantees that the Obama administration would like to extend to the industry is in part based on the hope that investments in new plants will put people back to work. If that's the case, greater investment in renewables may be a better bet. The inconsistency between the highly centralized vision of the power grid inherent in nuclear energy, and the decentralized nature of distributed generation has yet to be discussed at the national level.

The record on the collective effectiveness of many of these interventions is inconsistent. Tax expenditures are largely a black hole. We have a rough idea of the resources devoted to many of them, but rigorous ex-post evaluations are rare. For example, how much extra domestic petroleum do we obtain each year for the $2–3 billion in revenue we forego on the two major oil subsidies? The funds devoted to energy research and development have accomplished successes, and past experience suggests that the return to the increased resources going to this arena are likely to yield a positive return (Weiss and Bonvillian 2009).

Perhaps the most vexing questions pertain to energy efficiency programs. The extensive literature review by Gillingham et al. (2004) is equivocal on the key questions about the overall cost-effectiveness of these interventions and the extent to which we can rely on efficiency improvements as a strategy to mitigate the effects of climate change. Still, there are better data here than for most energy programs, and they suggest that investing in efficiency has a substantial payoff.

We should not lose sight of the accomplishments of energy policy over the last few decades. Policy has helped to stop the increase in the overall energy use of the U.S. economy. We have dramatically increased the production of renewable energy, albeit from a small base. And total oil use appears to have peaked, at least for the short term.

Finally—for how long can we afford all of these programs? Chapter 5 estimated the yearly cost of our energy interventions to be about $42 billion. Investments in energy policy are at risk of being an early victim of the almost inevitable cuts that will occur as we reduce our historically large budget deficits. One potential answer lies in creating other sources of revenue, including energy taxes or cap and trade systems. Those market-based instruments will be the focus of the next chapter.

Chapter 8

Policies Shaped by Taxes and Market Mechanisms

> The question is: What can we, as citizens, do to reform our tax system? As you know, under our three-branch system of government, the tax laws are created by: Satan. But he works through the Congress, so that's where we must focus our efforts.
>
> **—Dave Barry (1997)**

> The hardest thing in the world to understand is the income tax.
>
> **—Albert Einstein (quoted by the IRS, 2007)**

This chapter will consider a variety of current and proposed policies intended to use taxes and markets to raise revenues and change energy-related behaviors. At the federal level, our area of focus, there are very few actual energy taxes. Most energy-related levies are crafted as fees, to be paid by users or beneficiaries of various federal services. The streams of revenue generated are allocated to sustaining an ongoing service, or creating an emergency fund. We'll discuss several examples of energy fees and taxes, including the federal Highway Trust Fund, Oil Spill Liability Trust Fund, and the Gas Guzzler Tax.

Do Americans hate taxes? The country's humorists (Einstein included) have long mined a deep vein of popular disgust with taxes that reflects, among other things, a tradition of skepticism toward government and frustration with the complexities of the income tax system. But the story is not one-dimensional. When given the opportunity to vote on tax-related issues, Americans will often support tax cuts, such as California voters' passage of Proposition 13 in 1978, slashing property taxes. But voters are not antitax automatons. They will also vote for tax increases. Americans regularly vote for school levies that raise their property taxes. In 2010, as the Great Recession hammered state budgets, voters in several states, including conservative Arizona, which enacted a one-cent increase in the state sales tax, passed measures increasing taxes.

The emphasis in this chapter, though, will be on policies designed to mitigate the externalities related to energy use. One element of the discussion is the debate over the best policy choice for mitigating carbon emissions: a carbon tax or cap and trade system. Both are intended to decrease such emissions by raising the cost of high-carbon energy sources or the activities that rely on them. The main difference is whether the strategy should rely on constraining the quantity of emissions through a cap or raising the price of emissions via a tax. Both are, at their essence, a means of taxation. And neither will be sufficient in isolation to reach the policy goals established in legislation. A variety of other market-constraining regulations will be needed, including renewable portfolio standards, feed-in tariffs, and other policies, many of which are being used at the state government level and abroad. All of these policy mechanisms are likely to raise the price of energy. Will Americans be willing to support such policies? Or will they consider them yet another act by a diabolical U.S. Congress?

Federal Energy Fees and Taxes

The federal government has a number of fees and taxes on energy. Two of these we will briefly consider are the gas guzzler tax and the oil spill cleanup fee. In the aftermath of the 1970's oil crisis, the concept of a tax on cars with low fuel economy gained traction in Congress and was included as part of the Energy Tax Act of 1978. This tax applies only to passenger cars. In the late 1970s the categories of "minivan" and "sport utility vehicle" didn't even exist and vans and other light trucks (e.g., pickups), were used almost exclusively for business purposes. The IRS collects this tax from manufacturers of vehicles that are found to have combined city/highway fuel economy of less than 22.5 miles per gallon. A sliding scale is applied with tax rates that increase with each one-mpg decrease in fuel economy, starting at $1,000 for cars with 21.5 mpg and increasing to $7700 for cars with less than 12.5 mpg.

The U.S. Treasury collected $172 million from the tax in 2007 (the latest year available) and just over $1 billion between 2000 and 2007. In addition to this tax, until 2003 luxury car buyers were also hit by a luxury car tax established in the Revenue Reconciliation Act of 1990, with what was initially 10 percent of the value of autos priced over $30,000, which also applied to expensive boats, aircraft, jewelry, and furs. This excise tax expired at the end of 2002.

The gas guzzler tax is a disincentive to manufacture low-mileage passenger vehicles, though its effects appear modest. An analysis of the Environmental Protection Agency (EPA)'s data on 1101 vehicles in the 2010 model year that received a fuel economy rating found that 441 were exempt from the tax because they classified as light trucks. Of the remaining 660 vehicles, 95 were eligible for the tax, all luxury cars.

One policy option for improving overall fuel economy would be to eliminate the tax exemption for light trucks. These vehicles typically weigh about 1,100 pounds more than a passenger car and have much worse fuel economy, yet are very popular. Although 2009 was an aberration due to the recession, with only 10.4 million vehicles sold, compared to 17.4 million in 2000, three of the top 10 vehicles were light trucks, including the long-time top seller, the Ford F-Series pickup (Lassa 2010). This is a change, as five of the top ten sellers were light trucks in 2000, and light trucks outsold passenger cars each year from 2000 to 2007. More expensive gasoline and the recession brought the market back to rough parity between cars and light trucks beginning in 2008. This policy option is not likely to be enacted as it would run counter to another ongoing policy initiative, to strengthen the health of U.S. automakers.

Oil Spill Liability Trust Fund and Related Fees

The Challenger Horizon oil spill has focused attention on the country's capacity to respond to such environmental disasters. U.S. law, under the Oil Pollution Act of 1990, states that "each responsible party for a vessel or a facility from which oil is discharged" is responsible for the cleanup costs and damages, including those costs incurred by state and federal agencies. (The responsible party is usually either the owner and/or operator of the leaking facility.) There are liability caps that vary depending on the type of facility; an offshore facility, not including the one U.S. deepwater port, is responsible for those costs plus $75 million in damages. These caps do not apply in cases where gross negligence, willful misconduct, or the violation of applicable federal regulations may be proved.

To simplify the process of filing claims, and to provide a ready source of funds for quick responses to spills, Congress created the Oil Spill Liability Trust Fund as part of the Oil Pollution Act of 1990. The maximum size of this fund is $2.7 billion, with revenues primarily generated by a per-barrel fee on oil produced in or imported to the United States. This tax, initially 5 cents per barrel, was in effect from 1992 to the end of 1994; it was then reauthorized by the Energy Policy Act of 2005, beginning again in April 2006. The Emergency Economic Stabilization Act of 2008 (Public Law 110-343) raised the rate to 8 cents per barrel up to 2016, and 9 cents in 2017; at the end of that year the tax will sunset. Fines and civil penalties against other polluters, plus recovered funds from previous spills, are also placed in the fund. Currently the fund has about $1.6 billion. For any one spill, however, only $1 billion may be used from the fund. In addition to paying claims, these funds may be used for research, as well as for federal agencies to have a ready source of cash to pay for immediate spill cleanup costs.

The structure of this fund and the legal framework created by the Oil Pollution Act spreads the financial cost of a large oil spill among the responsible party, the industry, consumers, and the government (Ramseur 2010). The current arrangement is already being criticized for providing too much protection to energy producers, violating the polluter pays principle. In the case of the recent Gulf spill, BP has stated that it will pay all damage claims, over and above the $75 million liability limit. But public outrage over BP's management of this well is likely to lead to congressional action to raise the liability limit for future spills, as well as to consider a raise in the per-barrel excise tax to support a larger trust fund. Although this will be an important step, a more strategic policy response would be measures to reduce our dependence on oil in the transportation sector. Let's consider how the federal gas tax could be part of that solution.

Federal Gasoline Taxes

The federal government first imposed a national gasoline tax as part of the Revenue Act of 1932 in response to the collapse in federal revenues resulting from the Great Depression (Jackson 2006). It increased very gradually from 1 cent per gallon to 1.5 and then 2 cents up to 1956. The Highway Revenue Act, a title of the melodiously named Federal Aid Highway Act of 1956, raised the rate to 3 cents per gallon to create a stronger funding stream to support an expansion of the federal highway system. Since then, the rate has increased but remains relatively low by international standards. As of 2010, the federal excise tax on gasoline is 18.4 cents per gallon, and 24.4 cents per gallon for diesel. This is about 6.5 percent of average U.S. gasoline prices for 2010; gasoline taxes from all levels total about 17 percent of the average gas price.

In addition, all fifty U.S. states levy some type of gasoline tax. As shown in Figure 8.1, as of April 2010 the combined amounts of local, state, and federal taxes vary from 26.4 cents per gallon in Alaska to 67 cents per gallon in California, with a national average of 47.7. Diesel is taxed similarly but at a higher rate, with the national average of the three combined taxes slightly higher at 52.2 cents per gallon.

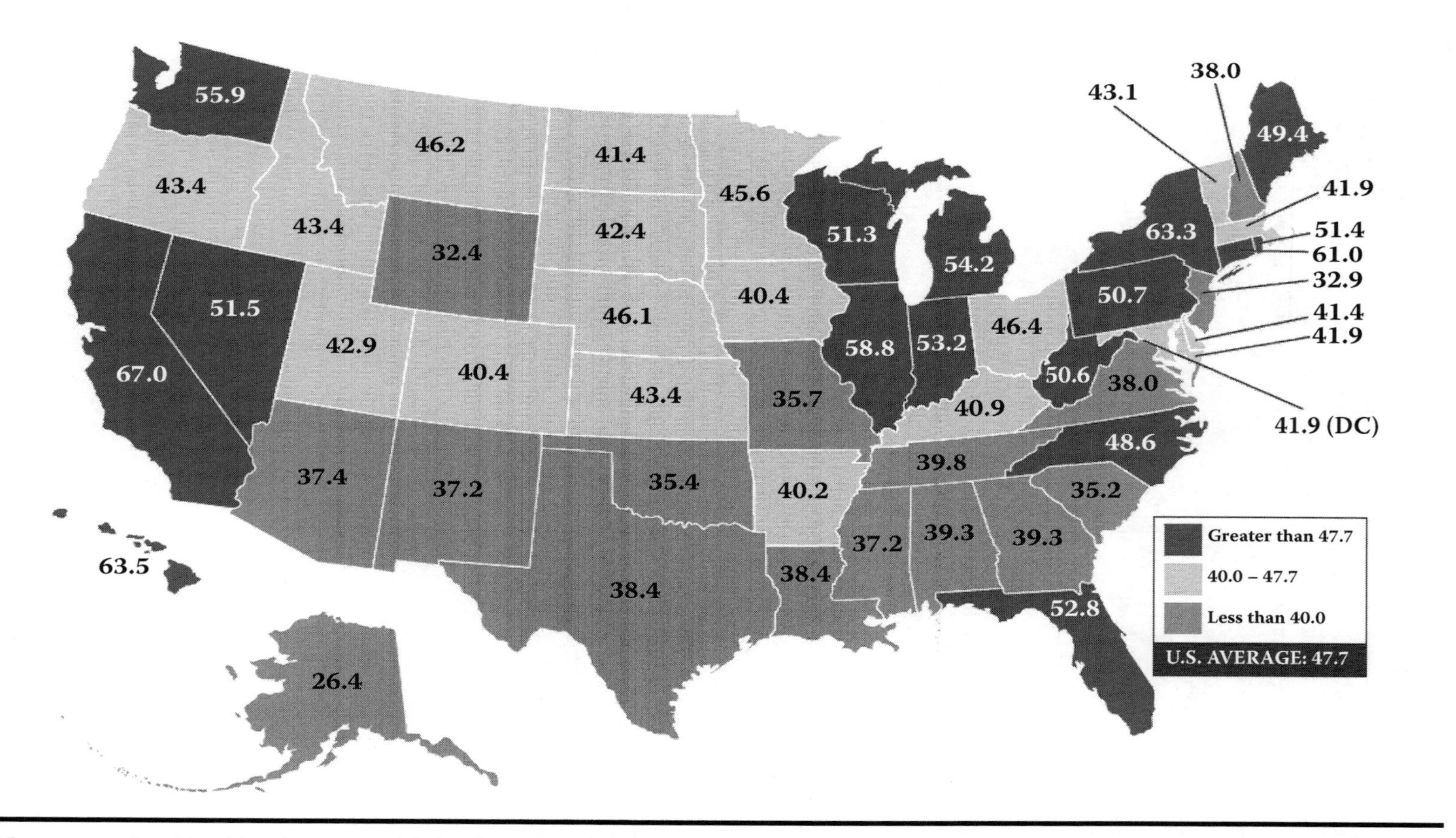

Figure 8.1 Combined local, state, and federal gasoline taxes. (From API [American Petroleum Institute]. 2010. Motor fuel taxes. American Petroleum Institute, http://www. api.org/statistics/fueltaxes/, accessed August 20, 2010.)

The federal gasoline tax is an excise tax. This is a tax applied to a specific good and is an indirect tax. The producer—in this case, the oil refiner—is expected to pass the value of the tax on to the purchaser or consumer, and then send the proceeds on to the government. Typically, energy excise taxes are collected per unit (e.g., per gallon), rather than ad valorem taxes, which are a percentage of the value of the item. This excise tax is considered a fee for users of the federal highway system, and most of the proceeds (15.44 cents/gallon) are placed in the federal Highway Trust Fund. This fund was created in 1956 under the Highway Revenue Act for spending on highway and bridge construction, maintenance, and other projects.

In addition, 0.1 cent per gallon is allocated to the Leaking Underground Storage Tank (LUST) Trust Fund. The final 2.86 cents per gallon is spent on mass transit through a special portion of the Highway Trust Fund. The "other projects" funded by this tax have been criticized for diverting funds away from highway and bridge maintenance, although most of these projects (particularly creating bike trails) do, in a broad sense, expand the capacity of the transportation system. Transportation museums and road-kill reduction programs are harder to justify (Coburn and McCain 2009).

Revenues from this tax have dropped each year since FY 2006, and totaled $30.4 billion in FY 2009. The long-term plateau and recent drop in U.S. vehicle miles traveled has decreased revenues. As a result, the Highway Trust Fund is structurally bankrupt, with tax income less than the legislatively mandated outgo to support road, bridge, and other projects throughout the country. Congress transferred $8 billion in general revenues to the fund in 2008, $7 billion in 2009, and $19.5 billion in 2010 through the Hiring Incentives to Restore Employment (HIRE) Act, HR 2847.

At present, basic principles of tax analysis suggest the federal gas tax is a significant policy failure (Pirog 2009). It does not generate sufficient revenues to pay for maintenance and investment in the federal highway system, so it fails as a user charge. This problem is going to become more acute as fuel efficiency increases thanks to stronger federal regulations, and electric vehicles gain market share. The tax has only a minor impact on gasoline prices and thus has little influence on consumption and sends, at best, a weak signal to motorists about the desirability of obtaining a more fuel-efficient vehicle. In contrast, gasoline taxes in Europe are several times higher than in the United States (Figure 8.2) and are designed to limit gasoline consumption and encourage consumption of diesel fuel instead. The higher taxes on diesel in the United States are intended to provide a differential "charge" to diesel trucks, which cause a disproportionate share of the wear and tear on U.S. highways.

If the gasoline tax is not working, what options are available for improving or replacing it? And what goals should be driving that policy choice? If the goals are to decrease fuel consumption, reduce traffic congestion, accidents and emissions, and encourage production and purchase of fuel-efficient cars, then a much higher tax intended to internalize the external or social costs of driving would be appropriate.

Estimates of the level of tax needed to properly reflect these social costs of driving vary widely; one recent estimate was $1.23 per gallon (Parry and Small 2009). Such a tax increase would decrease driving, reduce the costs of maintenance in the system, and generate significant funds for deficit reduction or income tax cuts and investments in alternative energy. From an economist's perspective, however, such a tax is not needed for conservation purposes. As oil becomes scarcer, the real or inflation-adjusted price of oil products will rise, consumption will decline, and production of more efficient options (e.g., high-mileage autos) will increase (Lazzari 2005).

But markets can move abruptly; sudden and prolonged oil shortages could be highly disruptive, and continuing reliance on imported oil is costly to the U.S. economy. One option floated by those concerned by the strategic impact of oil is a variable gasoline tax that would effectively

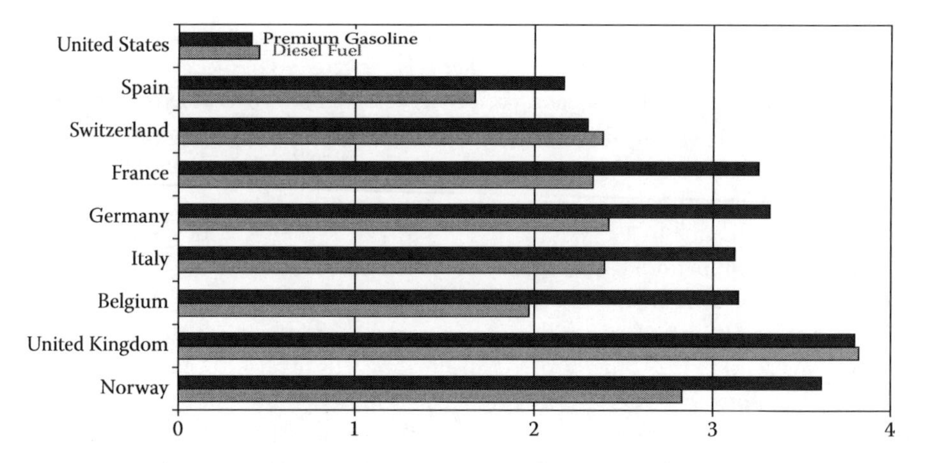

Figure 8.2 1999–2007 Average tax rates for premium gasoline and diesel fuel in Western European countries and the United States (nominal U.S. dollars per gallon). (From EIA (U.S. Department of Energy, Energy Information Administration). 2009c. *Annual energy outlook 2009 with projections to 2030: Electricity demand.* Global Oil Watch, http://www.globaloilwatch.com/reports/EIAEO09.pdf, accessed June 28, 2009).

set a price floor at a level that would significantly reduce driving, about $4 per gallon as of 2008 (Friedman 2008). If set high enough, either a gas tax or price floor would raise considerable revenues, roughly $100 billion for each $1 increase in the gas tax (Mankiw 2006). This is enough that the impacts of the tax on households and the overall economy would need to be carefully considered. This would be a highly regressive tax, so some form of rebates to low-income persons would be needed.

Finally, two other options are to introduce a tax on driving itself, with households paying a fee per mile, or introduce a much more extensive reliance on toll roads throughout the country. A driving tax would be a classic Pigouvian approach to internalizing the social costs of driving and could easily be implemented with current wireless technology, using smart cards, transponders, and potentially electronic odometers read by satellite. This would enable time of day or congestion pricing to enable higher charges to be levied for driving at rush hour. But this would be a highly intrusive policy. Data about each driver's daily activities would be generated, and establishing sufficient safeguards to prevent abuse would be essential.

Direct User Charges

Tolls, as a direct user charge, are effective at generating funds for ongoing maintenance and investment in long-run capacity. Their impact as an energy policy is nuanced—by easing congestion and aiding traffic flow they can actually *encourage* driving. Although tolling has long been part of the U.S. highway system, the budget pressures on state governments have led more and more states to privatize or lease important roadways.

California led the way with State Route 91 in Orange County, 10 miles of privately financed express lanes within a freeway. This was the world's first completely automated toll road. There are no toll booths; tolls are paid via electronic tolling using a required transponder on the car dashboard. Congestion pricing is used, so that tolls at peak rush hour reach $10, compared to the $1.20 toll off-peak. Initially built by a private consortium, and leased to them by the state, the road is now owned by the Orange County Transportation Authority. The most controversial element of the initial plan was the non-compete clause that prevented the state from providing upkeep to the other, non-tolled lanes of SR-91 for its entire 30 mile length; it also prohibited building mass transit near the road. Complaints over this arrangement were a factor in the road's sale to the Authority.

The best-known instance of highway leasing or concession was the state of Indiana's $3.8 billion, 2006 75-year lease of the Indiana Toll Road to a private investment company named Statewide Mobility Partners. This is a joint venture of two foreign firms, Cintra Concesiones de Infraestructuras de Transporte SA of Spain and Macquarie Infrastructure Group of Australia. This road is on a heavily used route between Chicago and the East Coast.

In the short run, it appears that the state was a big winner, since by at least one estimate the value of the road was closer to $1.5 billion (Bary 2009). The firm is on the hook for the cost of maintenance and improvements, and although tolls have increased for trucks and cash-paying vehicles, further increases are limited under the lease agreement. Meanwhile, the state is using the proceeds to fund a major highway improvement plan called Major Moves. With use and toll revenues down thanks to the recession, the road has been losing money, and its financing will need to be renewed in 2015. The recession-related credit crunch has limited the viability of similar projects. But 75 years is a long period. The road may yet prove to be a successful investment.

Whether it is good public policy is another question. An increased reliance on tolls is likely as states struggle to pay for highway maintenance—at a cost of up to $1 million per mile of freeway. But tolls have their own set of issues. Government entities managing toll facilities face conflicting pressures: political pressures not to raise the tolls, and countervailing pressure from the bond market, which wants to make sure that tolls are high enough to pay back borrowed funds. Compared to the gas tax, they are relatively costly to administer. Although transponder technology is very slick—tolls are instantly charged to a credit card—they also raise the violation rate, since cars without transponders can drive right through without stopping. Management of violations is costly and a political headache.

Tolls also create principal-agent problems. Structuring a contract for a 75- or 99-year lease, which effectively privatizes a road, is a daunting challenge. Without careful management of ownership and rights it is a constant temptation to neglect non-toll alternatives, to force drivers to use the for-pay alternative. And in poll after poll, respondents have been found to dislike tolls as much as they do taxes.

Of all of these options, tolling is the most likely to be widely implemented at the federal level. The G.W. Bush administration encouraged toll projects, and the return of a conservative administration to the White House could mean an increased number of tolls on interstate highways. One current constraint is that federal law allows states to implement tolls on portions of the federal interstate highway system, but only when the resulting revenues are to be used on that roadway.

Tolls are regressive, and while common in the eastern United States, remain rare west of the Mississippi. The likelihood of gas or driving taxes being implemented appears low. Privacy concerns will limit widespread use of driving taxes, while a large increase in the gasoline tax is unlikely during the country's likely slow and painful recovery from the Great Recession. However, it could form an element of a future deal combining tax increases and spending cuts to limit the growth in the national debt.

The Carbon Challenge

The most urgent challenge in the energy policy arena is how to craft a policy to reduce greenhouse gas emissions. In Chapter 2 we summarized the case for action to reduce emissions of carbon dioxide and other greenhouse gases (GHG). CO_2 concentrations in the atmosphere are projected to rise between 535 and 983 parts per million by 2100, with average global temperatures likely to rise between 1.8 and 4.0°C by the end of this century. In the United States, energy-related activities with fossil fuel combustion are responsible for about 94 percent of these emissions, making this environmental problem also an energy policy problem.

Technically, CO_2 is a *fund pollutant*; the environment has significant capacity to absorb all that we produce. However, it is generally described as a *stock pollutant*; it is accumulating in the atmosphere as a result of continuing emissions. As of 2010, there was an estimated 3,000 gigatons of CO_2 in the atmosphere—and that amount is rising faster than it can be absorbed or sequestered.

Worldwide, about 30 tons of carbon is added to the atmosphere each year. This amount includes about six tons from the United States. Emissions anywhere in the world add to this stock. So action to limit emissions in any one country, in any one year, will have minimal effects on the long-term stock. And there are significant unknowns about the environmental effects of increasing GHG concentrations, particularly the thresholds at which the environment (oceans, forests, and the atmosphere) can no longer absorb additional emissions and the levels at which severe environmental consequences could occur abruptly. We cannot be sure about the pace at which emissions need to be reduced in order to limit the risk of severe consequences.

The nature of this challenge limits the range of possible policy responses. To reduce additions to this stock, global cooperation is needed over an extended period of time, measured in decades. Action by the United States, still the world's second-largest emitter of carbon, is essential to create a policy space within which rapidly growing emitters such as the BRIC (Brazil, Russia, India, China) countries will be willing to agree to long-run limits on their own emissions.

The main options for reducing the country's carbon emissions are regulatory fiat, through command and control regulation, or raising the price of carbon, or a combination of both strategies. The two approaches for raising the price of carbon are a carbon tax, applied to energy sources at some point in the chain of production and consumption, or so-called cap and trade programs. These place an emissions cap on some portion of emission sources, and provide allowances to emitting firms to pollute up to that level. Firms that do not require all of their allowances may sell them to other, less efficient firms under certain conditions.

It is important to note that these policy options (carbon tax or cap and trade) are likely necessary but not sufficient to meet the country's energy policy goals—limiting future emissions at a reasonable cost while enabling economic growth and providing energy security and the continuing range of energy services Americans want, including household utilities and mobility. Most analysts agree that continuing investments in energy efficiency and innovation will be needed, along with supportive policies such as renewable portfolio standards that encourage the transition to renewable energy.

Carbon Tax vs. Cap and Trade

There is a huge literature on the relative merits of using a carbon tax versus a cap and trade design, to reduce carbon emissions (a small sample includes Helm 2005; Tietenberg and Lewis 2009; CBO 2008b; Freeman and Kolstad 2007; Ramseur and Parker 2009). We will summarize that literature and briefly discuss the particular approach included in the 2010 Kerry-Lieberman

American Power Act, beginning with the favored approach of most academics and policy wonks: the carbon tax.

A carbon tax is conceptually relatively simple. Policy makers determine the social cost of carbon, and apply a tax on either emissions or inputs, to force emitters to internalize those social costs. The tax would likely be expressed in dollars per million BTUs, which would reflect the relative carbon content of various fuels. This will increase the costs of fossil fuels and their products—gasoline, and coal- and natural gas–generated electricity, in particular—and encourage both production of nonfossil fuel substitutes and investments in efficiency. Fuels with relatively higher emissions per BTU will face a higher tax; this means that coal will be taxed more highly than petroleum or natural gas.

The tax gives cost certainty to producers and consumers. Its implementation, while complex, is manageable. It can be implemented most easily "upstream" at points of production, such as oil and natural gas wells and coal mines, although "midstream" sources, such as electric utilities and oil refiners, are also an option. There are about 1,374 coal mines, 641 coal-fired power plants, about 150 oil refineries and 530 natural gas processors, plus a couple of hundred oil and natural gas importers (Ramseur and Parker 2009). Uses of carbon-based fuels that don't lead to combustion would be exempt from the tax.

Of course, we don't know many of the parameters needed to manage such a tax. Recall from Chapter 7 that a pollution tax should be set at the point at which the marginal benefit of the tax is equal to the marginal cost. But we don't know about the shapes or slopes of these curves, there is wide disagreement about the social cost of carbon (Tol 2008a), and we don't know, exactly, how emissions will increase over time and what their effects will be. So it is difficult to estimate what tax rate would be required to prevent a given level of environmental damage.

What Is the Most Important Variable?

The most important variable influencing the rate of carbon tax recommended in various proposals is the discount rate applied. Higher discount rates applied to future streams of revenues and costs reflect reduced concern about the cost of today's emissions in the future. These will result in lower estimates of the needed tax. A lower discount rate that more closely equates emissions now with future damages will result in a higher estimate of the necessary tax rate.

An assumption made by analysts is the best decision rule in the face of uncertainty. Since our understanding of the science of climate change and the atmosphere is limited, though improving, some advocate a relatively low tax rate that increases over time—which is also consistent with a higher discount rate scenario. If we apply a precautionary principle approach in the face of such uncertainty, a higher tax would be needed to forestall the risk of abrupt and devastating change.

One option is to set an emissions target, and determine a tax rate and schedule required to meet it (Ramseur and Parker 2009). An advantage of the carbon tax is that it would allow relatively quick feedback loops, by monitoring emissions and adjusting the tax rate up or down to set the economy on a path toward the goal.

An advantage suggested by economists is that it is likely to be more economically efficient than a cap and trade system if, as is likely, the marginal benefits curve is relatively flat, and the marginal cost of abatement is relatively high. Since the effects of any one year's reduction in emissions makes relatively little impact on the overall stock, the marginal benefits of any one year's abatement is relatively low. The benefits occur over time (Pizer 1999). A consistent, known tax rate prevents the problem presented by the volatility of the price of allowances to pollute. Spikes in the price of allowances can force firms to purchase them when the price significantly exceeds the marginal benefit of the extra unit of mitigation, which imposes a significant burden on the economy. Finally, revenues from carbon taxes reduce environmental externalities, but can also be substituted for more economically distorting taxes such as income or capital gains taxes, thus creating a "double dividend" (Green, Hayward, and Hassett 2007).

The Primary Disadvantages of a Carbon Tax

First, it is a tax, and therefore in the United States a toxic form of public policy. This makes it politically hazardous to any legislators that support it. Even under the best of conditions, it could be difficult to set the tax at a rate optimal from an environmental perspective, due to concerns about its economic and social impacts. It is also a more visible form of intervention than a cap and trade system (Stavins 2003). Trying to call it a fee or a pollution charge won't disguise the reality that it's a tax. But a populace shocked by the devastation from the Challenger Horizon oil spill may be ready for such a tax.

Many countries and political jurisdictions in the United States and elsewhere have introduced carbon taxes, including the Canadian provinces of Quebec and British Columbia, the city of Boulder, Colorado, and several European countries. Voters in Boulder approved a Climate Action Plan tax in 2006 on residential, commercial, and industrial electricity users, with the proceeds to fund energy efficiency programs. The Boulder City Council has the authority to set rates within an approved range. The tax is expected to generate about $1.6 million in 2010. Despite the tax and substantial efforts to improve energy efficiency, the city's carbon emissions declined only 1 percent between 2006 and 2008 (Simon 2010).

Sweden is one of the few countries that have a comprehensive set of environmental taxes (Speck 2008). The country levies both an energy tax and a carbon tax, with the latter focused on taxing energy consumption. The national carbon tax is now 1.01 Swedish kroner per kg of CO_2. No tax is levied on electricity generators that participate in government-approved energy efficiency programs. Homeowners pay the full tax, but industrial users pay only a 50 percent rate. Renewable fuels, including biomass, are exempt. Much of Sweden's energy use is for winter heat, and the tax has encouraged use of biomass for combined heat and power projects.

In each of these instances, the jurisdiction's underlying political culture is much more accommodating to the use of tax policy to change behavior than is the case in the United States. Even in Sweden, CO_2 taxes were implemented as part of a grand deal to lower income taxes.

The Cap and Trade Option

Carbon tax options have another major weakness: although they set a price on carbon, they do not set a limit on emissions. Cap and trade systems first establish a yearly cap, or limit, on the emissions from included sources, then enable the owners of those sources to obtain allowances or permits that give them a limited right to emit those pollutants up to a certain level, typically on a per-ton basis per year. This requires an investment in data gathering by the government so that emissions levels from these sources are known.

The level of the cap is set so that it declines over time, along with the number of available permits. Firms find the easiest and cheapest initial means to lower their emissions, and regularly compare the cost of investing in emissions reductions to the costs of purchasing permits. Over time, many firms will invest in R&D and technology that will enable them to continue to meet lower emissions limits. In years when spurts in production or other factors are likely to push them over the cap, they purchase allowances to cover the difference. This allocates permits to the producers that value them most highly. Producers relinquish their allowances to the government at the end of each period. Companies with emissions in excess of their available allowances are fined.

This approach has proved remarkably effective in several cases, with the best known the U.S. acid rain program that created a system for trading permits for sulfur dioxide. Most assessments of the program believe that it enabled emitters (power plants) to meet the program's emissions caps much more efficiently than would have been the case under regulation (Stavins 2007).

A political advantage of cap and trade is that to voters and consumers, it is a relatively opaque process. Even though the end result is an increase in the price of various energy sources (particularly electricity, natural gas, and gasoline), there is no specific "tax" at which to point responsibility. But that opacity comes at the cost of complexity. As with a tax, policy makers would need to carefully consider which sources would be covered by the cap, as well as how to establish the number of initial permits or allowances, and allocate them.

In the economist's ideal world, all permits would be auctioned by the government. This would enable a market price for the allowances to quickly be set, and generate significant revenues. But other options have been used, including lotteries and administrative or political allocation based on prior levels of production and emissions. In the real world of politics, too many permits have often been allocated, and too many of these allowances have been provided for free—often on the basis of current power output—to lower the cost of the program to current emitters.

This is a windfall benefit for firms receiving these allowances, and it acts as a disincentive to new firms that could enter the market with new, cleaner technology. It complicates the growth of a market for trading allowances and thus neuters the "trade" side of cap and trade. This is a common criticism of both the initial phase of the European Union's Emissions Trading System, or ETS (which covers large industrial and energy sector emitters) and of the American Clean Energy and Securtiy (ACES) Act passed by the House in 2009. Implementation of the ETS was particularly difficult due to overallocation of free permits to industry, underallocation to the power sector, poor initial data on emissions, and caps that differed by country (GAO 2008a).

A Market-Based Approach

Cap and trade systems are a quintessential market-based approach, since they literally create a market for permits. But the recent performance of financial markets and those who participate in them has left voters and policy makers very skeptical, even cynical, about the vulnerability of such markets to manipulation. This has made the issue of how to structure a carbon market more complex.

Prices for emissions permits in various programs (including both the acid rain program and the EU's Emissions Trading System, or ETS) have been volatile, subject to influence from both the number of available permits and other factors (CBO 2008b). Since spikes in the cost of emissions allowances would impact energy prices, markets would need to be carefully designed to limit manipulation. Some type of ceiling would be required, perhaps by requiring the government to release more permits when it is reached. This, of course, would break the cap by allowing higher emissions.

The potential global market for carbon has been estimated at $2 trillion (Chestney 2010). Such a market would be critically important to its participants. Regulations will be needed to provide a framework for both the primary market for exchange of permits, and a secondary market. Market participants would wish to have market smoothing instruments at their disposal, particularly derivatives such as futures, options contracts, swaps, and forward sales (Pew 2010).

Compared to more esoteric derivatives, these "plain vanilla" instruments are relatively safe, and major exchanges in New York, Chicago, London, and other cities have had extensive experience with them. Still, creating the framework for these markets will be a difficult political task. Policy makers will need to decide what body would regulate them: the Securities and Exchange Commission, the Commodity Futures Trading Commission, or the Federal Energy Regulatory Commission (FERC). The Waxman-Markey bill passed by the House in 2009 gave the FERC such regulatory authority. Also, who would be permitted to trade allowances? Would unregulated over-the-counter exchanges be allowed, or would on-the-record trading with a counterparty be required? The latter would enable regulators to gauge the value of outstanding trades and overall level of risk in the market. What controls would be placed on derivatives? To what extent would either market be open to global trading?

Advantages and Disadvantages

A further complication is that the allowances traded in cap and trade programs typically have a shelf life. They can be designed to expire or be submitted to the regulating body at the end of each compliance period. But the programs are more economically efficient when the permits may be "banked" or retained for use in a later time period, or "borrowed" and used earlier than intended (Tietenberg 2006). One reason is that the allowances may be used as a bridge prior to bringing online major investments that will increase efficiency and lower emissions.

One advantage of a cap and trade policy would be the system's ability to build on the ongoing emissions markets around the world, including the EU's Emissions Trading Scheme, and the Regional Greenhouse Gas Initiative and Western Climate Initiative in the United States. International agreements may most easily be organized around quantities rather than prices and tax rates (Helm 2005).The growing market for offsets is also a factor. Offsets are emissions reductions produced by specific projects, such as investments in renewable energy.

Under a cap and trade policy, a regulated firm could be allowed to purchase offsets domestically or globally, rather than purchase an allowance on the open market. As we will discuss in Chapter 9, regulating this offset market will be a serious challenge. A cap and trade policy would likely be easier to harmonize globally than a carbon tax, although it is potentially feasible for a global deal to be reached that would include a roughly similar tax by all participants. But ensuring that each country monitored and enforced the tax similarly and consistently would be a heroic proposition (CBO 2008b).

Choosing the Best Option

Is there a clearly superior option? The short answer is no. Both a carbon tax and cap and trade share several challenges. To be effective, both will need to be broadly applied to as many sources of emissions as possible, including the agriculture, commercial, and residential sectors, will need to cover the most important GHGs (methane, nitrous oxide, and several other gases), and will need to be carefully designed to sensibly allocate the revenues generated (Cleetus 2009). Revenues could be rebated to U.S. households and businesses and used to support energy R&D and investments in energy efficiency. Both will impact the economy and have distributional impacts on U.S. society, although as discussed earlier, the size of such impacts has generally been estimated to be low, at or less than 1 percent of gross domestic product (GDP). But the global financial meltdown has made a cap and trade system, once a clear favorite, a more difficult alternative to sell to policy makers.

One result is that the Kerry-Lieberman American Power Act under consideration by the Senate in June 2010 did not attempt to create a unified cap and trade program. Instead, refiners of transport fuels and power generators would have separate allowance programs. Chapter 11 will include a more detailed analysis of this complex legislation.

Renewable Portfolio Standards

Chapters 5 and 7 described many of the policy instruments designed to directly or indirectly influence the pace at which the United States expands its renewable energy capacity. At the federal level these have primarily been tax incentives. Probably the most important nonfederal policy response since the late 1990s has been the steady introduction of renewable portfolio standards (RPS). As we briefly discussed in Chapter 5, an RPS is a requirement for electric utilities to obtain a specified proportion or amount of their production from renewable sources by a given date. Usually that percentage is based on retail sales of electricity; the goals typically start at a low level and increase over time. Utilities must document their compliance with the standard on an annual basis. Not all approaches have equal benefits; Table 8.1 points out the wide differences in the advantages and disadvantages of renewables.

Most state plans also allow suppliers to meet the requirement through purchase of renewable energy certificates (RECs). A REC is a financial instrument that represents one megawatt-hour

Table 8.1 Summary of Advantages and Disadvantages for Renewable Energy

Technology	*Typical Levelized Cost (Cents per kWh)*	*Advantages*	*Disadvantages*
Wind	4–5	Wide availability, scalable up or down, proven technology, global acceptance	Difficult to site; intermittent, efficient storage as yet not available; typically 20–40% efficiency
Photovoltaics	20–40	Ubiquitous resource, silent, long lifetimes, easy and quick to install	Very expensive, intermittent, some site limitations
Biomass	4–9	Relatively easy to locate, large availability of resource	Greenhouse gas and particulate emissions, site limitations
Hydropower	4	Additional benefits of flood control, agriculture and recreation; can be inexpensive	Has large land, water, and ecological impacts; limited availability of new locations
Geothermal	5–6	Can be inexpensive	Limited availability, depletable, little known environmental impact

Source: Komor, P. 2004. *Renewable energy policy*. New York: iUniverse.

Table 8.2 Renewable Portfolio Standards, by State and Goal Year[a]

10 to 14%	*15 to 19%*	*20 to 24%*	*25% and Over*
Maine: 10%, 2017	Arizona: 15%, 2025	Washington, DC: 20%, 2020	Illinois: 25%, 2025
Wisconsin: 10%, 2015	Missouri: 15%, 2021	Delaware: 20%, 2019	Minnesota: 25%, 2025
North Carolina: 12.5%, 2021	Montana: 15%, 2015	Kansas: 20%, 2020	Nevada: 25%, 2025
	Washington: 15%, 2020	Maryland: 20%, 2022	Ohio: 25%, 2025
	Rhode Island:16%, 2019	New Mexico: 20%, 2020	Oregon: 25%, 2025
	Pennsylvania: 18%, 2020	Vermont: 20%, 2017	West Virginia: 25%, 2025
		Massachusetts: 22%, 2020	Connecticut: 27%, 2020
		New Jersey: 22.5% , 2021	Colorado: 30%, 2020
		New Hampshire: 23.8%, 2025	California: 33%, 2010
		New York: 24%, 2013	Hawaii: 40%, 2030

[a] Three states have nonpercentage goals: Iowa, 105 MW; Michigan: 10% + 1,100 MW by 2015; Texas: 5880 MW by 2015.
Source: DOE, 2009.

of electricity created from renewable energy. As of May 2010, the EU, several countries including Sweden, Japan, Australia, and others, and 29 states plus the District of Columbia have renewable portfolio standards, as shown in Table 8.2. The states of North Dakota, South Dakota, Utah, West Virginia, Virginia, and Vermont have renewable portfolio goals.

The goals of most RPSs are to support investments in renewable energy and increase the proportion of renewables in the jurisdiction's electricity supply. They are a demand-oriented policy designed to influence technology choices. In the long-term they also aim to reduce greenhouse gas emissions by encouraging the transformation of the U.S. energy grid from its reliance on production of electricity from polluting fossil fuels, particularly coal. RPSs benefit politically from being perceived as an economic development strategy supporting investment and high-wage jobs (Rabe 2006).

An RPS is both a regulatory and a market intervention. Most include regulatory "sticks," or penalties that may be applied in the event that suppliers, including investor-owned utilities (IOUs) and other electricity distributors, do not meet a yearly standard. In practice, however, most of these sticks don't pack much sting, since most RPSs also have a variety of escape clauses, such as cost caps, that let these utilities off the hook. These utilities, however, have incentives to purchase the lowest-cost renewable power available, either directly from renewable suppliers or through RECs. One intention of the gradual ramp-up of the goals is to allow time for supporting systems to develop and competition to occur between renewables suppliers.

Meeting Requirements with Wind Energy

In most cases, RPS requirements in the United States have been met with wind energy. Along with the federal incentives mentioned in previous chapters, RPSs appear to have had a significant impact on the expansion of U.S. wind energy capacity.

Wiser et al. (2007) concluded that over half of wind capacity additions between 2001 and 2006 were influenced by state RPS requirements. This supports the findings of Menz and Vachon (2006) who found that RPSs were effective at encouraging development of wind energy. About 10,000 megawatts of wind capacity was installed in 2009, and total U.S. wind capacity is now about 35,600 megawatts (AWEA [American Wind Energy Association] 2010).

The Energy Information Administration (EIA) projected 2009 wind additions to be 35 percent of the increase in all U.S. generating capacity, second only to natural gas (42 percent) (EIA 2010). But additions in subsequent years are projected to drop drastically, due to the uncertainty over the federal incentive framework and the state of the financial markets.

Wind energy presents some challenges, however. It tends to peak at night (when loads are low) and the best sites are often located far from population centers, requiring extensive investment in transmission. It is more intermittent than most other power sources and thus is somewhat more complex to manage. Utilities selling wind energy may invest in additional back-up production (often *spinning* reserves, turbines literally spinning at less than full output that may be ramped up if needed) or purchase *shaping services* from utilities or entities with base-load power reserves, to accommodate the variable output from wind. But wind has a significant benefit—very low operating costs because no fuel is required.

In the United States, RPSs are a state-level intervention; therefore, it is no surprise that they vary widely. As Table 8.1 suggests, the most obvious differences are in terms of the percentage goals of the policy and the year by which that goal is to be attained.

Other fundamental differences include (1) the types of renewable energy accepted under the standard, (2) acceptable geographic boundaries for those sources, (3) what is considered *new* generation, whether different sources (usually solar, due to its higher cost) face different tier requirements, (4) the types of utilities to which the requirements apply (some states give smaller utilities lower goals), and (5) differing exemptions, duration, enforcement mechanisms, and regulatory bodies, among others (Wiser et al. 2007). Most states accept biofuels, biomass, geothermal, landfill gas, hydro, solar, solar thermal, tidal, wave, and wind power sources. But a sizable number accept municipal waste, a few include power from fuel cells, energy efficiency, and even waste tires.

Impacts of These Programs

What have been the impacts of these programs? This is a difficult question to answer. Some states have met or made considerable progress toward their standards. Texas met its initial goal of 2,000 MW by 2006, and by early 2010 accomplished its 2025 goal of 10,000 MW. California's initial goal is 20 percent by 2010, and 33 percent by 2020 by an Executive Order of Governor Schwarzenegger. But in 2009, its three major investor-owned utilities overall met only 15 percent of their residential power demand with renewables; one utility, San Diego Gas & Electric, is only at 10.5 percent. Although the state has had a flurry of wind investments, it is unlikely to meet the 2010 goal.

One consequence was for the state's Public Utilities Commission in December 2009 to authorize purchase of RECs to meet the requirement—the most likely source of which will be wind farms in Oregon and Washington states. Although RPSs have existed in the United States since

Iowa's early effort in 1983, most have been implemented since 1997. In the energy arena, with the long lead times required to obtain capital, project approval, and bring a new generation source online, it is fair to ask if these programs as a class have existed long enough to fairly evaluate their impact and success. Moreover, the distinctions between the programs present considerable methodological challenges.

In terms of their impacts on electricity prices and the environment, and overall effectiveness, the literature on RPSs is divided. Michaels (2008) panned the programs, concluding, "State RPS programs are largely in disarray, and even the apparently successful ones have had little impact." Wiser and Barbose (2008) were more positive, concluding that compliance with RPS goals set by early adopters of the policy has been significant. They found that RPSs overall had minimal impacts on electricity rates, however, and as of 2006, 9 of 14 states for which data were available reached 95 percent of their compliance target.

Measuring Performance

The easiest metric to apply to the programs is a simple one: has the United States begun to rely more on renewable energy? Department of Energy (DOE) data show that nonhydro renewables (primarily wind power) increased at a compounded annual average of 12 percent per year from 2000 to 2008. But even with that high growth rate, nonhydro renewables as of the end of 2008 still amounted to 3.8 percent of installed electricity capacity and 3.1 percent of generation in the United States (DOE 2009g). This appears to be due to lower natural gas prices, continuing increases in natural gas generation capacity, the overall growth in electricity demand, and possibly weaknesses in the policies themselves, including compliance waivers and cost caps (Cory and Swezey 2007).

Despite these problems, the RPS is one of the few levers available to policy makers, particularly at the state level, to encourage the transformation to renewable energy sources. They have the advantage of not requiring direct appropriations, and their impacts on employment also continue to resonate in state capitals desperately looking to boost job growth. Since state-level funded programs for renewable energy are likely to be on the budget-cutting block for several years to come, the RPSs are likely to remain a key element of U.S. energy policy.

Prospects for National Application

What are the prospects for a national-level RPS? Legislation for such a national standard has been introduced in Congress on several occasions and a combined efficiency/renewable energy standard of 20 percent by 2020 was included in the House energy bill (H.R. 2454, the American Clean Energy and Security Act) passed in 2009.

Crafting such a policy will be a challenge. It would have to clarify which renewable resources would be acceptable, which utilities would be covered, penalties for noncompliance, details on the national market for RECs to be created, what federal agency would administer the program, and how the program would mesh with existing state RPSs. Because southeastern states are less endowed with wind and solar resources than those in the central and western United States, policy makers from the region have often opposed a national RPS, arguing it would force their utilities to purchase RECs from Western suppliers. The alternative is increased investment in biomass facilities in those states. A national RPS would also require complimentary policies guaranteeing continuing investments in transmission capacity. Extending transmission lines will be necessary in order to bring online sources that are at an increasing distance from population centers.

The development of renewable energy sources faces a huge number of obstacles, over and above their significant capital and installation costs (Sovacool 2008b). For small projects, such as a rooftop solar installation, a key is to simply be allowed to connect the new system to the electricity grid. Most state laws now mandate such access, although the terms vary.

A second threshold is net metering, which allows electricity to flow back and forth from the grid. When generated power is running from the household onto the grid, the power meter will run backward, generating a credit. Again, laws in 43 states plus Washington, DC, provide for net metering, which was also mandated nationally by the 2005 Energy Policy Act, but only for publicly owned utilities. The laws contain many limitations. For example, Washington and many other states limit both the amount of power utilities may be required to buy, and the price paid for the power—usually a rate below that of the market. There are rational reasons for this—utilities need to be able to sell enough power to cover the costs in their rate base, for capital, and so on. But a rate higher than the market would provide a significant incentive to invest, both for distributed generation systems and utility-sized installations.

Feed-in Tariffs

Electricity rates designed to encourage the development of renewable energy sources are known as feed-in tariffs (FIT). The FIT is probably the most successful policy worldwide at encouraging investment in renewable energy sources. FIT policies typically include mandatory grid access to qualifying projects, plus a guaranteed rate per kWh for an extended, fixed period, often 20 years. As of early 2009, 61 countries and jurisdictions, including California, Vermont, and Washington, have implemented FIT policies (REN21 2009).

The FIT Experience in Germany

The most well-known FIT was created in Germany in 1991 and revised in 2000 and again in 2004. The 2004 law, the German Renewable Energy Sources Act (Erneuerbare-Energien-Gesetz, or EEG), provides a high guaranteed payment that differs by energy source, with PV solar provided a significantly higher tariff than wind or other options (Munoz et al. 2007). Germany, an often cloudy country slightly smaller than Montana, installed 3,800 MW of PV in 2009, about half of total world installations, compared to about 500 MW in the entire United States (Gipe 2010).

Each year the tariff declines by a set amount, known as the *degression rate*. This declining rate is intended to encourage rapid implementation and innovation. Overall rates paid to renewable producers are 6 to 8 times the market price of power; solar in 2004, for example, was .54 euros per kWh. As a result of the EEG, Germany dramatically increased its renewable electricity generation from 3.5 percent of total generation in 1990 to 15 percent in 2010 (IEA [International Energy Agency] 2008—dep ren). The rapid decline in the cost of PV and continuing high subsidies led the German Parliament to again cut tariffs and increase the degression rate in 2009; additional increases in the degression rate were likely in 2010.

The law has been broadly supported by German society; its advocates point to the beneficial effects on employment through support for domestic wind turbine firms, plus installation and maintenance of the new systems, as well as benefits to the environment. The face of Germany is being changed as solar panels pop up on urban buildings and larger installations dot the countryside; the new German Reichstag or Parliament building in Berlin is a shrine to renewable energy thanks to its solar panels, geothermal power, and sleek design. But critics have attacked the subsidies built into the program, which are paid by German households through slightly higher electricity bills, as well as the fact that solar PV, in particular, is a high-cost option for abating greenhouse gases (Frondel, Ritter, and Schmidt 2008).

Despite these concerns, the passage of a U.S. FIT is a longtime goal of renewable energy advocates. One reason is that a national FIT would mesh smoothly with a U.S. renewable portfolio standard. In tandem, this policy duo would greatly support the growth of renewables. The financing and planning of such projects would be eased since they would have a guaranteed revenue stream (Cory and Couture 2009; Sovacool 2008b).

Conclusion: Role of Interventions in Energy Policy

Market-oriented environmental policies are not a panacea for delivering the United States to either energy independence or a cost-free road to a lower-carbon future. But some form of carbon pricing will be needed to lessen the U.S. reliance on fossil fuels, reduce our carbon emissions, and give the United States the legitimacy to be able to negotiate a global climate treaty. The United States must think and act ever more globally about its energy policies and their ramifications; that will be the focus of our next chapter.

Chapter 9

International Cooperation on Energy Policy

> A tremendous challenge for human society is to move beyond the tangible pressures experienced today and to manage global resources with future generations in mind. ... Continued dialogue and consensus building within the international community, between the public and private sectors, and within society at large is needed to advance sustainable energy policies at the national, regional, and global levels.
>
> **—Goldemberg, Johansson, Reddy, and Williams (2001)**

International economic and geopolitical problems, policies, and a variety of international "actors" have a significant influence on the crafting of U.S. energy policy. Among the international problems that American policy makers have had to deal with recently are high energy prices, the failure to arrive at global agreement on limiting greenhouse gas emissions, the rise of a newly assertive China, and armed hostilities between nations over energy transportation and distribution. These problems are at the complex nexus between the energy arena and U.S. foreign policy. The goal of policy makers is often more to manage than to expect to solve such problems. Managing them requires careful and painstaking diplomacy with all of these actors—diplomats from nation-states, representatives of international and supranational organizations, multinational companies, and nonstate actors or nongovernmental organizations (NGOs). Many of the decisions that influence U.S. energy policy directly or indirectly are made through cooperative arrangements with these countries and organizations.

In this chapter we will consider the nature of the international arena, the major global energy issues, and the cooperative arrangements in the energy and environmental sectors among the United States, other countries, and regional and international organizations and forums. Crafting a global deal over climate change is one of this generation's preeminent challenges. The near-collapse of the Copenhagen summit was a potent reminder of the differences in philosophy, governance, and economic status among countries, all of which are barriers to reaching such an agreement.

As we will discuss below, nation-states tend to support cooperation and collaboration when it suits them. Fortunately, there are several successful examples of agreement on important global environmental issues that illustrate how countries may craft suitable policies and institutions when it is in their—our—collective interest. But climate change is a daunting problem, due in part to the magnitude and complexity of the issue, the weak nature of global governance, and the unwillingness of many important countries to engage in bargaining. To better understand why several of the key players in climate change negotiations appear so intransigent, it is essential to start with a brief overview of world energy and its linkages to the global political economy.

Global Energy: Sources, Consumption, Inequities

The basics of the global energy challenge may be briefly summarized. First, the most important fossil fuel sources are scattered unevenly across the globe. A small number of countries control a high proportion of the global reserves and production of oil, natural gas, and coal, as shown in Table 9.1. The prices for these energy sources are set in opaque ways that mix market forces

Table 9.1 Top 5 Global Producers and Consumers of Oil, Natural Gas, and Coal

	Reserves		*Production*		*Consumption*	
Petroleum						
Reserves and production: million barrels	Saudi Arabia	262.3	Saudi Arabia	10,782	United States	9,498
	Canada	179.2	Russia	9,790	China	7,831
Consumption: MBD (all 2008)	Iran	136.3	United States	8,514	Japan	4,785
	Iraq	115.0	Iran	4,174	India	2,962
	Kuwait	101.5	China	3,973	Russia	2,916
Natural Gas						
Reserves: TCF, 2009	Russia	1,680	Russia	23,386	United States	23,195
	Iran	992	United States	20,377	Russia	16,799
Production and Cons: MCF, 2008	Qatar	892	Canada	6,037	Iran	4,201
	Saudi Arabia	258	Iran	4,107	Japan	3,572
	United States	238	Norway	3,503	U.K.	3,388
Coal						
Reserves and Production: Million tons	United States	238,308	China	2,761	China	1,406
	Russia	157,010	United States	1,007	United States	565
Consumption: MTOE (all 2008)	China	11,4500	India	490	India	231
	Australia	76,200	Australia	325	Japan	129
	India	58,600	Russia	247	South Africa	103

Sources: EIA, IEA, BP.

and taxation, and (particularly for oil) often create sizable economic rents for those producers; that is, in the short run they can enforce prices much greater than the price needed just to bring them to market. They have a powerful incentive to keep the world reliant on these sources, even though dependency on oil revenues, in particular, is often toxic to a country's political and social development (Karl 1999). And from the perspective of consuming countries, overreliance on foreign oil and gas is often perceived as a serious issue of national security as well as a drain on national finances.

Second, consumption of these energy sources is highly unequal, but this is changing fast. The developed countries consume about 55 percent of all petroleum, and half of their natural gas, but are only 20 percent of world population. This is reflective of long-standing patterns of unequal social and economic development. About 1.5 billion people on the planet still lack access to electricity.

This pattern of consumption is changing as the result of globalization, increasing urbanization—half the world population now lives in cities—and the rapid economic growth of Brazil, China, India, and other countries. Most scenarios project slow growth in energy consumption in the north, and continued strong growth in the energy needs of these countries. They believe they will need increasing amounts of energy to literally fuel that growth, and will need to import much of it. China and India are each expected to meet over two-thirds of their expected oil demand in 2030 through imports.

Finally, there is recognition by most nation-states of the problem of climate change, the role of fossil fuels in contributing to the problem, and the need for collective global action. Greenhouse gas emissions don't recognize borders; most countries (other than the United States and China) are relatively small contributors to the problem (giving them little incentive to act on their own), and uncoordinated national climate change programs are unlikely to slow the growth in global emissions (Barrett 2009; Roeglj et al. 2009).

The still-developing countries view the problem as having been created by, and primarily the responsibility of, the developed world, represented by the Organisation for Economic Co-operation and Development (OECD) countries. They loathe sacrificing their own development, which they see as essential to raising the standards of living of their citizens and their stature in the global community. This fundamental difference in interests is a serious challenge to collective action on climate change.

States, Realists, and Idealists

Effective action against the causes of climate change will require the creation of global institutions adequate to the task of implementing and sustaining a comprehensive framework of policies. This is likely to include global market frameworks for offsets, monitoring emissions, and enforcing agreements to act. One of the main challenges to creating such institutions is the weak nature of global governance.

To understand the causes of that weakness, it's necessary to start with some of the basic concepts of international relations used to describe the international system. Nation-states are sovereign entities—that is, within a fixed geographic territory, they are the supreme authority to set and enforce laws. But sovereignty has eroded over the past few decades, in both the economic and political arenas. Most states can no longer tax, spend, promote protectionist policies, or abuse their citizens with impunity. Their wealthier citizens and business interests will move away, global credit markets will raise the states' interest rates, the World Trade Organization will penalize them, or the strengthened International Criminal Court will take action. The struggle over sovereignty is certain to impact the creation of a climate change regime that could create vast flows of revenue and require consistent and difficult enforcement of limits on the use of fossil fuels. States will be reluctant to hand over sovereignty to a global regime that could attempt to constrain policies viewed as vital to national development.

As sovereignty is changing, so is the very nature of the global system. Throughout most of the period since the creation of the nation-state system in the mid-1600s, states and their leaders have viewed the global system as anarchic. Since the planet lacks a world government with power to enforce mutual security, states sought to survive and to further their national interests through the use of diplomacy, backed by military and economic force, and often engaged in balance-of-power tactics to keep the peace. This realist framework is based on deep suspicion of the motives of other countries. It is still useful as a means of comprehending the motives of state behavior and the causes of interstate conflict.

Important international developments over the past sixty years are poorly explained by this concept for several reasons. First, states are no longer the sole actors important to global affairs. At one end of the continuum, the creation of international and supranational organizations such as the United Nations (UN) represents a significant break with the state system. These organizations have a legal status and decision-making authority that transcends control by national governments. At the other end, nonstate actors such as NGOs, multinational corporations, and terrorist organizations now have a powerful global influence and often create the agendas to which nations and global institutions must respond.

Second, the institutions of collective security and economic and social development created after World War II, including the United Nations and World Bank, reflected a more *idealist* notion emphasizing collaboration and common interests, not competition. The United Nations Charter is a powerful statement of our need to work together as a global community on matters of collective security and development. Although the UN system has its flaws, it remains a vital set of institutions for identifying issues of global concern and organizing cooperative action.

The struggle between the realist and idealist forces in world politics influences the nature of organizations with a regional or global reach. Most countries believe they need to protect their national interests, yet in some circumstances they will pursue cooperation. So although a supranational organization such as the UN or European Union may be endowed with a legal standing separate from their members, financing and decision-making rules often favor stronger member countries. The UN, for example, has a one-country, one-vote General Assembly with little power except to approve the organization's budget. The real power is held in the Security Council, which is dominated by its permanent members: the United States, China, France, Great Britain, and Russia. And national sovereignty is jealously guarded, so the power of most supranational organizations is carefully limited.

Regimes and Global Environmental and Energy Governance

When a public policy problem crosses state borders, international action is required to manage it. Such problems may be local, regional, or global. Often such problems involve three or more parties and are thus *multilateral*. Several of the globe's most intractable issues, from peace and security, to economic and social development and environmental protection, require worldwide multilateral action.

Scholars of international affairs use the concept of regimes to define the frameworks within which such action occurs. Typically, a regime begins when an international meeting focuses attention on a problem; over time countries negotiate a treaty or agreement defining the problem and needed action, and create an institution for carrying it out. The regime includes the international law, principles, norms, rules, and decision-making procedures that apply to the endeavor (Krasner 1982). The regime is thus a framework for cooperation and decision making. By setting principles and rules, the regime helps to limit self-interested behavior on the part of participating countries (Karns and Mingst 2004).

There are hundreds of such regimes worldwide, including at least 150 significant multilateral regimes with an environmental focus (Young 2010; Chasek, Downie, and Brown 2010). Additional examples include the Convention for International Trade in Endangered Species (CITES); the Montreal Protocol and subsequent actions to reduce the release of CFCs (chlolofluorocarbons) and other ozone-depleting substances into the atmosphere; and regimes that govern the oceans, Antarctica, the atmosphere, and space (Joyner 2005).

The concept of the international environmental regime is important since coordinated global action on climate change is likely to take place through such a regime. The initial attempt to do so through the UN Framework Convention on Climate Change (FCCC) and the Kyoto Protocol, has not been successful, as we will discuss ahead. What attributes are associated with successful regimes that may inform the design of a successor to Kyoto? We will return to that question at the end of the chapter.

Global Energy Regimes and Regional Energy Institutions

There are, however, only a small number of international regimes in the energy arena. The two most important are the atomic energy regime, with primary oversight provided by International Atomic Energy Agency (IAEA), and the International Energy Agency's (IEA) emergency response regime, which focuses on oil.

The IAEA works to both encourage the safe global development of nuclear energy, and to ensure that nuclear materials intended for energy use are not diverted to military purposes. It is responsible, under the Treaty on the Non-Proliferation of Nuclear Weapons (NPT), for administering international safeguards and providing verification. The gradually increasing number of nuclear states, and concerns about weapons development by North Korea and Iran, continue to present serious challenges to the IAEA and the global nonproliferation framework. The IAEA and its former director, Mohamed ElBaradei, were awarded the Nobel Price for Peace for their work in 2005.

The small number of formal international energy regimes reflects the reality that key energy sources are traded commodities, and that the markets for these commodities, though imperfect, have functioned reasonably well. They are also very large markets in monetary terms, and beyond the control of any one government. Trade in energy commodities generally occurs within the framework of world trade, including World Trade Organization rules. While the international energy institutions and organizations described in the following text fulfill useful tasks, most are limited to improving the performance of energy markets. It is notable that one important energy policy challenge—the volatility of energy prices—has proved to be beyond the reach of energy policy makers.

International cooperative institutions exist for each of the energy policy components. They include organizations representing the production and consumption of coal, liquid fossil fuels, natural and manufactured gas, nuclear energy, electricity and its many means of generation, transportation, and distribution. International cooperative organizations are also important in the entire gamut of existing and proposed renewable energy resource industries. Still more cooperative organizations represent energy conservation and efficiency on a global scale. In the following text we provide a brief overview of some of the international energy institutions and forums with which the United States cooperates.

International Influences on U.S. Energy Policy

The energy policy of the United States has long been influenced by external events and policy decisions made outside of the nation's borders. Because its domestic energy policy is directly influenced by these global economic and geopolitical policy-shaping forces, the United States participates in a wide variety of regional and global energy-related associations and forums. U.S. energy and environmental efforts are implemented through such cooperative venues as: (1) formal, multilateral commitments negotiated with the advice and consent of Congress, (2) the use of executive agreements without congressional participation, and (3) by administrative action exercised through the nation's membership in regional and global organizations formed to enhance collaboration and cooperation on a single policy area such as energy, security, and the environment (Caruson and Farrar-Meyers 2007). This section will describe U.S. participation in a sample of these difficult but often highly effective global associations, organizations, institutions, and forums.

Four of the organizations with which the United States cooperates and which have been identified by the government as being most influential in the field of energy policy are the International Energy Agency (IEA), the Asia Pacific Economic Cooperation Energy Working Group, the North American Energy Working Group, the International Energy Forum (GAO 2007b).

The International Energy Agency

The United States is a founding member of the Paris-based International Energy Agency. The IEA is an independent, intergovernmental organization established by the Organisation for Economic Co-operation and Development (OECD) in 1974 to provide energy policy guidance and cooperation for its 16 founding-member countries; that number has since grown to 28. IEA was formed during the tumultuous years of the oil crisis of 1973 and 1974 in order to coordinate international cooperation during that and any future energy supply emergencies.

As its membership has grown and the global energy condition has changed, so has the mission of the organization. Today, the IEA supports international cooperation in three major policy areas: energy supply security, energy issues relating to economic development, and energy and the environment. One of its major responsibilities is global coordination of responses to oil supply emergencies, in concert with member country governments with significant oil stocks. Within its three major policy areas, the organization supports work on climate change policies, energy market reform, and collaboration in energy technology among members and the rest of the world (IEA 2010).

Table 9.2 Cooperative Initiatives in the APEC Energy Working Group

Short-Term Programs	*Long-Term Programs*
Monthly oil data sharing	Public and private sector responses to energy investments
Maritime security cooperation	Trading and trading data in natural gas
Real-time energy emergency information sharing	Nuclear power data
Oil supply emergency response information	Energy efficiency programs Renewable energy development and information sharing on hydrogen, methane hydrates, and clean fossil energy

Source: APEC (Asia-Pacific Economic Cooperation). 2009. APEC Energy Working Group. http://www.ewg.apec.org/documents/ProgressReport-EGEDA.pdf (accessed January, 2011).

Asia Pacific Economic Cooperation Energy Working Group

The Energy Working Group (EWG) unit of the twenty-one-nation Asia-Pacific Economic Cooperation (APEC) was established in 1990 to support continued regional economic development and limit the environmental damage from the use of fossil fuels (APEC 2009). The mission of the EWG is carried out by four expert groups: Clean Fossil Energy, Efficiency and Conservation, Energy Data and Analysis, and New and Renewable Energy Technologies. A separate Task Force on Biofuels was recently formed to begin sharing information on biomass energy. The EWG was also the first APEC unit to form a public and private sector cooperative body: the EWG Business Network (EBN). This agency advises members on energy policy issues from an industry perspective; it also promotes dialogue between the private and public sectors.

In 2000, APEC adopted the EWG's recommendation for an Energy Security Initiative (ESI) as a vehicle for preparing the region for potential petroleum supply disruptions. The short- and long-term cooperative actions included in the initiative's agenda are displayed in Table 9.2.

An example of how APEC and other international organizations and forums are cooperating on energy policy is the joint initiative on petroleum-base fuels presented at APEC's 2009 annual conference in Santiago, Chile. The mission of this Joint Oil Data Initiative (JODI) is to collaboratively collect and share international oil supply and consumption information. Partners in JODI include APEC and its energy working group, Organization of Oil Exporting Companies (OPEC), the IEA, the International Energy Forum, and the UN Environment Program.

North American Energy Working Group

Canada and Mexico are the two major suppliers of imported oil and petroleum products to the United States; together they provide nearly 30 percent of the country's oil imports (Table 9.3). Maintaining a viable North American energy market is, therefore, of great importance to all three countries.

International cooperation on energy issues in North America are facilitated through the efforts of two initiatives: the secretariat-level North American Energy Working Group (NAEWG), and

Table 9.3 Petroleum Statistics for Canada, the United States, and Mexico, 2008

Petroleum (thousand barrels/day)	*Canada*	*United States*	*Mexico*
Total oil production	3,350	8,514	3,185
Crude oil production	2,596	4,950	2,791
Consumption	2,261	19,498	2,128
Net exports/imports (–)	1,089	–10,983	1,057
Exports to the United States	2,493	n/a	1,302
Refinery capacity	1,969	17,594	1,540
Proven reserves (billion barrels)[1]	178,092	21,317	10,501

Source: Russell, D. 2009. *North American energy relationships.* Draft paper for discussion. The Pembina Institute (Alberta, Canada) and the International Institute for Sustainable Development (Winnipeg, Canada).

[1] As of January 1, 2009.

the executive-level North American Security and Prosperity Partnership (SPP). The NAEWG was formed in 2001 to promote communication and collaboration on energy matters of common interest. The SPP was formed in 2005 as a trilateral program for increasing the security and enhance the prosperity of the United States, Canada, and Mexico through cooperation on the key issues of energy, transportation, financial services, technology, and business. Additional collaboration takes place on protection of the environment, agricultural trade and food supply security, and disease prevention and protection. The NAEWG and SPP help to facilitate energy cooperation through three broad initiatives: market facilitation, energy technology, and clean energy.

The U.S. president, the prime minister of Canada, and the president of Mexico issued a joint declaration in August 2009 that signaled their continued commitment to joint aggressive action on climate change and clean energy. Highlights of the August 10, 2009, declaration are presented in Box 9.1.

International Energy Forum

The International Energy Forum (IEF) membership consists of the energy ministers and secretaries from IEA and OPEC countries. Together, the IEF membership accounts for more than 90 percent of the world's total oil and gas supply and demand countries. The group also includes key players like Brazil, China, India, Mexico, Russia, and South Africa. Through the forum and its associated events, ministers, and energy industry executives meet to hash out solutions to increasing important problems of maintaining global energy security. The IEF's global energy research, data collection, and discussions are administered by a permanent staff (Secretariat) based in Riyadh, Saudi Arabia (IEF 2010).

Two of the IEF's more important initiatives are the International Energy Business Forum (IEBF) and the JODI, both of which are also administered in the Saudi center. The IEBF was established in 2004 as a means for fostering discussions between countries' energy ministers and the chief executive officers of leading energy companies. IEBF sessions are held in association with

BOX 9.1 NORTH AMERICAN 2009 JOINT DECLARATION ON ENERGY AND THE ENVIRONMENT

Continuing their commitment to trilateral North American collaboration and cooperation on sustainable development, environmental cooperation, and clean energy research, the elected leaders of the U.S., Canada and Mexico announced their plans to continue to honor trilateral cooperation programs begun in 2000 and 2005.

They announced support for a global goal of reducing global greenhouse gas emissions by at least 50 percent below the 1990 level by 2050, with the world's developed countries reducing their emissions by 80 percent or more by 2050. Selected energy related points in the joint declaration include agreements to work together to:

- Implement mid-term and long-term goals to reduce national and North American GHG emissions.
- Implement a proposal by Mexico to set up a Green Fund to support mitigation and adaptation activities.
- Build capacity and infrastructure for future cooperation in emissions trading.
- Develop climate friendly and low-carbon technologies, including North American electricity Smart Grid.
- Phase down use of HFCs (a potent greenhouse gas).
- Protect each country's forests, wetlands, croplands and other carbon sinks.
- Reduce transportation emissions.
- Reduce GHG emissions in the oil and gas sector, and promote best practices in reducing emissions and the venting and flaring of natural gas.

(From White House, Office of the Press Secretary. 2009. North American leaders' declaration of climate change and clean energy (August 10)).

annual IEF sessions, thereby ensuring that the concerns of energy industry leaders are available for IEF ministerial deliberations. The mission of the IEBF is defined as:

> The IBEF provides a platform for industry leaders to register and debate their views and concerns with a wide audience of the world's key policy-makers. It is a unique opportunity to address sensitive subjects under the impartial cover of the IEF umbrella which would otherwise be left unsaid or dealt with less effectively on an ad hoc and bilateral basis. (IEA 2010, n.p.)

The JODI was established by the IEF in 2001 in response to a request of six international organizations interested in oil data: APEC, Eurostat, IEA, OLADE (the Latin American Energy Organization), OPEC, and the United Nations Statistics Division (UNSD). Participants were concerned over a lack of data transparency in oil markets. That lack of transparency was seen as a cause of excessive fluctuation in oil prices. JODI became a permanent unit of the IEF in 2005;

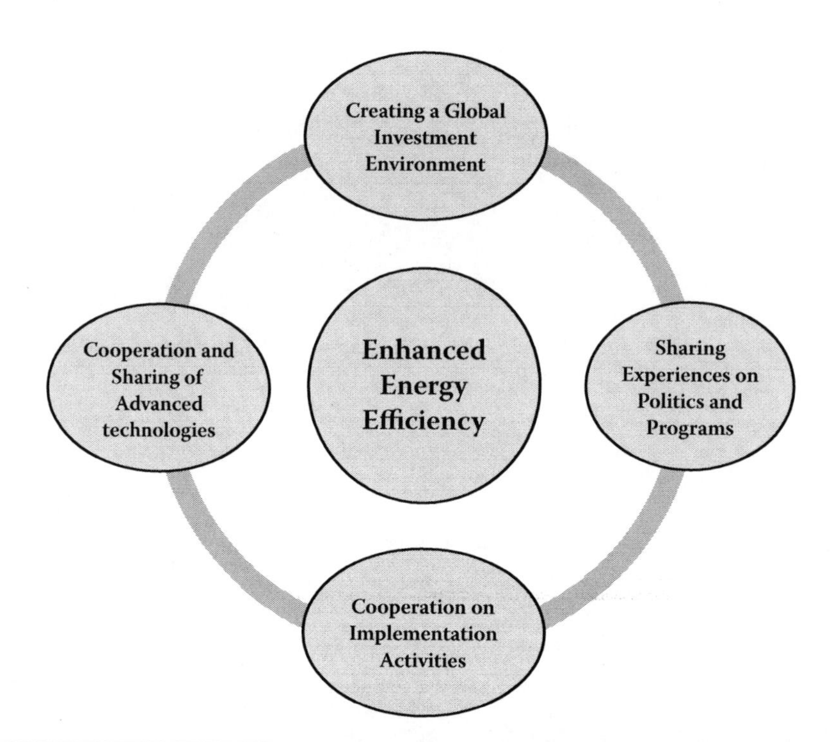

Figure 9.1 Levels of cooperative activity in energy efficiency policies. (Adapted from Energy Charter. 2008. *About the charter.* Energy Charter website, http://www.encharter.org/index.php?id=7&L=0 accessed August 20, 2010.)

it regularly reports on global crude oil production and demand, crude oil imports and exports, petroleum stocks on hand, and global demand for a variety of petroleum-based products.

Other International Energy Organizations

In addition to these four major international energy collaboration and cooperation organizations, the United States plays an active role in many other large and small global institutions. Typical of these are the United Nations Economic Commission for Europe (UNECE), the Energy Charter Secretariat (ECS), and European Union (EU) cooperation. Formed in 1947 to promote pan-European economic integration, the UN agency has added a number of energy and environmental issues to its work agenda. These include the Energy Efficiency 21 Project (EE21) and the Energy Efficiency Investment Project Development for Climate Change Mitigation (Energy Charter 2008) as shown in Figure 9.1.

There are several other organizations that seek to influence energy policy and decisions in other regions. For example, the ECS, an association of fifty-one nations across Eurasia plus the European Communities, was established in 1998 to promote energy security through more open and competitive energy markets. The ECS focuses on intergovernmental cooperation on energy questions covering the entire energy supply chain from exploration to end-use and all energy products and related equipment.

Finally, the European Union and EU member countries are engaged in many interlocking efforts in the arenas of energy security, energy efficiency, and conservation.

The European Union has adopted an energy and environmental policy that calls for an almost carbon-free electricity generating system by the year 2050, through the "20-20-20" program. The European Union's goals are to reduce carbon dioxide emissions by 20 percent and to have a 20 percent share of renewable energy by 2020.

Meeting Europe's Growing Demand for Electricity

European demand for electricity is expected to require additional power production of close to 6,300 TWh by 2020 (terawatt hours; 1 terawatt = 1,000 gigawatts). While improvements in energy efficiency are expected to reduce this growth in power consumption somewhat, EU administrators are looking to advances in renewable primary energy sources. Renewable energy is projected to supply from 500 to 1,200 TWh or around 8 to 12 percent of the projected additional demand. Most of this contribution is expected to be from efficiency improvements made to existing hydropower plants and from wind and some solar power. There is also a major drive underway globally to install new nuclear power generating facilities.

Global Cooperation on Greenhouse Gas Reduction

Global efforts to reverse damage to the environment resulting from global warming have resulted in two related energy agendas. The first, pursued primarily by states and corporations, is finding and developing the primary energy resources that can substitute for petroleum and coal and maintain economic development. The second and more critical is global action to limit the amount of greenhouse gases discharged into the atmosphere from the continued use of these fuels until affordable clean energy is available everywhere.

Of the six most critical greenhouse gases (GHG) being released into the atmosphere by human activity, by far the largest—and hence, most harmful—is carbon dioxide (CO_2), which is almost exclusively the result of burning fossil fuels. Therefore, limiting the emissions of carbon is of paramount importance for energy policy makers everywhere. Shares of this and other greenhouse gases are shown in Table 9.4. As the data indicate, the very high global warming potentials of several synthetic gases such as hydrofluorocarbons make it imperative that they be included in global GHG reduction strategies.

The Kyoto Protocol

International agreement on the dangers of climate change and resultant global warming was achieved at the 1992 UN Framework Convention on Climate Change—the Earth Summit—held in Rio de Janeiro (UNCED 1992). Global agreement on the need to reverse global warming by reducing greenhouse gas emissions became a reality at the conference with approval of a goal of stabilizing GHG concentrations in the atmosphere at "a level that would prevent dangerous anthropogenic interference with the climate system" (Pew Center on Global Climate Change 2010b).

Negotiations on what was to become the Kyoto Protocol began in Berlin the next year with the first planning session of an ad hoc group of representatives from interested countries (the Conference of the Parties, or COP). That conference produced what became the *Berlin Mandate,* an agreement on establishing an open-ended, ad hoc group charged with negotiating a formal agreement on a plan to deal with global climate change (Depledge 2000).

Table 9.4 Shares of Greenhouse Gas Emissions from U.S. Sources, 2006

Greenhouse Gas	*Major Sources*	*% of U.S. Total GHS Emissions*	*Global Warming Potential*
Carbon dioxide	Fossil fuel combustion, nonenergy use of fuels, iron and steel production	85%	1
Methane	Landfills, natural gas and petroleum production, agriculture, coal mining	8%	21
Nitrous oxide	Agricultural soil management, transportation, animal wastes	5%	310
Synthetic gases (HFCs, PFCs and SF6)[a]	Substitution of ozone-depleting substances, electric power transmission and distribution, and aluminum production	2%	140 to 23,900

Source: Stephenson, J. 2009. Climate change: Observations on the potential role of carbon offsets in climate change legislation. March 5, 2009 testimony before the U.S. House of Representatives subcommittee on energy and environment. Report GAO-O9-456T. http://www.gao.gov/new.items/d09456t.pdf (accessed January 25, 2010).

[a] Hydrofluorocarbons, perfluorocarbons, sulfur hexafluoride.

After a series of informal conferences, a compromise agreement was reached on how to achieve the desired reduction in GHG emissions. This agreement was presented at the 1997 Framework Convention held in Kyoto, Japan. The goal spelled out in the Kyoto Accord was a 5.2 percent reduction in human-caused greenhouse gases from the 1990 base year by the industrialized countries, with further cuts to be negotiated later. The reduction was to occur from 2008 to 2012, with each party agreeing to a specific target reduction in emissions. Methods of complying with the Protocol are exceptionally flexible. For example, developed countries can make cuts at home or abroad; they are not held to a timetable; the cuts can be by actual cuts or through accounting measures. Countries can also pick which combination of the six greenhouse gases to target.

Companies taking action outside of their national borders can do so through joint implementation (JI) with firms in other developed countries (i.e., joint ventures); or, they may finance climate-neutral ventures in developing countries through the Clean Development Mechanism (CDM). There were no financial sanctions for not complying; the only financial support was for projects by developing countries. These and related advantages and disadvantages of the Protocol are summarized in Table 9.5.

The long process of gaining acceptance by developed nations (Annex I Parties in the agreement) began after Kyoto. Together, these nations contributed 55 percent of global GHG emissions. But the U.S. Senate unanimously passed a resolution in 1997 stating it would not support any agreement that did not include developing countries (Non-Annex I parties).

The United States withdrew from negotiations in early 2001. The Bush administration stated that the agreement was flawed by its failure to include developing countries and that the policies

Table 9.5 Strengths and Weaknesses of the Kyoto Protocol

Strengths	*Weaknesses*
The Protocol has been signed by virtually all nations, making it the first truly global agreement on the environment.	Only a minority of nations committed themselves to concrete GHG-reduction targets, resulting in a small base of concrete actions to reduce GHG emissions.
The Protocol established the first global administrative structure for dealing with climate change.	Rapidly rising GHG emissions in developing and transition economy nations are not regulated by the Protocol.
The variety and flexibility of means for implementing controls of GHG emissions—carbon cap and trade, joint implementation (JI), and Clean Development Mechanism (CDM)—potentially make the Protocol highly cost effective.	Incomplete participation and different reduction targets resulted in unlevel playing field; firms with constraints have a competitive disadvantage compared to firms in nations without constraints.
The Protocol includes specified binding targets and timetables for reductions in GHS emissions in the industrialized states.	The Protocol's targets are only a small portion of the GHG reduction needed to halt global climate change.
	The Protocol only sets GHG emissions targets; technical measures are ignored.
	No penalties for noncompliance are provided in the Protocol.
	Carbon trade, JI, CDM, and carbon sinks can eliminate domestic action by making it possible to comply by accounting measures.
	Developing countries cannot afford the high costs of the administrative and technical load required.
	The 2008–2012 time horizon was too short.
	Too much emphasis on emission-burden sharing instead of encouraging new economic opportunities.
	The framework of regulation for private companies has unspecified or ambiguous rules.

Source: Wijen, F., and K. Zoeteman. 2004. *Final report of the study* Past and future of the Kyoto Protocol. Tilburg, the Netherlands: Tilburg University Globus Institute for Globalization and Sustainable Development.

needed to attain its ambitious goals would damage the U.S. economy. After the U.S. withdrawal, Russia, which contributed 17 percent of global GHG, was needed to sign for the agreement to be ratified. Russia signed in 2005, and ratification of the Protocol by 141 countries was achieved in February 2005, seven years after its introduction.

BOX 9.2 LACK OF OVERSIGHT HURTS TRADING IN EUROPEAN CARBON EMISSIONS

Hungary's Ministry of Environment and Water sold 800,000 Certified Emission-reduction credits (CERs) under the Kyoto Protocol's Clean Development Mechanism (CDM), the system that lets developing countries sell the credits they earn for their reductions in greenhouse gas emissions to developed countries. Industrialized countries can used the credits as a way to meet the Kyoto Protocol reductions they are required to make in their carbon emissions. The problem with the sale was that the CERs were already used by firms in Hungary to offset their emissions. Offset credits that have been used have no value on European carbon exchanges; the EU requires that one credit must equal one ton of CO_2. Double-marketing of the credits defeats the purpose for which they were designed: to cut carbon emissions.

The Hungarian credits were reportedly purchased by a Japanese buyer. Firms in Japan can use already used credits to be ready when Japan implements its own carbon trading plan. Moreover, Japan will be able use the recycled CERs to meet its own national carbon-cutting commitments. Somehow, however, the Hungarian credits resurfaced at a carbon credit exchange headquartered in Paris. When the European Commission discovered the recycled credits, the trader was forced to temporarily halt trading and the price for CERs dropped precipitously.

The price for CERs recovered somewhat shortly afterward, but remained weak for some time. One global banker was reported as saying that there were signs that the recycling effort had hurt investor confidence, and that prices are likely to stay low.

The Protocol expires in 2012. Coming up with a replacement system that restricts double-counting could make even more trouble for the infant cap-and-trade system. Hungary says it will continue to sell recycled credits, but will attach stricter rules so that they do not show up again in Europe.

(From *The Economist*. 2010. The wrong sort of recycling.)

The Kyoto Protocol has become the primary mechanism for international cooperation on efforts to reduce greenhouse gas emissions (Bartsch, Müller, and Aaheim 2000; Depledge 2000; Grubb, Vrolijk, and Brack 1999; Grubb et al. 2003; Gupta 2001; Gupta and Tol 2003; Oberthür and Ott 1999; Wijen and Zoeteman 2004). Although not required to make cuts, developing countries that signed the Kyoto accord were required to report the extent of their GHG emissions. One of the challenges with the CDM—double counting of credits for emissions reductions—is discussed in Box 9.2. The UN Framework Convention on Climate Change (FCCC) remains responsible for the preparation of periodic reports on the state of climate change.

Cooperating on Cap and Trade Agreements

Improving and extending the collection and disposal of CO_2 is the target of most of the economic, scientific, and technical research on ways to reverse global warming. One of these methods is the global cap and trade system of reducing CO_2 emissions through carbon offsets. This is a relatively new approach to reducing the costs of limiting GHG emissions only became possible with Protocol ratification in 2005.

Cap and trade agreements benefit both industrialized and developing nations. Organizations in industrialized nations may purchase offset allowances from low-carbon-emission developing countries; developing countries are then able to use those payments to fund their own carbon reduction programs.

Cap and trade programs begin with setting specific limits on carbon emissions from regulated organizations—the cap. Then, regulated organizations are given or sold a set of "allowances," which are permits to emit a set amount of GHG. Organizations that invest in facilities to reduce their emissions below their cap can then sell their excess allowances to other regulated organizations. Companies purchase unused allowances because they are less costly than investing in emissions reduction equipment or facilities.

The offset program was established in 2005 under guidelines provided in the Kyoto Protocol (Hornsby, Summerlee, and Woodside 2007). The Protocol encourages investments by industries and governments to reduce the emission of all greenhouse gases. Developed nations agreed to cap the greenhouse gas emissions generated by their industries. The businesses are then given or sold emissions allowances based on their performance. The allowances do not have to be used by the business; instead, they can be bought and sold on the open market. This allows the nation to achieve its carbon reduction targets by the purchase of benefits of carbon reduction actions taken by firms in other countries (UNFCCC 2005).

Carbon offsets refers to a market-driven program to reduce or remove CO_2 and other greenhouse gas emissions from the atmosphere by using the removal or reduction of emissions at one site or activity to compensate for the emissions taking place at another site or activity (Schmidt 2009; Stephenson 2009). The carbon offset market thus created for trading in unused or excess allowances is receiving growing attention in developing and developed countries alike.

The cap and trade system has become the market's major contribution in the fight against climate change and, in one form or another, is being implemented in most of the world. Asia-Pacific Economic Cooperation (APEC), one of the last regional associations to fully embrace the system, announced in late 2009 that it had completed a draft energy trade and investment action plan to reduce or eliminate barriers to energy trade and investments.

The Voluntary Carbon Offset System

The system for carbon offsets exists in two forms: a voluntary and a compliance form. The voluntary form is the method used in the United States, which did not sign the Kyoto accord. The United States does not have binding limits on carbon emissions; thus, U.S. businesses are permitted, but are not required, to participate in the cap and trade system.

Two examples of how the voluntary system has been adopted in the United States are the Regional Greenhouse Gas Initiative (RGGI) and the Western Climate Initiative (WCI). The RGGI is a cooperative made up of 10 northeastern U.S. states that allows utilities to apply offsets toward their compliance target of a 10 percent cut in emissions between 2009 and 2018; the Western Climate Initiative is a similar cooperative effort that includes the states of Arizona, California, New Mexico, Oregon, Utah, and Washington, together with the Canadian provinces of British Columbia and Manitoba.

The Compliance Carbon Offset System

The compliance system was established by countries that have ratified the Kyoto Protocol. This system took its final form in the European Union, where regulated organizations (entities) must meet mandatory limits on carbon emissions. In the European Union, the distribution of allowances is controlled by the EU Greenhouse Gas Emission Trading System (EU ETS). Carbon offsets are generated under a CDM established by the Protocol. To qualify as an acceptable offset CDM project, four criteria must be met.

The countries that signed the Kyoto agreement are permitted to control and distribute cap and trade program allowances according to the needs of their own economy. The Protocol permits allowances to be sold, given away, or auctioned by the countries that signed the treaty. They must be:

- **Additional:** They must be programs or projects that would not have happened without the offsets.
- **Quantifiable:** It must be possible to measure the real or projected reductions.
- **Permanent:** The greenhouse gases they keep out of the atmosphere won't be released later, such as when a replanted forest is cut.
- **Real:** They must be verified by third-party inspectors.

Reducing Carbon Emissions: REDD and Carbon Sinks

Reducing or mitigating the release of carbon into the atmosphere can occur through many different processes, as shown in Table 2.3 describing Pacala and Socolow's (2004) stabilization wedges. From a global perspective, all are important. The global pattern of emissions differs from that of the United States and other developed countries, however.

Worldwide, about 20 percent of emissions are generated by land use changes, particularly deforestation. Deforestation is particularly damaging since normally forests and jungles are vast absorbers of carbon dioxide, and cutting them down releases CO_2 into the atmosphere. As a result, preventing the destruction of global forests, particularly in the tropics, is vital. This problem is (awkwardly) termed LULUCF (Land Use, Land Use Change and Forestry), while the Reducing Emissions from Deforestation and Forest Degradation program (REDD) is the set of efforts that seek to prevent deforestation, and support reforestation and sensible farming practices.

The broad goal is to protect carbon sinks that naturally collect and store CO_2. Geological carbon sinks are Earth's oceans and porous underground rock formations; biological carbon sinks include forests and grasslands (Goodale et al. 2002; Grace 2004; Fung, Doney, Lindsay, and John 2005). *Afforestation* is planting trees where forests have not existed for more than 50 years; *reforestation* is replanting forests after timber harvesting or destruction by such natural disasters and forest fires. Some other agricultural actions taken after 1990 such as returning tilled lands to grazing are also included.

The roles of carbon sinks, REDD, and emissions trading under the Kyoto Protocol remain controversial subjects (Bettelheim and D'Origny 2002). The Protocol allows credit for both afforestation and geosequestration (underground storage of CO_2) sinks for carbon credits under the JI plan of the developed countries and the CDM of the developing countries. Important advantages of REDD projects, however, include their relatively low cost, along with subsidiary benefits for biodiversity and economic development (Blackman 2010). But several serious concerns exist, including how to create a reliable framework for REDD project payments and how to monitor such programs.

Evaluating Carbon Offset Programs

The Government Accountability Office (GAO) has evaluated the European Union's implementation of the carbon offset program to determine what aspects of their required compliance system might work in the United States (GAO 2008a). Four challenges have been identified: First, the carbon offset concept is complex and can involve a wide variety of activities. For example, carbon reduction offsets can be achieved by a variety of means. For instance, they might involve:

1. The reduction of greenhouse gases from the capture of methane from landfills or coal mines
2. A focus on the sequestration of carbon dioxide in underground formations or the planting of tree farms and other forestry or rangeland activities
3. The avoidance of greenhouse gases through installation of renewable energy facilities such as wind farms, geothermal plants, solar arrays, or other renewable programs

Because of this complexity, there is mounting evidence that many projects approved in the European Union have not met all four of the requirements. Another aspect of this challenge is that buyers and sellers, as well as the regulatory bodies in the various Kyoto Protocol signatory nations, often have different interpretations of what constitutes a valid carbon offset.

A second challenge is establishing criteria for evaluating the long-term certainty of offsets. Two characteristics of offsets have to be considered. Proving the validity of the additionality requirement has been particularly problematic. The carbon or other gases to be removed must be measured against a base of what emissions would have occurred if the offset project was not implemented. Adding to the difficulty is that no single correct method for establishing additionality has been agreed upon.

The third challenge facing Congress are the economic and environmental trade-offs that must be considered when using offsets as a regulatory program for limiting GHG emissions. Instead of making an effort to reduce their own GHG emissions, firms opt to purchase less-expensive offsets that might or might not achieve the promised reductions. The ability to purchase offsets in place of reducing their own emissions may also delay investments in research and development of new technologies and fuels.

Finally, a mandatory GHG limitation scheme in the United States simply may not achieve the emissions reductions desired. The European Union has found it nearly impossible to measure the overall ability of its CDM offset program to reduce global GHG emissions. In addition, a number of offsets that do not pass the test of additionality have already crept into the market; others have been counted more than once. If more of these nonqualifying offsets are allowed, overall emissions are likely to increase rather than decline. As a result of these difficulties, opinions remain mixed on whether a compliance scheme should be a part of any domestic U.S. emissions reduction system (GAO 2008a).

The Verdict on Kyoto

The evidence suggests that the Kyoto framework has not been effective at slowing the growth of GHG emissions. Monitoring of emissions has revealed most of the world falling short of its GHG reduction goals, and in fact, global emissions have increased. Of the major countries covered under the Protocol, only Germany, Russia, and the United Kingdom met the primary GHG stabilization goal, and that accomplishment is likely more the result of economic factors than the effectiveness of the Kyoto regime (Betsill 2010).

The experience of Canada, one of the early countries to ratify the Protocol, provides another example; GHG emissions grew by 35 percent during the 15 years after 1990 (Hakes 2008). For the United States and Australia, the two developed countries that did not ratify the Protocol, emissions over the same period grew by 19 percent and 54 percent, respectively; Australia ratified the Protocol in 2007. Since none of the developing countries have reduction targets, China and India, the two largest developing nation economies, represent a clear indication of what might be expected in future global emission control efforts. Emissions in China, which did not attend the Kyoto conference, rose by 137 percent; India's rose 22 percent.

Why was Kyoto unsuccessful? There is a profusion of literature on this topic, but the concerns expressed by several authors are germane. Barrett (2009) suggests that a global climate treaty must meet several conditions. These include broad participation, inclusion of major emitters (the United States, the European Union, Japan, Russia, China, and India), means to ensure compliance, and the capacity to sustain both participation and compliance as parties implement difficult changes in their energy policies. Kyoto met none of these conditions.

A key phrase in the climate change regime to date, going back to the Rio Conference in 1992, is for countries to pursue "common but differentiated responsibility" for reducing emissions, implying that the level of appropriate response is contingent on level of development. But the precise meaning of the phrase is open to interpretation (Hultman 2010).

Prins and Raymer (2007) suggest that the problem was with the overall approach, in expecting a top-down treaty of vast scale and scope to create a successful market for carbon. The successful program against chlolofluorocarbons (CFCs) created confidence that environmental action at a global scale was possible, but that program took on a relatively tame, technical, and limited problem. Substitutes for CFCs were not hard to find. But climate change, argue Prins and Raymer, is a definitive example of a "wicked" problem.

In a classic paper, Rittel and Webber (1973) distinguished between tame and wicked problems, with the latter lacking a definitive formulation, no clear boundaries with other related problems, no clear criteria for knowing when the problem is solved, and no clear, unambiguous set of solutions. Working with such problems is an exercise in patience and persistence since they require the development of deep understanding of interrelated social, technical, economic, and political systems, how to develop consensus around a problem definition and goal, and how to move the interlocked systems in a common direction. The wicked nature of the climate change problem suggests that a top-down strategy will not be sufficient. A more comprehensive approach will be required that gives individuals, countries, corporations, and others incentives to engage in decarbonizing activities of many kinds.

The Copenhagen Accord

The Kyoto Protocol on preventing climate change and stopping global warming expires in 2012. Over the last several years, discussions and informal negotiations have continued on the issue of what should succeed this agreement. On December 19, 2009, representatives from 193 nations formally recognized the Copenhagen Accord, the latest development in the global climate change battle. The Accord is an indication that the world is slowly becoming more committed to doing something about the relentless march toward environmental disaster resulting from global warming. It included a nonbinding agreement between the United States, China, India, Brazil, and South Africa to keep global temperatures from rising more than 1.5 degrees Celsius—a compromise from the 2 degrees Celsius called for—and more financial and technical help (initially, $30 billion) for the poor countries from the rich countries.

On paper, the Accord looks promising; it spells out key agreements on important elements of the climate change problem. In reality, however, the Accord has no teeth. A Copenhagen Protocol to replace the Kyoto pact did not emerge from the Denmark session of the UNFCCC. The Accord does not include specific commitments to cutting greenhouse gases, nor does it contain any penalties for not meeting stated country goals for GHG cuts. From a diplomatic perspective, the Accord is not a formal FCCC agreement, and the 193 parties technically only "took note" of it. What it is instead is an agreement to think about the problem some more, and to continue to work on coming up with an action plan that all developing, transition, and developed countries can agree on—an ambitious objective, indeed.

One of the problems facing negotiators at the December 7–18, 2009, Copenhagen climate change conference (the 15th Conference of the Parties) was its sheer size and complexity. The conference site could only accommodate a third of the more than 45,000 official delegates signed up for the conference, and tens of thousands of unofficial protesters showed up attempting to disrupt the proceedings. The next official conference on climate change was scheduled for Mexico City in November 2010.

Negotiations during the Copenhagen conference did not go smoothly and few participating nations went away from the conference happy. As one observer noted (Goode 2010, 12):

> [P]oor nations, led by those from Africa, repeatedly threatened to walk out. They temporarily halted the formal talks at the beginning of the conference's second week because they objected to an attempt to move away from having twin negotiating tracks—one aimed at extending the 1997 Kyoto Protocol climate treaty and the other focused on crafting a pact that would include the world's top polluters, the United States and China. The U.S. refused to sign the Kyoto agreement; China skipped Kyoto entirely. … Unanimity was [only] preserved by having all 193 nations take note of—rather than formally adopt—the Copenhagen Accord. U.N. officials argue that "taking note" has the same effect as adopting the accord.

The Accord did salvage a useful agreement. The Cancun Conference in late 2010 built on that modest progress. Participating governments agreed on several actions, including a Green Climate Fund, and improvements in measurement and reporting of mitigation actions, among others. But the way forward for global climate negotiations is far from clear.

Challenges to Global Action on Climate

What are the prospects for global action on climate change? The current outlook is highly uncertain. There are reasons for optimism. International action has led in the past to significant successes, as noted above. The issue is already on the international agenda, and on the policy agenda of major emitters, including the United States. There is considerable agreement that action is in the best interests of most countries, and can be implemented at a reasonable cost.

There is general agreement on the part of many observers that an agreement should include several attributes. In particular, it should do the following (Hultman 2010; Olmstead and Stavins 2009; Stern 2009):

- Set a specific goal, or goals (in terms of global CO_2-e emissions, or the maximum rise in global temperature to attain by a particular year, or both)
- Include all of the largest emitting countries (both industrialized and developing) and as many countries as possible overall [as] participants
- Include all of the most important GHG
- Integrate a process for systematic and reliable reporting of national emissions
- Rely on market-based approaches, including a framework for trading of carbon credits that should build on the relative success of Kyoto's Clean Development Mechanism
- Include means to finance and support massive investments in research and development of renewable energy technologies
- Provide substantial support to both the REDD program to reduce deforestation, and for extensive adaptation programs

There is also growing acknowledgment that any new global agreement will not be sufficient on its own. Substantial national efforts and binational and multilateral partnerships will be needed to support a profusion of programs (Joyner 2005). Yet many areas of disagreement remain. Should the agreement be comprehensive and multilateral? Some observers, such as the authors of the Hartwell Paper (Prins et al. 2010) are skeptical that such an agreement is either achievable or desirable.

Should any agreement build on the Kyoto Protocol, or start from scratch? Should it reflect the Kyoto division of the world into Annex I/non-Annex I countries—a developed/developing world split—or aim for a more comprehensive framework? What provision should be made for compliance? What is needed to raise confidence in the effectiveness of offsets? These questions remain ripe for negotiation.

But reaching an agreement will take time and considerable political will, in part because of the magnitude of the task. To hold global atmospheric GHG concentrations to 500 ppm CO_2-e (and thus a temperature rise to about 2°C) will require about a 50 percent cut in global emissions by 2050, relative to 1990; a yearly reduction in emissions of over 3 percent per year would be needed between 2020 and 2050 (Rogeli et al. 2010). Cuts between now and 2020, however, would make the challenge less daunting.

Chasek, Downie, and Brown (2010) examine in detail the many obstacles to creating strong and effective environmental regimes, including several pertinent to climate change. Among the most important are the unequal costs of adjustment, and links to important economic and social activities and interests. A strong treaty that influences fossil fuel use will have some level of economic impact. Developed countries, particularly relatively inefficient ones such as the United States, will have to adjust their consumption to higher prices; countries that rely on fossil fuel production will have to find other sources of growth. Each country that aims to ratify the treaty will face powerful domestic constituencies opposed to it. And the sheer complexity of the topic and the empowered status of developing countries that are large emitters of carbon will make negotiations tough.

A fundamental challenge for climate change policy is that many leaders in the global south view global institutions with considerable suspicion, which was only strengthened by their lack of involvement in the drafting of the Copenhagen Accord. The International Monetary Fund (IMF), World Bank, and World Trade Organization are considered by many in the global south to be elements of a system that has helped to sustain patterns of global inequality, and the governance structures of these agencies are effectively controlled by the developed countries (Karns and Mingst 2005; Ballesteros et al. 2009). Effective global action on climate change is likely to require the creation of a global regime with strong institutions, but so far, essential countries such as China and India have shown little willingness to accept the creation of such a regime. In addition, two other factors—the need to rectify wide disparities in global energy consumption, and the reality that projected growth in emissions comes primarily from developing countries—gives those countries considerable bargaining leverage. The path to an agreement will be arduous.

Conclusion: Influences of Global Cooperation on U.S. Energy Policy

This chapter provided an overview of global energy issues and institutions. We summarized key trends and examined regional and global energy institutions, and global regimes that aim to manage environmental problems associated with energy use. The chapter emphasized the ongoing process of negotiating an agreement to slow the advance of global climate change. Although the obstacles to reaching such an agreement appear daunting, all parties understand the seriousness of the challenge, and there are many ongoing efforts to improve the efficiency of energy use worldwide, discover new energy sources and technologies, and to make its availability more equitable.

Chapter 10

Policies for a New Energy Future

> Moving towards clean energy is about our security. It's also about our economy. And it's about the future of our planet. And what I hope is [that] the policies that we've laid out—from hybrid fleets to offshore drilling, from nuclear energy to wind energy—underscores the seriousness with which my administration takes this challenge.
>
> **—President Barack Obama (March 31, 2010)**

> The tragedy unfolding on our coast is the most painful and powerful reminder yet that the time to embrace a clean energy future is now. Now is the moment for this generation to embark on a national mission to unleash America's innovation and seize control of our own destiny.
>
> **—President Barack Obama (June 15, 2010)**

Despite the profusion of U.S. energy policies we have documented in this text, the country still lacks a comprehensive and clear energy strategy. We have followed a blunderbuss approach that subsidizes primary energy sources and various technologies indiscriminately, while our most potentially useful interventions are supported only intermittently and with a lack of long-term commitment.

There are three problems in the energy sector and economy driving the need for a new strategy. The first is climate change. Policy elites from the scientific community, government, business, including the energy sector and many citizens support decisive action to begin to lower the country's carbon emissions. The second is energy security. The national security establishment and many ordinary Americans are nervous about the country's reliance on petroleum imported from regimes that are either unstable or actively hostile to us. Third is a combination of problems with an economic focus. In the aftermath of a painful recession, it is unclear what segments of the economy will drive economic growth, and clean energy and energy investments are perceived by many elites as one possible answer. Related to this is that the country's lack of a clear strategy is making life much more complicated for the corporate leaders whose investment decisions finance energy infrastructure. Without a clear carbon regime, energy decisions may be driven on a state-by-state basis and at a lower level than would be optimal.

With these concerns in mind, it is clear that any energy policy of the United States aimed toward 2030 and beyond must be crafted around programs to achieve several objectives. First, we must have a secure, affordable, and adequate supply of primary energy for all sectors of the economy. Second, we must reduce our almost exclusive reliance for primary energy on burning fossil fuels that pollute the atmosphere by emitting climate-changing greenhouse gases and carbon particulates. Third, we must engage global leaders to continue to work collaboratively toward a global climate change treaty. Fourth, policy makers must continue to support the current high levels of investment in energy research and development. This will help us develop and deploy the technology that will allow the country to maintain our current level of energy services with less primary and secondary energy. Fifth, we must diversify our energy portfolio and reduce our reliance on imported petroleum, while decreasing our carbon emissions. Finally, we must minimize the impact of these policies on persons and households least able to pay.

What mix of policies is most appropriate to accomplish these objectives? And how likely is such a set of interventions to obtain the popular and political support needed to make them law, in a highly partisan era? This chapter will suggest a set of possible policies and examine several of the challenges they face, including their popular support and political viability over the medium term. We will also consider in some detail a few of the difficult policy choices Congress faces as they craft such legislation. These include the role of nuclear energy, expanded access to federal lands, and research and development of carbon sequestration technology.

In his Oval Office address on June 15, 2010, President Barack Obama called for legislative action on an energy bill that would "end America's century-long addiction to fossil fuels," decrease our need for imported oil, and accelerate the transition to a clean energy future. He is not the first president to call for action to reduce our reliance on imported oil: each of his previous seven predecessors made similar statements (Milken 2010). But none of these presidents faced the grinding effects of the worst environmental disaster in U.S. history. Passing an energy bill that reduced the country's need for petroleum and moved the country toward renewable energy would be a significant legislative achievement. It would help remove the stigma of corporate and governmental failure that may forever link his legacy with the disastrous effects of the Challenger Horizon spill.

The Energy Options Portfolio

A range of energy options is available to policy makers, as shown in Figure 10.1. The general strategies that could be pursued include supply-side measures, such as expanded domestic production of fossil fuels, including oil and gas, increasing renewables production, or investments in nuclear power. Demand-side measures include conservation, efficiency investments, energy taxes, and cap and trade systems that would raise the cost of primary and secondary energy sources.

From this range of options, decision makers—primarily members of Congress—must enact a portfolio of energy policies that will work together to accomplish the policy objectives described above. The elements of such a portfolio could include:

- Policies that will create a direct or indirect price for carbon, and generate revenue used to support consumer rebates and energy programs.
- A national portfolio standard and feed-in tariff.

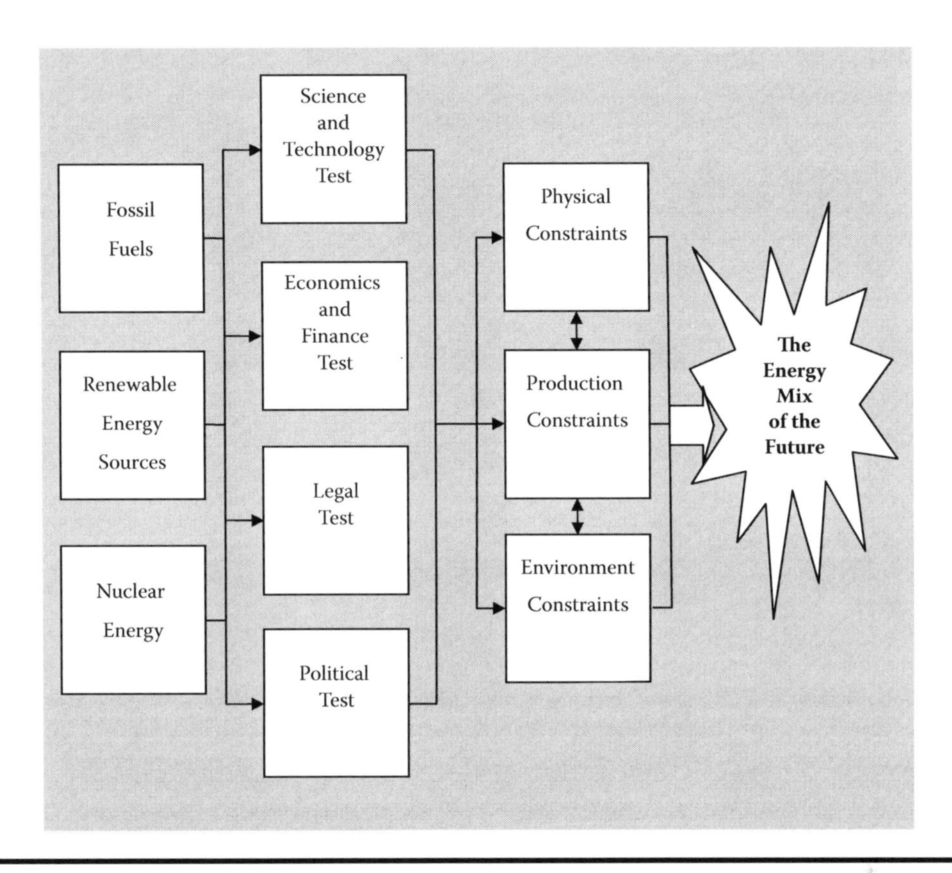

Figure 10.1 Constraints in crafting the energy mix of the future. (Adapted from Galston, W. A. 2006. Politics and feasibility: Interests and power. In *The Oxford handbook of public policy*, ed. M. Moran, M. Rein and G. E. Goodin, 543–556. Oxford, U.K.: Oxford University Press; Immergut, E. M. 2006. Institutional constraints on policy. In *The Oxford handbook of public policy*, ed. M. Moran, M. Rein and G. E. Goodin, 557–571. Oxford, U.K.: Oxford University Press; Bobrow, D. B. 2006. Social and cultural factors: Constraining and enabling. In *The Oxford handbook of public policy*, M. Moran, M. Rein, and G. E. Goodin, 572–586. Oxford, U.K.: Oxford University Press.)

- Increased support for energy efficiency programs and investments, including infrastructure such as transmission systems for modernizing the electricity grid.
- Continued support for research, development, and deployment in the energy sector.
- A phase out of most energy subsidies.
- Appropriate regulatory action.

Here we will provide a brief sketch of how the interventions described and analyzed in Chapters 7 and 8 could work with these policies. Similar proposals have been suggested in considerable detail by several other authors, including Furman et al. (2007), the Committee on America's Energy Future (2009), and The Pew Center for Global Climate Change (2006).

Either a carbon tax or cap and trade system would be an appropriate means to give large producers or users of fossil fuels price signals about the carbon content of those fuels. This would create incentives to use less energy, find more efficient production methods, and upgrade their equipment. On balance, a carbon tax is the best approach due to its ease of administration and relative lack of complexity compared to the more cumbersome requirements of a cap and trade system. Initially, such a tax could be set at a rate roughly equal to the official estimate of the current social cost of carbon, about $20 in current dollars, and set to increase by about 5 percent per year. This would raise about $100 billion per year (Interagency Working Group 2010; Parry 2009b).

Carbon Taxes on Fossil Fuels

A carbon tax will hit coal harder than petroleum due to its higher carbon content. A Congress intent on truly reducing U.S. oil consumption as the primary strategy of reducing imports would also raise the gasoline tax gradually but significantly over an extended period, as suggested by Mankiw (2006). This would allow individuals to adapt and firms to respond by improving the fuel efficiency of the automobile fleet.

Another strategy for reducing imports is increasing domestic production; however, that is not viable in the long term given the country's limited reserves in relation to its consumption. In combination with current regulations in place raising corporate average fuel economy (CAFE) standards, improved mass transit, and (eventually) longer battery life in battery electric vehicles, we are close to having the policy tools and technology to cut considerably our use of oil for transport. Revenues from such a tax would be rebated to households to lower their income taxes, protect low-income households from the effects of higher energy costs, and also be used to fund the efficiency and R&D programs discussed in the following text.

The electricity sector would require policies influencing both the demand side and supply side. The United States is capable of considerable improvement in building and industrial energy efficiency. McKinsey estimated that a comprehensive efficiency program could reduce energy use by 23 percent by 2020 compared to the business-as-usual case (2009). A holistic program of demand reduction would be needed aiming to improve technology and change behavior by people and firms, in the residential, commercial, and industrial sectors.

Creation of a national renewable portfolio standard would create a nationwide incentive for investors and utilities to add renewable energy in place of other generation options, particularly coal. In combination with a feed-in tariff, this would encourage the addition of distributed generation and business and household investment in solar, wind, and small hydro projects. Additional investment in transmission and an upgrading of the national electricity grid will be needed, as well as a gradual move to smart meters. This will allow distributed generation to be more seamlessly added to the generation mix, and allow actual time of use pricing for households and businesses that more closely reflects the true marginal cost of electricity.

Continuing the present high levels of investment on R&D in the energy sector would improve the likelihood that breakthroughs could occur in large-scale energy storage, batteries for battery electric vehicles (BEVs), photovoltaic efficiency, carbon sequestration, long distance transmission of electricity through direct current, and several other important technologies. In the long run, the likely secondary impacts on the U.S. economy through such investments would be sizable.

Finally, one source of funding for this portfolio of interventions would be the gradual elimination of other energy subsidies, including the various tax credits for wind and solar energy. Repeal of these tax benefits over time would rationalize the energy system by removing incentives to invest in and consume fossil fuels, and provide consistent price signals to utilities and investors

(Sovacool 2008). The EPA would be required to supplement such a regime with rigorous data collection on emissions of CO_2 and other greenhouse gas (GHG), plus regulation of mobile and source point emitters as needed under the Clean Air Act.

Passage of such a program would put the United States in a stronger position to negotiate a global climate change treaty that included developing nations. A national carbon tax could be harmonized with similar taxes in other countries (Nordhaus 2007). It would likely enable the country to meet a goal of reducing U.S. greenhouse gas emissions to 17 percent below 2005 levels by 2020, and make it possible to be 80 percent below 2005 levels by 2050. It is consistent with a goal of limiting the increase in atmospheric CO_2 levels to 550 ppm.

What would be the cost of such a program? While dependent on the specific rates of carbon tax and gasoline tax assessed and the level of funding for efficiency and other initiatives, such a program would be likely to reduce gross domestic product (GDP) by less than 1 percent of GDP per year by 2030.

The Challenges Facing Policy Makers

What are the barriers to implementing this set of policies? There are several, including politics, the inertia of the U.S. energy system, energy supply issues, and the environmental and behavioral barriers to renewables and energy efficiency.

Challenge I: Politics

Is this portfolio politically viable? As of early 2011, the probability of the U.S. Congress passing such a package approaches zero. Passage of a carbon tax or cap and trade plan through a more conservative Congress will not happen without a national crisis. Policy proposals that raise the price of gasoline or power would be framed as growth-busting energy price hikes that will hurt hardworking Americans. Attempts to cut oil and gas subsidies would be framed as a tax increase and meet bitter opposition. Given the intense pressures to cut federal spending, energy politics will be a free-for-all in which energy interests will engage in intense lobbying to retain favored programs. Tighter regulation will remain a favored policy tool and will create tension between Congress and the Obama Administration.

A new approach to energy politics was evident in President Obama's 2011 State of the Union address, in which he set a national goal of producing 80 percent of U.S. electricity from "clean" sources by 2035 (Obama 2011). "Clean" was defined as renewables, nuclear, clean coal and natural gas. Although this stretches the definition of "clean" beyond recognition, it might be broad enough to attract bipartisan support for a bill supporting R&D and incentives for such production that will also generate green jobs. Biofuels and support for electric vehicles could be included in this policy proposal, as actions to lower U.S. reliance on imported oil and generate jobs may be a winning formula at a time of slow growth and high geopolitical tension.

There was a policy window open for a more ambitious bill open in mid-summer 2010. The Challenger Horizon oil spill reminded Americans of the problems associated with our reliance on oil. A May 2010 poll found Americans (particularly Democrats and independents) more concerned about environmental protection than about encouraging growth in U.S. domestic energy supplies (Jones 2010). A Stanford University survey found support for tougher regulations on emissions, plus subsidies for renewables, and energy efficient autos, appliances, homes and buildings (Krosnick 2010). The results of this poll also suggest that there is an "issue public" of citizens passionately concerned about climate change, supportive of strong government action, and likely to lobby in its behalf. Despite the costs, many voters seem willing to continue to support continued energy subsidies and efficiency-mandating regulations.

But any policy mix that may raise the cost of energy faces a difficult fight. Consistent with many other surveys, the Stanford results found most respondents opposed both taxes on electricity (78 percent) and gasoline (72 percent) to reduce consumption. One observer (Pielke 2010, 59) has concluded that an "iron law" of climate policy is that "when environmental and economic objectives are placed into opposition with one another…it is the economic goals that win out." Although the policy arena is too unpredictable to validate such a law (who would have predicted passage of a national health insurance law in mid-2008?) the politics of energy will remain very difficult.

Challenge II: The Inertia of the Energy System

Even with relatively strong legislation that aims to shift the country's energy strategy, the United States is likely to be reliant on fossil fuels for many decades. The existing energy infrastructure is a vast and costly system, combining physical assets (electricity generating stations, oil refineries, gasoline stations) with market and regulatory mechanisms at the national, state, and local levels. Moving this system toward reliance on renewables will take decades, and is likely to start at the margins—with each individual decision by a utility or state about new sources of generating capacity. This highlights both the importance of the renewable portfolio standard (RPS) as a policy choice influencing those decisions, and the reality that the burning of fossil fuels will be a part of any energy policy. Much of the petroleum used for transport will continue to be imported. The immediate needs, then, are for Americans to use less carbon-based fuel, and to continue to vastly improve the efficiency of the machines—autos, generating plants, and prime movers—that will continue for some time to be powered by oil, natural gas, and coal.

This transition is not going to be easy. Analysts predict that there will be at least a 20 percent increase in global demand for energy in 2030, although growth in U.S. demand will be less. The U. S. Department of Energy (DOE) predicts growth of nearly 17 percent in the gigawatt electrical generating capacity from 2009 to 2030. Table 10.1 displays prices and demand for electrical energy for the years 2007, 2009, and projections for 2030. Average prices per kilowatt hour of electricity are projected to rise only moderately over this same period, from 9.1 cents in 2009 to 10.4 cents in 2030. Note the continuing role expected to be fulfilled by coal in the country's generation mix.

Challenge III: Energy Supply

We examined in detail the state of U.S. and world primary energy supplies in Chapter 1. Few serious energy analysts deny that the world is running out of oil; reserves of natural gas are high in the United States due to recent shale gas finds, but could dwindle rapidly if substituted en masse for coal for electricity generation; coal is plentiful but dirty and expensive to convert for transportation uses. The relationship between these energy sources and the viability tests facing new and nontraditional energy sources, and development and production constraints is displayed in Figure 10.2. Increasing production of any of the current major primary energy categories faces severe limitations, including:

Table 10.1 Advantages and Disadvantages of Renewable Energy

Energy Source	*Typical Cost Per kWh (¢)*	*Advantages*	*Disadvantages*
Wind	4.0 to 9.3	Widespread availability, can be easily scaled to meet local requirements	Often difficult to site; wind speed is variable and intermittent
Solar: Photovoltaic Solar: Thin-film Photovoltaic Solar: Thermal	33.5 to 39.4 18.0 to 23.0 11.0 to 14.0	Although variable, exists everywhere; silent; long life of equipment; no cost for fuel Least-expensive solar	Very expensive, difficult to site large facilities; storage batteries needed for night and cloudy days Day use only
Biomass	7.5 to 9.0	Known technology; supply generally widespread	High energy needed to produce; contributes to air pollution; expensive
Hydropower	4.0 to 11.3	Although installation expensive, operation inexpensive	Few sites left for new facilities; impacts air, water, and ecology
Geothermal	5.0 to 6.0	Inexpensive and reliable	Limited number of sites available; unknown effects on natural environment

Sources: Komor, P. 2004. *Renewable energy policy.* New York: iUniverse; GAO 2007a; Bailey, R. 2009. Energy futures: A quick guide to alternative energy. *Reason* (June).

- **Physical limitations:** Fossil fuels and other resources are not evenly distributed around the world, and all are exhaustible.
- **Production limitations:** Natural, economic, scientific, or social limitations can curtail growth in production. The continuing fight to expand drilling for oil off the California coast and in the Arctic Wildlife Refuge are examples of how these forces constrict petroleum production expansion.
- **Financing limitations:** The more than 30-year halt in construction of new nuclear energy–powered electricity generators reflects market choices about the high cost of such power and consumer and regulatory concern about its safety and long-term effects.
- **Environmental limitations:** Air pollution, geophysical degradation, limited water supplies, global warming caused by greenhouse gases and microparticulate emissions are among the major constraints limiting use of existing primary energy sources (Moriarty and Honnery 2009).
- **Political limitations:** Politics as usual results in business as usual, and that means trouble for the economy and for the environment. A good example of politics as usual is the recent U.S. Senate rejection of a bill to reduce subsidies to the oil and gas industry (Gaudiano 2010).

Figure 10.2 Possible focus points for new energy policy. (Developed from materials from Obama-Biden energy policy platform, 2008, Barackobama.com. 2008. *Barack Obama and Joe Biden: New energy for America*. Organizing for America, http://www.barackobama.com/pdg/factsheet_energy_speech_080308.pdf, accessed January 31, 2009.)

Challenge IV: Overcoming Environmental Barriers to Renewables

Earlier in the text we discussed the many external costs associated with fossil fuel use. But renewable energy sources come with their own set of advantages and disadvantages. Their general advantages are that they do not require combustion, are inexhaustible, and expand the available portfolio of energy supply options. But they generate their own externalities, and are sensitive to a variety of political, economic, scientific, or social constraints.

Environmental and land-use limitations have the capacity to reduce the overall potential of renewable energy. Hydro, wind, and solar sources used for electricity generation have a lower power density than their fossil fuel counterparts; it takes many acres of wind turbines, solar panels, or even mirrors for a concentrated solar facility to equal the output of a 500 MW combined cycle gas turbine plant. This is due to the relatively low conversion efficiency of photovoltaic panels, and size requirements to optimize the power output from wind turbines. Such turbines need to be tall (60 to 100 m high) since wind velocity increases with distance above the ground; they also must be placed about 5 blade-diameter lengths apart to minimize wind wake.

Siting issues with wind turbines and concentrated solar power can be politically explosive, as discussed in Box 10.1. Moreover, all renewable energy sources can have potentially serious environmental impacts when implemented on a large scale (Abbasi and Abbasi 2000; Moriarty and Honnery 2009). Wind can impact bird and bat populations; concentrated solar power disturbs

BOX 10.1 CONCENTRATED SOLAR POWER: OPPORTUNITIES AND CHALLENGES

There is a rough consensus among energy analysts that to accomplish important energy goals, including reducing the risks of climate change, fundamental breakthroughs in new energy technologies are needed. Without them the U.S. and world energy systems will decarbonize at a slow but steady pace—likely too slow to prevent climate change. But promising new technical solutions face many hurdles between the lab bench and cost-effective implementation. After the technical issues have been conquered, investors must surmount planning, regulatory, practical and competitive challenges that may leave once-promising solutions to join the ranks of other "almost" winners—such as the Betamax video player or Segway personal transporter. One promising energy technology, concentrated solar power (CSP) may be at risk of joining this list.

In locations with intense year round sunlight—the Mojave Desert, Southern Spain or North Africa—various methods may be used to collect solar energy on a large scale (*Economist* 2009d). One method is to use acres of mirrors that are precisely calibrated to reflect sunlight onto a "power tower" containing a fluid, which boils from the heat and drives a turbine, creating electricity. Another is to use acres of parabola shaped mirrors that focus sunlight onto tubes containing a fluid, which is collected, and the heat exchanged to water, which then boils and drives the turbine. CSP has great advantages: it operates at a relatively large scale, is nonpolluting, may easily be combined with a small natural gas plant (for 24/7 power production) and the generated heat may be stored (one option is to heat molten salt). A few such plants are operating in the U.S., and many others around the world. An optimistic 2005 study projected that CSP could produce 5% of global electricity by 2040 (Arringhoff et al 2005).

But concentrated solar plants have drawbacks. They require significant volumes of water for cooling, in typically dry desert environments. They require large areas. The recently approved 370MW Ivanpa project in San Bernardino County, California, will occupy 3,500 acres, or 5.5 square miles, of desert habitat managed by the U.S. Bureau of Land Management (BLM 2010). Since development will disturb native plants and animals, this has led to many legal challenges to CSP projects. Finally, they remain relatively costly, and are at risk from a surprising source, the rapidly improving efficiency of photovoltaic panels (Kanellos and Prior 2010).

(From *Economist, The.* 2009d; Kanellos, B. and B. Prior. 2010; Aringhoff, R., G. Brakmann, M. Geyer, and S. Teske. 2005; BLM [U.S. Bureau of Land Management] 2010.)

desert habitat; hydropower causes extensive damage to riparian areas and fish populations. Overall, however, these pale in comparison to the external costs of petroleum, coal, and even natural gas.

The renewable source generating the most concern is biomass. Population growth and ongoing climate changes may adversely affect the production of useable energy from renewable biomass sources. So the potential for sustainable bioenergy contribution to the 2030 demand picture could close to zero. Increasingly, agriculture experts estimate that only a limited amount of biomass energy can be harvested globally without cutting food production or worsening climate change (Field et al. 2008). With present and foreseen technology, cellulosic biomass may never be a feedstock for liquid fuels. Estimates of the energy return on energy invested (EROEI) of corn ethanol have varied, but a consensus figure is 1.3:1, a very limited return on the resources invested in manufacturing it. A recent paper (Felix and Tilley 2008) found that conversion of switchgrass to ethanol takes as much or more energy to produce than it actually produces.

Challenge V: Energy Behaviors and Increasing Energy Efficiency

The potential for improving U.S. energy efficiency is vast. A consensus estimate is that energy efficiency investments could reduce the country's energy use by 25 to 30 percent over the next 20 to 25 years (Ehrhardt-Martinez and Laitner 2008). Processes for transforming primary energy into useable secondary sources are inefficient, as are many end uses, such as residential heating and cooling and (especially) transportation. But efficiency improvement opportunities are spread across an estimated 100 million U.S. locations; they require up-front investment in new technology and a willingness to engage the task (McKinsey 2009). Often this last issue is the most challenging, as the relatively small and hard-to-measure savings accrue in the future, while the investment cost of the new devices plus the time and attention needed for their installation are incurred in the present.

In Chapter 6 we discussed a number of the specific barriers to investing in energy efficiency, including the principal-agent problem and prohibitive cost of new investments, and the desire of households for a quick payback on their investment. Efficiency subsidies in the form of loans and tax credits, and efficiency standards, can be used to surmount some of these barriers. Introduction of electricity pricing closer to its marginal cost will also provide consumers with an incentive to invest in efficiency.

Making the Tough Choices

The portfolio of energy policy choices we propose in this chapter omits several policy options that present policy makers with particularly difficult dilemmas. These include the appropriate role for nuclear power and for carbon sequestration, or so-called clean coal, what to do about the potential movement of the high-carbon industry to countries with less-restrictive carbon regimes (*leakage*), and access to public lands and the continental shelf. We will briefly consider each of these below.

Nuclear Power

Perhaps the most controversial energy policy dilemma is whether the United States should continue to offer subsidies for the development and construction of new nuclear power plants. As discussed in Chapter 7, U.S. policy since the 2005 Energy Policy Act has been to expedite regulatory approval of new plants and provide sizable subsidies for their construction and power production in the form of loan guarantees and tax credits.

Proponents of such policies argue that nuclear is a relatively low-carbon energy source, that it is a well-known and improving technology with, overall, a strong safety record. Moreover, they argue that its levelized cost of power (i.e., total cost based on constructing and operating a generating plant over its economic life in inflation adjusted, present value terms) is competitive with natural gas and wind, and that it offers a reliable source of baseload power.

Opponents emphasize the still-unresolved issue of what to do with waste nuclear fuel. Storage in casks onsite is now the default option. However, the risks of accidents and terror attacks against nuclear facilities and the risks of nuclear proliferation from increasing reliance on nuclear power make this solution more problematic than ever. Private investors have been reluctant to support nuclear energy without a combination of federal insurance and subsidies, largely due to the very high cost of construction, now $8 billion per reactor in the United States in current dollars (although that could go down with standardized reactor designs). Decommissioning at the end of the plant's useful life is also costly. Some critics also cast doubt on whether the supply of uranium is adequate to support a strong push toward nuclear power, although supplies are now adequate to meet expected demand for the next 200 years.

Conflict over Two Underlying Issues

The underlying conflict about nuclear is over two issues. The first issue is the future direction or vision of the U.S. electricity system. An expanded role for nuclear energy implies the continuation of a highly centralized, corporation-dominated utility sector potentially resistant to too much distributed generation. A nuclear-dominated utility needs to sell its power to pay back bondholders; expanded micropower production by businesses and homeowners eats into that revenue.

Nuclear power is likely to do better under traditional rate-of-return regulation, rather than in a competitive environment that favors the lowest-cost producers. Does this mean that a carbon price will be in place that could give nuclear a more equal footing with fossil fuels? A 2008 CBO study concluded that a $45 per-metric-ton CO_2 charge would be sufficient to make nuclear cost competitive with coal and gas, even without subsidies (CBO 2008b). With such a charge in place, nuclear generators could be strong competitors even in a competitive electricity market, and squeeze out renewables unless wind and solar continued to receive federal and state subsidies, improved their overall efficiency, and reduced their cost.

The second issue is the up-front cost of nuclear energy; $8 billion or, at best, even $6 billion, should costs decline with standardization of design, is a "lumpy" and substantial investment, particularly when combined-cycle gas turbine plants may be built at a fraction of the cost. Smaller, common-design manufactured reactor designs are one possible answer.

The most recent Energy Information Administration (EIA) analysis of generation cost highlights the difficulty of the choices facing utilities and society (Table 10.2). Natural gas generating plants are the cheapest to construct to begin with, but face the risk of fluctuating natural gas prices. Nuclear has very high up-front costs, lower fuel costs, but considerable fixed maintenance costs. This table does not show energy efficiency as an option, but investments in combined heat and power and other efficiency measures often have a quicker payoff than investments in generation. So should society invest in a relatively small number of nuclear units, or use those resources to fund a combination of efficiency, renewables, and natural gas, especially with the expansion of U.S. natural gas reserves? The latter is likely a better option, although investment in sufficient nuclear capacity to maintain its current 20 percent contribution to the electricity supply may be prudent.

Coal and Carbon Capture and Storage

The abundance, relatively high energy content, and low cost of coal make it an excellent source of fuel for electricity generation. But coal is full of impurities and carbon. Each ton burned releases 5,720 lbs. of CO_2 into the atmosphere (assuming a carbon content of 78 percent), along with

Table 10.2 U.S. Projections for Electricity Prices and Demand, 2007–2030

	Year		
Item	*2007*	*2009*	*2030*
Average end-use price[a]	**9.1**	**9.1**	**10.4**
Residential	10.6	10.8	12.2
Commercial	9.6	9.3	10.6
Industrial	6.4	6.3	7.4
Electricity generation[b]	**4,190**	**4,938**	**5,181**
Coal	2,021	2,121	2,415
Oil	66	57	60
Natural gas	892	815	1,012
Nuclear	806	831	907
Hydro and other[c]	374	555	758
Net imports	31	17	28
Capability (gigawatts[d])	**996**	**1,050**	**1,227**
Coal	315	331	360
Oil and natural gas	448	458	563
Nuclear	101	104	113
Hydro and other	131	157	191

Source: EIA 2009h

[a] In 2007 cents for kilowatt hour.
[b] In billion kilowatt hours.
[c] "Hydro and other" includes conventional hydroelectric, pumped storage, geothermal, wood and wood waste, wind, solar, other biomass, municipal waste, purchased steam, and miscellaneous.
[d] One gigawatt equals one billion watts.

mercury, radon, and other pollutants. The ubiquity of coal as an energy source, and its high carbon emissions, have led to intensive research into techniques for extracting that CO_2 from the emissions of coal combustion and finding ways to compress the gas, transport it, and sequester it underground, to prevent it from entering the atmosphere. This technique is known as carbon capture and storage (CCS). Possible storage locations are primarily geologic and include depleted oil and gas reservoirs, salt formations, the deep ocean, and coal seams. Compressed CO_2 may also be used to assist enhanced oil recovery as well as production of coal bed methane. Estimates of potential reservoir space worldwide range from hundreds to thousands of GtC, while global emissions are roughly 7 GtC per year (Herzog and Golomb 2004).

The CCS process was included by Pacala and Socolow (2004) as one of their 15 stabilization wedges, due to coal's importance as a contributor to climate change. CCS includes a variety of technologies for extracting CO_2 from various industrial and power generating technologies, including natural gas plants. Currently over 80 research projects are underway worldwide to improve the various stages of the technique (extraction, compression, transport, and storage) and test whether the approach is commercially viable for particular applications. The technique has proved successful in small-scale pilot tests, but none at scale. The DOE program includes the FutureGen project, additional R&D, and seven regional partnerships.

Sequestration is a classic "eat your cake and have it too" strategy. It would allow us to enjoy the benefits of relatively cheap electricity from coal, without having to be concerned about at least some of its environmental effects. But does it make sense? Under current technology, the Department of Energy estimates that CCS adds about 75 percent to the cost of electricity from a new plant using pulverized coal. The extra cost represents, in part, the additional fuel required (25 to 40 percent more) because CCS is an energy-intensive process. The injected gas would need to be stored for hundreds or thousands of years. The relative safety of the technique is one issue, although reservoirs are capped and in some cases the CO_2 physically binds to the storage medium (e.g., coal), reducing the risk of a large-scale, catastrophic release.

The largest potential reservoir is the deep ocean, which already serves as a sink for atmospheric CO_2. However, some data suggest that the oceans are turning slightly acidic at current levels of absorption. An unresolved question about large-scale applications of CCS is what proportion of CO_2 would not be captured, and thus enter the atmosphere. Finally, who would assume liability for terrestrial storage sites? In 1986 a naturally occurring release of CO_2 from the depths of Lake Nyos in Cameroon led to 1,700 deaths. The widespread use of CCS would inevitably lead to industry requesting government involvement to insure against such risks.

With present technology, use of CCS in conjunction with coal or natural gas electricity production is uneconomic. Further research, improvement, and deployment of the technology is needed to find means to lower the cost of the various components of CCS. Issues of liability will need to be clarified, and a carbon tax or cap and trade equivalent would bring it closer to commercial viability. Overall, however, this technology makes little sense as a long-term solution as long as other cheaper and cleaner options are available. It could help ease the transition to a lower carbon future but to date the label "clean" coal is a stretch. Accordingly, adoption of a cap and trade system or a carbon tax in the United States will raise production costs for energy-intensive industries.

Access to Public Lands and the Continental Shelf

The United States currently holds about 2 percent of global oil reserves, but consumes about 23 percent of world production—19.5 million barrels per day (mbd) of the world's 85.5 mbd. Of this amount, the U.S. imports about 11.1 mbd, which includes both crude oil and refined products. An available short-term strategy is to reduce our proportion of imports, from the current 57 percent, which costs roughly $300 billion per year, by increasing domestic production. Current U.S. proved reserves of crude oil were 19.1 billion barrels at the end of 2008 (EIA 2009i). In addition, the United States has an estimated 127.28 billion barrels of undiscovered, technically recoverable oil, of which 41.4 billion barrels are on land (BDEMRE 2009), and 85.88 billion barrels on the outer continental shelf (MMS 2006). Total world crude oil proved reserves are estimated at about 1.3 trillion barrels. Oil producers and national governments around the globe regularly seek to expand these known reserves. In the United States, this often means greater political and economic pressures to expand access the public lands and the continental shelf.

One option for expanding domestic oil production would be to open the Alaska National Wildlife Refuge (ANWR), particularly Area 1002, to oil and gas development. Under federal law, this is currently prohibited. ANWR lies about 100 miles east of the Prudhoe Bay oil field on the north slope of Alaska. The U.S. Geological Survey estimated in 1998 that Area 1002 held between 5.7 and 16.0 billion barrels of oil, with a mean estimate of 10.4 billion barrels, of which 7.7 billion barrels lie in the federal portion of the 1002 zone (DOE 2008). The EIA estimated that about 10 years would be required to begin oil production in the area, which would then peak about 10 years later at between 510,000 to 1,450,000 barrels of oil per day. At current rates of use, this is between 6 and 17 percent of daily U.S. oil production. But drilling for oil in ANWR has long been opposed by a broad coalition of environmentalists and many ordinary Americans, who believe the Earth should be left undisturbed. Polls on the issue over the last decade found support for drilling increased along with the cost of oil.

This issue has long been a political football. In July 2008, Republican Congressman John Linder of Georgia joined 84 of his fellow members of the House of Representatives in sponsoring what he described as energy legislation that would lower American's energy prices and move the United States toward energy independence. House Bill 6566, then known as the *American Energy Act*, was a Republican energy bill that brought together a number of legislative proposals into a single combined bill with these sections:

- It would have repealed bans on oil drilling in the U.S. deep continental shelf and the Arctic coastal plain (the location of the Alaska Wildlife Refuge).
- It would have provided tax incentives to all Americans who add to energy efficiency, whether it is through fuel-efficient automobiles or business operations.
- It would have also repealed the section in the Energy Independence and Security Act (EISA) that blocks development of coal-to-liquid technology.

The United States could also further expedite development of oil production on other federal lands and the outer continental shelf. But the Deepwater Horizon oil spill greatly complicated the process of reaching any policy agreement on offshore drilling. A lengthy moratorium on new drilling is likely while the causes of the spill are investigated; eventually new regulations and additional safeguards will be put in place. The cost of an offshore exploratory well already exceeds $100 million, and this is likely to increase. Whether the aftermath of the spill will unravel one of the assumed "grand bargains" in constructing an energy bill—support for expanded offshore oil and gas production in exchange for tighter controls on emissions and support for renewables—is unknown.

Drilling for oil offshore is not the only issue clouding the public lands access question. A number of offshore wind farms have also been proposed, the largest of which is the wind power project proposed for construction off Cape Cod, Massachusetts. Details of that proposal are discussed in Box 10.2.

Greenhouse Gas Leakage Problem

Greenhouse gas leakage occurs when production of products with high GHG emissions moves to countries not participating in a global climate regime, or such countries simply increase their production while the output of companies in the regulated nations declines (Frankel 2009).

BOX 10.2 SITING ISSUES: THE CAPE WIND PROJECT

The old joke in the real estate industry is that there are three important factors about a piece of property: location, location, and location. A similar reality confronts the energy sector. The process of converting primary energy sources into secondary ones, particularly electricity, requires structures and facilities that are large, obtrusive, and create pollutants or other external costs. There is a reason they are usually located far from population centers. Transmission lines are notoriously difficult to build in part because very few people are enthusiastic about having the structures passing through or even near their property. It is understandable that property owners and residents would exclaim "Not In My Backyard!" in reaction to proposed construction of such facilities. The need to accommodate such NIMBY concerns is an inescapable process in a democratic society.

Wind farms are relatively benign compared to other electricity generating options and some people, including the authors, find them strangely compelling. Obviously, they must be located where there is a steady wind, and most such locations are remote.

One exception is an area just south of Cape Cod, Massachusetts, on Horseshoe Shoal in Nantucket Sound, which on maps of suitability for wind energy stands out as the best location for a wind energy facility in the northeastern United States. Recognizing this, in 2001 an energy firm, Cape Wind Associates, filed plans to develop a 486 MW offshore wind farm with 130 3.6 MW wind turbines. The controversy over this proposal continues, 9 years later and even after Interior Secretary Ken Salazar announced federal approval of the project in April 2010. The project generated opposition for several reasons (Associated Press 2010). It would be visible from the shore of the Cape and portions of Nantucket Island, prized scenic locations, as well as from the compound of the Kennedy clan in Hyannis Port. Local fishers argue that the shoal would be lost as a fishing ground and force them to go much farther out to sea. It is also near shipping lanes. Local Indian tribes objected, stating that the facility would obstruct views and disturb tribal burial grounds. Proponents argued that the United States needs renewable energy, and the project would create badly needed jobs. The offshore location is also a benefit as prevailing winds increase in the afternoon at times of peak load. The nature of much of the opposition—the relatively wealthy landowners on the Cape and Nantucket shores—also drew attention. Most polls showed Massachusetts residents overall in favor of the project, with opinion on the Cape more closely divided.

Although the project has received federal regulatory approval, it faces other hurdles. Developers must sign contracts to sell the power likely at twice the current rate in Massachusetts, or around 20 cents per kW, and then raise the $2 billion required for construction (Wald 2010).

Vulnerable industries include steel, glass, basic chemicals, pulp, and paper. Since less-developed countries often have less-advanced technology and less-robust environmental regulation and compliance, leakage could lead to the unintended consequence of increasing worldwide emissions of GHG.

Leakage is perceived as a threat to the economic competitiveness of developed countries and a potent incentive to cheat on international climate agreements or avoid participating altogether. To limit the effects of leakage, several different types of policies have been proposed. One set covers the production of goods. Tariffs could be raised against nonparticipant countries, or importers of goods from nonparticipating countries could be required to buy allowances to cover the associated emissions.

On the fuel side, a secondary form of leakage could occur if the costs of coal, or possibly, oil, decline as a result of a widely adopted climate agreement. Nonparticipating countries could increase their use of such fuels, making their goods more competitive. Or they could increase production of fuels from "dirty" sources, such as tar sands or heavy oil, that would still be competitive against more highly taxed fuels in participating countries. Again, tariffs or allowances could be assessed to equalize costs to consumers.

The federal government, through Section 526 of the Energy Independence and Security Act of 2007, has already prohibited federal agencies from procuring synthetic fuels or fuels from nonconventional sources, unless an analysis of their life-cycle GHG emissions finds they are less than emissions from conventional oil (Frankel 2009). This could make U.S. government purchases of gasoline refined from Canadian oil sands or Venezuelan heavy oil illegal.

The leakage problem creates a dilemma for the global community by pitting collective action on climate change against the relatively liberal world trade regime. Inclusion in the American Clean Energy and Security Act passed by the House in 2009 of a tariff or "border tax" provision, plus a provision providing allowances to trade-sensitive industries, generated concern about the potential for the bill to start a trade war. From a European perspective, both the United States and developing countries are benefiting from failure to take action on climate, while they have already put in place a system to limit carbon emissions (Frankel 2009). Any climate bill emerging from the Congress soon is likely to contain at least a weak border tax clause, both for domestic political reasons and to serve as a goad to advancing global climate negotiations.

What Are Sensible Policy Criteria?

The portfolio of policies outlined in this chapter represent one set of approaches to meeting the policy objectives we described. Starting with different objectives would lead to a different set of policies. The policies we describe also reflect our judgments about the criteria upon which to base some decisions, such as between cap and trade and a carbon tax. In our view, the relative ease of administration of a carbon tax is a decisive factor. For others, the fact that price rises under a cap and trade system are less transparent makes it a more viable alternative from a political perspective.

There is also a broader issue for which the notion of applying policy criteria is relevant. Congress has markedly increased funding of DOE research and development programs for energy efficiency, renewable and environmentally friendly energy, and energy infrastructure projects by public and private agencies and organizations. At this stage of the energy policy dilemma there is some justification for investing in research on nearly every scheme that holds some promise for weaning us away from our dependence on oil and coal. The policy analysis challenge is how to choose between the many options available. The literature of energy policy is rife with passionate justifications for devoting more research and development money on one or more solutions to all the world's energy problems. These range from arguments promoting biomass conversion to ethanol, wind, wave, sunlight, geothermal, algae, or other renewable energy (RE) resources; using natural gas, "clean coal," or in the far distant future, hydrogen or more nuclear energy for generating electricity and eliminating the current use of "dirty" coal and petroleum products. But moving away from a scattershot approach to policy means consistently applying clear criteria to our energy policy choices.

In practice, this is difficult to do. Once granted, subsidies generate economic rents to producers that are difficult to take away, unless they have a built-in termination date. Thus many oil and gas subsidies have staying power, since they were enacted in the days before congressional PAYGO rules forced a clear estimate of subsidy costs and required new subsidies to have built-in sunset dates. Creating coalitions to pass legislation means the preferences of many members must be accommodated. And we don't know what technologies will succeed, either in the laboratory or the competitive world of consumer choice.

In answer to this challenge, Holden (2006) suggests a series of viability tests that should be applied before any large-scale adoption of new or nontraditional energy sources. The most important of these is that large-scale application of the energy component must be possible; it must pass the test of science and applicable technology. Holden suggests:

> What we must know is what are the uses [for each energy source] and who are the users. We must know the gains and losses *and therefore, the interest likely to be affected, activated, or neutralized politically*, in any set of imagined decisions. From this, we are likely to have some idea of the likelihood of feasibility and viability. (Holden 2006, 879)

Holden's model also suggests two other criteria. The production limitations test for expanding production first involves determining whether anyone has the wherewithal and the will to invest the very large sums needed to bring the energy alternative to market. There are, for example, very few corporations willing to invest the very large sums for construction of a nuclear generating facility.

The legal test restricting production expansion includes defining the rights and obligations of all parties and institutions in the value chain. This includes more than the active participants in the research, development, and operation of the energy system; it also includes the regulatory agencies that oversee the energy product and the many, diverse social, environmental, and political institutions with a stake in the success or failure of the source.

In addition, several broader, social criteria deserve attention in the process of crafting energy policy. One is distributional equity—how would the policy affect the distribution of wealth, income, and opportunities in society? This issue does not receive much consideration in energy debates as discussions of supply, environmental effects, and national security dominate. But given the substantial data suggesting that U.S. society has become more unequal over the past 30 years, any energy policy changes we consider should include a careful analysis of the effects on low-income persons and households. Sovacool (2008b) suggests a national systems benefits charge be applied to electricity bills to create funds that could be used in part to lessen the impact of higher energy prices.

Conclusion: The Recurring Issue of Local Control

In addition to the issues described above, the question of local control must be resolved. At present the system is weighted in favor of utilities and regulators, although local forces (as in the Shoreham nuclear plant case) also have had victories. We see the battle between Big Energy and NIMBY (not in my backyard) proponents as frustrating and fruitless for both parties. What is needed is a planning process that goes beyond public hearings—often after energy facility siting proposals have already been announced—and involves citizens in long-term planning over the mix of generation and its location. Long term, we should aim to lessen the top-down dynamic of energy policy and encourage a more participative process for policy development.

For a period in the late 1970s there was an optimistic belief that alternative energy offered a route to a new society, one free of the grid and corporate control (Glover 2006). When oil prices collapsed and federal support for renewables in the 1980s declined, belief in this model also largely disappeared, although a surprising number of people remain off the grid. The relatively high cost of solar and wind for homeowners and the rise of wind, solar, and other forms of renewable energy have brought them firmly into the mainstream. The rise of climate change as an issue, plus lower long-term costs for renewable energy options is likely to renew the passion around renewables and encourage people to take action.

The last of these criteria is sustainability, in at least two senses. But the importance of this criterion makes it an appropriate theme for our final chapter.

Chapter 11

Aftermath of the Gulf Oil Spill: Prospects for Policy Changes

> The profound changes of recent decades and the pressing challenges of the twenty-first century warrant recognizing energy's central role in America's future and the need for much more ambitions and creative approaches....The staleness of the policy dialogue reflects a failure to recognize the importance of energy to the issues it affects: defense and homeland security, the economy, and the environment. What is needed is a purposeful, strategic energy policy, not a grab bag drawn from interest-group wish lists.
>
> **—Timothy E. Wirth, C. Boyden Gray, and John D. Podesta (2003)**

As we were completing the initial draft of this book in mid-2010, the Deepwater Horizon oil spill in the Gulf of Mexico was still not under control. The spill became the largest ever in U.S. history, causing immense damage to beaches, wildlife, and marshes, despoiling a large portion of the Gulf of Mexico's ecosystem, and wreaking economic havoc on the Gulf states. The long-term damage to the Gulf, its people, and wildlife will be severe.

Who, or what, was responsible for this mess? Lawsuits, congressional hearings, and exposés in the press will gradually clarify the series of events that led to the disaster. They are likely to identify human and systemic factors, including poor decisions driven by aiming to save relatively small amounts of money, and poor regulation, caused in part by a tradition of chummy familiarity between oil companies and their federal regulators. The deeper and less obvious factors contributing to the disaster may, however, not be aired and discussed quite so openly. Was it a failure of our national energy policy? You bet. Can we turn that policy failure around? We had better.

Policy Failure and the Gulf Oil Spill

One such factor is the very American assumption that technology can quickly and relatively painlessly solve problems. That technology has not been up to the task, so far, of stopping the leak from the ruptured oil well, which is over one mile beneath the surface of the Gulf and releasing oil with enormous force. Granted, the problem is very difficult. Yet the risk of such a failure and an honest assessment of the company's capacity to fix it should have influenced decisions about both regulatory standards and well design, but did not (Barstow et al. 2010). Improved technology is widely seen as a potential cure-all for the energy dilemmas faced by this country, but the Gulf spill suggests that we need to temper our optimism that it can solve them easily.

A second and more basic factor, in our view, is the fundamental American belief that we have the right to use as much energy as we want, limited only by what we can afford. An expression of this trait is our insatiable desire for mobility and by extension the oil that makes it possible. In his book *Dave Barry Hits below the Beltway*, humorist Dave Barry offers a revised and painfully funny version of the U.S. Constitution that includes Amendment XIII (Barry 2001, 42):

> If any citizen wishes to drive a "sport utility" vehicle with the same weight, fuel economy, and handling characteristics as the Lincoln Memorial, then nobody shall have the right to stop that citizen, because this is America, dammit!

Mobility, via large and comfortable vehicles, is perceived as an American birthright, and we have organized our lives around it. Sunday drives, drive-in theaters, burger joints, drive-up tellers and liquor stores, suburbs, commuting, driving vacations—all rely on the availability of cheap oil. There is a lot of oil left on the planet, but it has become more costly to find and develop. The easy oil is mostly exploited; what is left is either under the control of unstable and often unfriendly countries, or deep under the sea.

One of the largest oil finds in decades is the Tupi field in the Atlantic Ocean about 160 miles off the coast of Brazil, east of Rio de Janeiro. This estimated 8 billion barrels of oil lie roughly 20,000 feet under the surface of the sea, with about 7,000 to 10,000 feet of ocean, 9,000 or more feet of rock and sand, and then another 7,000 feet of salt in between (Barrionuevo 2007).

Developing this field is an element of Brazilian oil giant Petrobras's five-year investment plan; the estimated cost of the project is $220 billion (Viscidi 2010). If this is the best of the oil that remains, we need to prepare now for a future with scarce and expensive oil. And a large portion of that cost will come in the form of degradation to the environment.

The Deepwater Horizon spill has torn the "green," environmentally friendly mask off the face of BP and helped to clarify the state of the oil business. In the United States, the greed of the oil industry is harnessed to continue to find and develop new sources of petroleum. It's a lucrative endeavor; oil companies such as Exxon-Mobil and others continue to be some of the most profitable companies in the world. Outside of the United States, the oil industry is dominated by national oil companies; these state-owned oil companies are motivated by a more complex mix of profit seeking and nationalism. Whether publicly or privately owned, the big players in the oil industry are able to exert their financial and political power to weaken regulations and retain subsidies.

Problems Regulating the Energy Industry

Regulating an industry with such concentrated political, economic, and social power is very difficult. But that is no justification for inept performance. The Gulf spill has exposed serious failures of U.S. federal regulation of the offshore oil industry. The Minerals Management Service (MMS) failed to demand adequate safety precautions on the well, and accepted shockingly amateurish safety plans from BP. And these are not the first serious problems found at the agency. Even more glaring is the multi-billion-dollar "mistake"—covered up for 6 years—when increased royalty rates to be paid to the federal government were omitted from new drilling contracts in 1998 and 1999 (Ivanovich 2006).

There are now 3,509 active oil platforms in the U.S. portions of the Gulf of Mexico, including 25 already drilling at a depth of more than a thousand meters. And BP is preparing to drill in the Arctic. How confident should we be in the design and safety of these projects?

Culpability for this disaster is shifting from BP to the MMS, the Department of Interior, to the federal government and President Obama. The spill has already shaken public confidence in government regulation, and is beginning to again undermine the very idea of government action, as occurred when Federal Emergency Management Agency (FEMA) failed after Hurricane Katrina (Rich 2010). This is a matter of grave concern for all of us.

Limited Policy Options

When a government action succeeds—think Apollo Program, and Neil Armstrong walking on the moon—all Americans share in that success. When it fails, spectacularly, we try to distance ourselves from that failure. We point an accusing finger at distant institutions, deny our own culpability, and become more cynical about government action. This is a very human reaction, but it's shortsighted in the energy arena. Our options for organizing this sector are limited.

Yes, conceivably we could nationalize BP's U.S. operations, but the entire petroleum sector? Even if we could financially, there is little evidence that national oil companies perform better overall than private ones, although in some cases (especially Norway's Statoil) they have consistently been more safety conscious. And the private sector majors (Exxon Mobil, Chevron, and ConocoPhillips in the United States) are an endangered species, with the national oil companies owned by Saudi Arabia, Venezuela, China, and other countries now controlling the lion's share of both global oil reserves and output (Viscidi 2010).

Another policy option is for the government to back away and, instead, hand even more control over to the energy industry through complete self-regulation. But that's not likely in the aftermath of the worst oil spill in U.S. history.

Instead, the federal government must improve its regulatory performance and raise public confidence that it knows what it is doing. That means changing the culture at the MMS, in part by hiring people who know the energy industry and won't hesitate to hold it, and themselves, accountable. It must also pass and implement policies to help the American people prepare for a future with more costly oil, increasingly limited use of other fossil fuels, and greater reliance on renewable and more sustainable energy sources.

We have concentrated on the oil segment of the energy sector in this last chapter mostly because the all-consuming disaster of the Deepwater Horizon oil spill screams for an energy policy that includes more active and more effective federal regulation of that powerful industry. But the problem demands a comprehensive, systemic response that goes beyond oil. In the Introduction we suggested a short collection of components that are necessary for producing a viable energy policy for the nation. Key elements in the policy planning hopper must continue to include expanding spending on research and development of alternative energy sources, encouraging greater conservation in the use of existing energy resources, improving the efficiency of energy distribution and consumption, and changing the way Americans think about and use limited energy resources, both to ensure existing supplies tide us over until viable alternative energy sources are available, and to cease polluting the global environment.

An equally important issue that our energy policy must address now is the undeniable effects that the continued use of coal and natural gas for electricity generation and oil to fuel our transportation system are having on the environment. What must be of particular concern is the contribution to global warming caused by greenhouse gas emissions resulting from the use of these fossil fuels to power U.S. households, commerce, and industry.

A comprehensive approach is needed because action on climate change demands it, and because moving away from petroleum as a transport fuel will test our energy system. Although hydrogen fuel cells are still a possible option, the most likely replacement for internal combustion engine and hybrid cars are battery electric vehicles (BEVs). But our electric utility sector may not be ready for rapid adoption of BEVs. We need to ensure that we have sufficient (and clean) generating capacity, plus an electric grid capable of handling the additional, more intermittent flows of power.

But our current grid is based on old technology. Relatively little data flows back and forth from houses and businesses to utilities to help them manage power demand and supply. The key usage data are read on your 1920's-era electric meter (Achenbach 2010). The vulnerable state of the national grid was exposed by the August 14, 2003, blackout that turned out the lights on 50 million people in eight U.S. states and the Canadian province of Ontario. We need to continue to invest in this infrastructure, by strengthening the transmission connections among the four major electricity regions in North America (Western, Texas, Eastern, and Quebec Interconnections), better linking new renewable sources to the grid, and accelerating the implementation of smart meters. Several pieces of energy legislation over the last five years have provided funds to accelerate this process, but it must continue.

Meanwhile, according to Energy Information Administration (EIA) data, 15 states plus the District of Columbia have fully deregulated their electricity sectors, and allow retail choice—consumers may choose their provider. Seven have suspended deregulation (in part to avoid another California-style deregulation debacle) and the remaining 23 retain some form of state-regulated, monopoly service provision. How this hybrid, partially deregulated system would react to the complexity and cost of climate change legislation—particularly a cap and trade system—in addition to the profusion of state-level renewable portfolio standards and other policies, is unclear. Will there be pressure to further concentrate the industry into fewer firms with a multistate reach? Following the financial crisis, will utilities and energy developers have access to sufficient capital to bring online the needed grid enhancements? And the renewables called for by renewable portfolio standard (RPS) requirements? And are our regulatory systems adequate to manage a more complex grid? The organizational structure of the industry going forward is a considerable unknown. As with the oil industry, we rely mostly on the private sector to provide the power we need, although with a stronger role for publicly owned entities. To bring about a cleaner energy future will require that we have regulatory systems and institutional models up to the task.

Transitioning to a New Vision

The mix of policies we outlined in Chapter 10 would gradually lead the United States toward a new energy future. In this vision, increased energy efficiency, and changes in land use, transportation, heating and cooling, and transitioning to carbon-neutral energy sources would lower our energy use and reduce our carbon emissions. It is also likely that we would choose to build smaller houses and cars, drive less, and fly less. This would ease the impact of energy use on the environment while providing much higher levels of energy services to people around the world who now lack them. It is probable that this will not happen fast enough to prevent global greenhouse gas (GHG) levels from continuing to rise, so we will also have to plan and budget for measures that will help us adapt to climate change.

But we believe such a future will still allow us to enjoy significant mobility and levels of energy services. One way forward has been suggested by Swiss researcher Eberhard Jochem, whose concept of the "2,000 Watt Society" envisions a world society with high levels of efficiently delivered energy services (Jochem et al. 2002). It is unlikely that the United States will ever be a 2000-watt society; in the United States we now consume about 12,000 watts per person per year. In Europe, per capita energy consumption remains at about 6,000 watts per year, while the average in Africa is close to 500 watts (Cascio 2005). But working toward the goal is a worthy task (Smil 2009).

Is this vision of a new energy future supported by a broad segment of the American public—a group sufficiently large, and politically strong and intense enough to be a political force? As we noted in Chapter 10, recent poll results are contradictory. Over 50 percent of Americans continue to be very or somewhat worried about climate change, a proportion that has changed little since 2008. Beyond the substantial core group of Americans focused on climate change (39 percent in a recent survey were characterized as "alarmed" or "concerned" about it), many voters are inclined to support change as long as it doesn't cost anything, particularly in the way of higher gasoline taxes or electricity bills (Leiserowitz et al. 2010).

Two elements of the new vision for energy in the United States are transitioning the transportation sector away from fossil fuels and changing the way that Americans use energy. Weaning the transportation sector from its addiction to oil and gasoline has been an idea that's been kicked around since the oil crises of the 1970s; there has been lots of talk but little lasting action. That may be changing, finally. The short-term solution has been to make cars and trucks more energy efficient. Increasing corporate average fuel economy (café) standards to 35 mpg or higher was a move in the positive direction, but the only way a permanent solution can be found is to come up with a clean, affordable, efficient, renewable new fuel. Corn-based ethanol has been tried, but it takes more energy to produce than it replaces and takes land from food production. Others solutions being experimented with include biodiesel produced from nonfood sources such as algae, and plug-in electric vehicles. Governments in the United States, Germany, and Japan are also investing in hybrid vehicles employing hydrogen-based fuel cell electric power. Three "seed regions" are in the works in the United States: Southern California, New York, and Washington DC. General Motors, Honda, Hyundai, Kia, Mercedes-Benz, Nissan, Toyota, and Volkswagen are all participants in the California Fuel Cell Partnership. The German, Korean, and Japanese firms are testing demonstration fleets in their home countries. GM has estimated that an investment of $100 to $200 million in infrastructure development would support nearly 15 million fuel cell electric vehicles (FCEVs) in Southern California (Simanaitis 2010).

The second major policy direction shift lies in the way policy makers view the energy industry. The United States and Europe are transitioning away from depending upon large, concentrated organizations managed by energy suppliers and instead focusing on a system of small, dispersed operations managed by users. Examples include the move to co-generation and small, locally sited natural gas–powered electricity generators. The attention being placed on improving energy efficiency and conversion to energy self-sufficient homes and commercial structures is another indicator of the transformation taking place in energy policy (Ōyama 1998).

Citizen Reaction

And how might Americans react to this new vision, and new reality? The 1970's oil shock was a blow to the national psyche. But we got over it. Much higher gasoline prices in 2008 were met with shock, alarm—and adaptation. Energy price events tend to be abrupt. There is a significant, but hard to quantify, probability that oil and gasoline prices will jump considerably in the next 10 to 20 years. Should we wait for it to happen, letting the companies controlling oil production gain the windfall, or begin now to steer our society away from its need for oil—a need so great that even President George W. Bush characterized it as "an addiction"? Similarly, should we wait for stronger and stronger evidence of climate change, or act now to forestall the clear risks it presents?

Making Policy Change Happen

What might move our stuck policy system to a different equilibrium that would allow both stronger regulation and coherent energy policies that would help us move toward such a vision? There are several possible changes that could lead to a more focused policy outcome—some positive for the country's long-term prospects, and some of more concern.

One might be increasing public recognition of the effects of climate change. A 2009 report provided powerful documentation about the impacts of climate change on several fundamental systems in the United States, including water, energy, transportation, agriculture, ecosystems, and health (Karl, Melillo, and Peterson 2009). It forecast worsening conditions across much of the country in the absence of concerted global effort. But so far not enough voters have noticed or been convinced. Many Americans have simply not seen enough evidence, and been moved by concern about climate change to take or support action on a new energy policy.

There is some evidence that they are willing to take action as consumers, but not in their role as citizens (*ScienceDaily* 2010). People are willing to shut off their electronics and turn off the lights, but relatively few are engaged in political action to mitigate climate change and environmental degradation. Whether the Gulf oil spill will lead people to become angry and get involved, as the 1969 Santa Barbara oil spill did, will be an important test.

The current Gulf oil spill is likely to make energy a more visible campaign issue. Yet in a recent nationwide poll, energy continues to lag far behind the economy, immigration, financial reform, and education as a concern for voters (*New York Times*/CBS News 2010). The same poll found respondents continuing to prioritize development of new energy sources over protecting the environment (49 percent to 29 percent). While an impressive 58 percent supported fundamental change in U.S. energy policy, once again a majority opposed increasing gasoline taxes to pay for needed changes.

The lack of engagement in energy issues should not be surprising, given the top-down nature of energy policy making and the background nature of energy phenomena. Most of us have neither the time nor inclination to get involved in energy issues as citizens, unless an energy facility is proposed near our home, or the utility wants an outrageous increase in rates. In most cases, opportunities for public input are pro forma. Perhaps the best goad to public involvement in energy would be implementation of smart meters and real-time pricing of electrical energy, forcing individuals and families to be confronted more directly with the results of their energy choices.

Taking Action

But not everyone is disengaged; there are many examples of people, nonprofits, and programs that are seeking to create systemic change in energy systems from the ground up. One example is the New Energy Cities program in the Northwest. This program combines community building, programs for improving energy efficiency in buildings and transportation, new infrastructure (such as distributed generation and smart grids), and innovative financing methods. This long-term effort is one of several similar programs nationwide.

A more powerful political force could be created by appealing to Americans' concerns about the safety, reliability, and cost of imported energy. A 2010 poll by pollster Frank Luntz found significant support among voters for policies that helped reduce oil imports, improve national security, and hold polluters accountable (Eilperin 2010). Luntz is well aware that fear can be a primary motivator of policy change, and the global situation is tense and volatile. Policies that reduce such tensions while improving our own economic situation could find favor across the political spectrum.

A wildcard in the long-term energy policy system is the financial status of the federal government. With large deficits continuing into the foreseeable future, an escalating federal debt, and increasing voter concern about the issue, the sustainability of federal energy programs is not guaranteed. As discussed in Chapter 10, some policy scenarios include continued high levels of funding for research and development, energy efficiency programs, and support (i.e., subsidies) for both renewable and nuclear energy. But coming up with the money to pay for these programs is still problematic.

In a speech on June 2, 2010, in Pittsburgh, Pennsylvania, President Obama committed both to leading the U.S. transition to clean energy and to substantial federal belt-tightening through a three-year freeze on discretionary spending—not including funding for national security—beginning in 2011 (Obama 2010b). The President's Fiscal Commission also is in the process of developing a set of proposals to ease the country's long-term fiscal crisis that will inevitably include a mix of budget cuts and revenue increases.

Fortunately for the U.S. Department of Energy (DOE), a freeze would fix energy spending at relatively high levels. But what would happen to such spending should President Obama's successor be a conservative politician more sympathetic to the oil and gas industry and skeptical about renewables? If we are serious about implementing a long-term strategy to reduce our dependence on oil and reduce carbon emissions, an equally robust funding strategy will be needed. So any sensible set of proposals would include generating such revenues through a cap and trade system or carbon tax.

Conclusion: Get Started, Get Involved, Be Heard

Solving our energy dilemmas and creating a more sustainable society and public sector by 2030 will require a combination of sustained individual effort and collective action through public policy. The steps individuals and families can take to reduce their energy use are well known; start with the simple and progress to the more ambitious—use compact fluorescent lights, turn off lights and electronics when not in use, take the bus, buy a fuel-efficient car and a smaller house. But for public policy to make a difference, it will be essential for people to get involved, raise their voices, and move the political system.

APPENDICES 3

Appendix A: Chapter Discussion and Review Questions

Introduction

1. Why is energy policy perhaps the most Janus-faced of all the policy arenas in American politics?
2. Why will climate change caused by humans have disastrous and irreversible effects on Earth's biosphere? Explain your answer.
3. What are the environmental conditions framing energy policy making described by Professor Michael McElroy?
4. What percentage of the world's oil supply does the United States use?
5. How do governments intervene in energy supply decisions?
6. Define energy in terms that make sense to you in light of your daily life.
7. Why does driving a large SUV influence the price you pay for the energy you use?
8. Describe the similarities in how useful energy is produced and distributed.
9. Who are the stakeholders in the energy policy-making process?
10. Review the list of questions that energy policy makers might want to ask. What questions would you add to this list?
11. What actions should citizens take to ensure society has enough of the right kinds of energy at prices everyone can afford?
12. What actions should citizens take to halt the greenhouse gas emissions that are blamed for global warming?
13. Has the United States always depended upon foreign sources of energy?
14. What will happen when the world runs out of oil?
15. What energy resources are likely to serve as a replacement for oil to power transportation?

Chapter 1: The Political Realities of Energy Policy

1. Why is energy policy such a complex and paradoxical issue?
2. Why is it that we make few decisions directly to buy or consume energy?

3. What is your answer to the question, "Where does electricity come from?"
4. Describe the path that coal makes as it moves from where it is mined to where it is used.
5. In what states are most U.S. coal deposits located?
6. Why does the United States rely so heavily on coal as an energy source?
7. Should the U.S. government and/or state governments own, operate, and maintain the country's energy infrastructure? Why or why not?
8. What percentage of the energy in the system is wasted overall as a result of inefficiencies in both generation and power transmission? How can this inefficiency be improved?
9. What percentage of greenhouse gas emissions comes from energy-related activities?
10. What is the meaning of the phrase, "The Earth's atmosphere is a common-pool good"?
11. What is the function of energy policy?
12. Why does the United States use so much energy?
13. Describe the U.S. energy system.
14. Discuss recent patterns in residential energy use.
15. What do we mean when we say that energy use is out of balance?

Chapter 2: Energy Policy in Transition

1. Describe what connection there might be between the dying BC forests and energy use.
2. Who was Joseph Fourier? What role did he play in global warming research?
3. Who was Svante Arrhenius? What role did he play in global warming research?
4. How much warmer is the Earth projected to be by the end of the decade under the existing rate of fossil fuel use?
5. How sure can we be about the forecasts of the results of our use of fossil fuels? Why is it important to continue to make forecasts of future energy needs?
6. What is *integrated resource planning* (IRP)? What role does it play in energy forecasting?
7. Describe why confusion exists over when the world will reach the peak in known reserves of recoverable oil.
8. Why did oil producing countries such as Iran, Kuwait, Saudi Arabia, United Arab Emirates, and Venezuela raise the estimates of their reserves and production capacity?
9. What is likely to happen when peak oil does occur and production begins to decline?
10. Why have emissions of carbon to the atmosphere caused by fossil fuel combustion increased so much since the 1990s?
11. What roles are China and India expected to have in future energy use and carbon emissions?
12. Describe the energy-related challenges facing U.S. policy makers and their likely responses.
13. Describe the "stabilization wedges" concept introduced by Pacala and Socolow in 2004.
14. What effect is the 2006 release of *The Stern Review on the Economics of Climate Change* having on energy policy making?
15. Discuss the problems facing the efforts of policy makers to set carbon pricing policies that are equitable, just, and effective.

Chapter 3: The Art and Science of Crafting Energy Policy

1. Name the individual stages in the stages model of public policy making and explain why it has some shortcomings.
2. Describe the classical liberalism tradition of public policy making.
3. Describe Adam Smith's view of economic behavior.
4. What did Carl Jung mean when he suggested that people have a shadow?
5. Describe the new conception of the state that emerged after World War II.
6. What role, if any, do the federal, state, and local governments play in setting and carrying out our energy policy?
7. What role does the U.S. Congress play in setting public policy? Is Congress's influence growing or declining? Why?
8. Describe Robert Dahl's pluralistic or group model of political decision making.
9. How does the Iron Triangle model of policy making differ from the policy network model?
10. Describe the role of social construction theory in public policy design.
11. How did the U.S. political system perform on energy issues in 2010?
12. Describe the different ways conservatives and liberals frame the energy policy issue.
13. What would our approach be to designing a new way of thinking about energy policy?
14. Describe the shifts taking place in the U.S. economy and how those shifts will impact energy policy.
15. Will the U.S. economy follow the stagnation pattern exhibited by the Japanese economy during the 1990s and into the new century? Why or why not?

Chapter 4: The Long Search for a Sustainable Energy Policy

1. Name the three comprehensive energy policy acts enacted since 1992 and explain why only three such laws were passed during the period.
2. What does America want from an energy policy?
3. What were the country's biggest energy concerns during the surplus years from 1945 to 1970?
4. What were the country's biggest energy concerns during the unexpected and unprepared for energy shortages decade of the 1970s?
5. What were the country's biggest energy concerns during the years of deregulating the energy industry?
6. Why did the federal government finally begin to think about adopting a comprehensive energy policy during the 1990s?
7. How did the link between our needs for energy and security of supply come to be so relevant during the early years of the twenty-first century?
8. Describe the events that led up to passage of the country's first really comprehensive energy policy bill during the administration of President George W. Bush.
9. What are the main focal points of the Obama/Biden energy program? How did the explosion of the deepwater drilling platform and subsequent oil spill disaster change that focus?
10. What federal agency is the largest consumer of fossil fuel energy? What are they doing to reduce that amount?
11. What are CAFE standards, and how do they impact fuel consumption?

12. What does the phrase "energy independence" mean? Is it a state that the country can arrive at easily?
13. Where does the United States get most of its imported oil? Are those exporters friendly allies with stable economies?
14. Why are American and European energy policy makers concerned over the energy requirements of China and India?
15. Why is it no longer possible to plan energy policy without including environmental policy?

Chapter 5: Difficulties in Achieving a Balanced Energy Policy

1. Name and define the six factors that make setting an energy policy so difficult.
2. The chapter grouped energy stakeholders into four different groups. Name and define each of these groups.
3. Which of the example questions that stakeholders might ask do you feel have the greatest impact on energy policy making? Why?
4. Why do some stakeholders focus on energy supply issues, while others focus on demand issues?
5. Why do some stakeholders focus on energy security issues, while others focus on energy and environmental issues?
6. Discuss the impact of special interest group lobbying and financial donations upon energy policy making.
7. Why is it so difficult to show direct links between campaign contributions and voting patterns?
8. What are the goals of U.S. energy policy?
9. What energy goals did President Barak Obama promote during his presidential campaign?
10. Why are energy policy interventions an imperfect means to accomplish important energy goals?
11. What did Lester Salamon mean when he described the tools used for government interventions?
12. How much did the federal government spend on energy programs from 2004 to 2009? On what types of programs were the funds spent?
13. How was government money used to promote energy efficiency?
14. How were federal funds used to encourage research and development on renewable energy resources?
15. Why is it important for us all that the federal government rethink and modify its commitment to energy subsidies?

Chapter 6: What's on the Current Energy Policy Agenda?

1. What were the six categories of programs in the Obama/Biden energy platform?
2. Describe how the fifteen Pacala and Socolow wedges are on the energy policy agenda.
3. Why is a climate bill on the energy agenda?
4. Why is lack of a clear carbon price a major impediment to energy sector investments?
5. Briefly define each of the eight additional energy issues that could be included in a discussion on energy policy.

6. What role is nuclear energy likely to play in the crafting of a new comprehensive energy bill?
7. Are we right to be concerned over the rapidly declining stores of oil? What role should this play in crafting our new energy policy?
8. In 2003, researchers Wirth, Gray, and Podesta identified two policy issues that must be addressed immediately. What were they, and what steps have been taken to resolve the issues since they were proposed?
9. What percentage of U.S. electricity is generated with renewable energy resources? How likely is this to be increased greatly in the near future?
10. The chapter noted that in drafting our new energy policy, policy makers must carefully consider the impact of each policy proposal on market structures, incentives, private investment, and costs to consumers. Explain why this must be done.
11. Why have many states so eagerly implemented renewable portfolio standards (RPS)?
12. What are the components that together make up the electricity Smart Grid?
13. What changes if any should be made in allowing private firms to have access to public lands for developing energy resources? What government agencies now monitor and control these public lands?
14. Discuss the issues related to allowing drilling for oil and natural gas on the outer continental shelf.
15. What are the pros and cons of opening of the Alaska National Wildlife Refuge to drilling for oil and natural gas?
16. Is carbon capture and storage viable? What is carbon sequestering?
17. What are biofuels? How can they help us achieve energy self-sufficiency?
18. Why is natural gas back on the energy policy agenda?

Chapter 7: Crafting Policy with Subsidies and Regulations

1. How does government intervene in the economics of the energy sector?
2. A number of questions were asked about these interventions collectively. Review those questions and make your own conclusions.
3. Why does government make these energy-related interventions?
4. Discuss the successes and failures in government interventions relating to energy efficiency.
5. What is the principal-agent problem, and how does it influence energy policy?
6. What do we mean when we say that energy consumption in the United States and throughout the world is highly inequitable?
7. Describe the different types of energy subsidies. What impact do they have on energy supply?
8. What is the role of the strategic petroleum reserve?
9. How do subsidies help or hinder nuclear power producers?
10. Why are loan guarantees for energy projects risky business for government?
11. What are tax expenditure interventions? How are they used to promote energy development?
12. Describe the practice of expensing exploration and development costs and intangible drilling costs.
13. Describe the major Obama energy-development theme discussed in the text.
14. Discuss the practice of federal regulation in the energy sector. What are its limitations?

15. What is the Energy Star program and why has it come under some criticism?
16. Describe the Low Income Energy Assistance Program (LIHEAP).

Chapter 8: Policies Shaped by Taxes and Market Mechanisms

1. Discuss the pros and cons of the two optional policy strategies: constraining the quantity of emissions through a cap, or raising the price of emissions via a tax.
2. What is the Oil Spill Liability Trust Fund, when was it created, and how does it work to help clean up after oil spills?
3. What is the Oil Pollution Act, when was it enacted, and what does it hope to accomplish?
4. Why are gasoline taxes different from state to state?
5. Why is the federal gas tax considered by some people to be a significant policy failure?
6. What options are available for improving or replacing the federal gasoline tax?
7. How effective are direct user charges (tolls) at decreasing road use?
8. What is the difference between a fund pollutant and a stock pollutant? Which is CO_2? Does it make any difference how it is treated?
9. Discuss the relative merits of using a carbon tax versus a cap and trade design.
10. Is the renewable portfolio standard (RPS) system a regulatory or a market intervention approach? Why?
11. How does wind energy function as a component of an RPS system?
12. What are some of the reasons why the states differ in the way they elect to fulfill their RPS portfolio requirements?
13. Has the requirement for more renewable energy under the RPS system raised the cost of electricity to homeowners?
14. Discuss the following statement: The development of renewable energy sources faces a huge number of obstacles, over and above their significant capital and installation costs.
15. What is the FIT system, and how is it used in Germany?

Chapter 9: International Cooperation on Energy Policy

1. Why are some nation-states unwilling to give their complete cooperation and collaboration to global efforts to solve energy-related climate problems?
2. Briefly describe the basics of the global energy challenge.
3. Explain why global institutions must work together to solve the energy-related environmental problem.
4. Explain how the regime concept may serve as a framework for energy and environment cooperation and decision making. What two such regimes exist today in the global energy arena?
5. In what ways do international institutions impact U.S. energy policy?
6. With which four international organizations does the United States often cooperate on energy policy issues?
7. How does the United States work with the North American Energy Working Group to resolved energy policy issues of concern to all its participants?
8. With which additional international organizations does the United States often cooperate on energy policy issues?

9. What are the six most critical greenhouse gas (GHG) emissions resulting from the burning of fossil fuels?
10. How is the United States working with international organizations to control these emissions?
11. Describe the role of the UN Framework Convention on Climate Change in gaining international cooperation on reducing GHG emissions.
12. Why was the United States reluctant to sign the Kyoto Protocol?
13. How does the voluntary carbon offset system differ from the compliance form? Why are there two versions of this program?
14. What is the Reducing Emissions from Deforestation and Forest Degradation (REDD) program? How does this program work as a way to control carbon emissions?
15. Describe the challenges facing Congress as it seeks to come up with a workable carbon offset program.
16. Why was the Kyoto Protocol unsuccessful?
17. What was the result of the 2009 Copenhagen Accord?
18. Comment on the following assertion: A fundamental challenge for climate change policy is that many leaders in the global south view global institutions with considerable suspicion.

Chapter 10: Policies for a New Energy Future

1. What three problems make it important for the United States to adopt a comprehensive energy policy?
2. The chapter introduced four objectives that should be included in any long-term energy policy. Name and comment on these objectives. Do you agree or disagree? Would you add or delete any?
3. What are some of the policies that are needed to achieve those objectives?
4. What has President Obama said about these objectives and the policies needed to reach the objectives?
5. Discuss the energy options mentioned in the chapter discussion on a policy portfolio.
6. What are the advantages and disadvantages of setting a tax on fossil fuel carbon emissions?
7. What are the political challenges facing energy policy makers?
8. How does the inertia among leaders in energy sector organizations hinder implementation of appropriate energy policy provisions?
9. Discuss the policy challenge associated with declining supplies of petroleum.
10. Discuss the following statement: "Siting issues with wind turbines and concentrated solar power can be politically explosive."
11. Why is increasing energy efficiency a promising short-term solution to the peak oil problem?
12. What decisions have to be made and actions taken before large-scale expansion of nuclear energy to generate electricity is a viable energy policy option in the United States?
13. What do we need to do to make carbon capture and sequestration a more attractive energy policy option?
14. What is the "leakage problem," and how does it affect any energy policy requiring global cooperation and collaboration?

15. Why are granting developers of fossil fuel energy resources greater access to public lands and permission for deepwater drilling in areas once barred from drilling for oil considered to be controversial energy policy options?
16. How does the local control issue affect implementation of many if not most possible energy policy options?

Chapter 11: Aftermath of the Gulf Oil Spill: Prospects for Policy Changes

1. What are some of the factors that have contributed to energy policy failures?
2. Do you agree or disagree with the quoted statement of Dave Barry? State your reasons why.
3. Do you support Brazil's decision to continue to open the Tupi field in the Atlantic Ocean near Rio de Janeiro?
4. What are some of the problems the U.S. government faces when trying to regulate the energy industry?
5. Knowing our need for oil and gas, do you feel the government should increase or decrease its regulation of the petroleum industry?
6. What stipulations and rules should be in place for oil companies to continue to drill in the deep waters of the Gulf of Mexico?
7. What is your view of a new energy future?
8. Is it realistic to think that the United States can go from a 12,000-watt society to a 2,000-watt society and keep its high standard of living for its citizens?
9. What can Americans do to reduce their energy use?
10. What are some examples of actions some citizens are taking now to reduce energy use and cut GHG emissions?
11. Describe the New Energy Cities program.
12. If you were asked to establish an energy policy for the United States, what would it include?

Appendix B: Timeline of Energy Policy Developments, 1950–2010

1950–1970	
December 20, 1951	First electric power generated from a U.S. nuclear reactor in Arco, Idaho.
June 26, 1954	First Russian nuclear power plant (APS-1) connected to the Russian power grid. The 5-MW plant was the world's first nuclear power plant to generate electricity for commercial use.
August 30, 1954	President Eisenhower signs the Atomic Energy Act of 1954, beginning the development of a civilian nuclear power program.
December 23, 1957	The Shippingport (Pennsylvania) Atomic Power Station, the world's first full-scale nuclear power plant, begins operations.
November 9, 1965	The nation's first major electric power blackout hits the Northeast.
September 23, 1970	Northeast heat wave causes a series of electric power brownouts.
1971–1980	
June 29, 1973	President Nixon establishes the Energy Policy Office to coordinate all energy policies at the presidential level.
October 6, 1973	Yom Kippur War breaks out in the Middle East; Organization of Oil Exporting Companies (OPEC) declares oil embargo, forcing first "energy crisis," on October 17.
November 7, 1973	President Nixon kicks off Project Independence to achieve energy independence by 1980.
December 4, 1973	The Federal Energy Office replaces the Energy Policy Office and is assigned responsibility for allocating limited petroleum supplies to refiners and consumers and controlling the price of oil and gasoline.

May 7, 1974	President Nixon signs the Federal Administration Act of 1974, creating the Federal Energy Administration to replace the Federal Energy Office.
October 11, 1974	President Ford signs the Energy Reorganization Act of 1974; the Atomic Energy Commission is abolished and the Energy Research and Development Administration, the Nuclear Regulatory Commission, and the Energy Resources Council are established.
December 22, 1975	President Ford signs the Energy Policy and Conservation Act, extending oil price controls, setting automobile fuel economy standards, and abolishing the national strategic oil reserve.
February 2, 1977	President Carter signs the Emergency Natural Gas Act of 1977.
April 18, 1977	President Carter announces first National Energy Plan in his first energy policy speech and calls for establishment of a cabinet-level energy department.
August 4, 1977	President Carter signs the Department of Energy Organization Act of 1977. The Federal Energy Administration and Energy Research and Development Administration are abolished.
October 1, 1977	The Department of Energy is established.
November 9, 1978	President Carter signs the National Energy Act, which includes the National Energy Act, the Power Plant and Industrial Fuel Use Act, the Public Utilities Regulatory Policy and Tax Act, and the Natural Gas Policy Act. The Energy Tax Act defines gasohol as a blend of gasoline and at least 10 percent alcohol by volume and establishes a subsidy of $0.40 for each gallon of ethanol blended into gasoline.
March 28, 1979	Accident at the Three Mile Island nuclear power plant.
June 20, 1979	President Carter announces a plan to increase use of solar energy and fund research.
July 10, 1979	President Carter proclaims a national energy supply shortage.
July 15, 1979	President Carter proposes $88 billion effort to increase production of synthetic fuels from coal and shale oil reserves.
June 30, 1980	President Carter signs the Energy Security Act, which includes six major components: the U.S. Synthetic Fuels Corporation Act, the Biomass Energy and Alcohol Fuels Act, the Renewable Energy Resources Act, the Energy and Energy Conservation Act, the Solar Energy and Energy Conservation Bank Act, and the Ocean Thermal Energy Conversion Act. The Energy Security Act provides loans to small ethanol producers and sets the first tariff on imported ethanol.

1981–1990	
January 28, 1981	President Reagan signs Executive Order 12287, providing for decontrol of refined petroleum products.
February 25, 1981	Major reorganization of the Department of Energy (DOE) to improvement and greater emphasis on research, development, and production.
May 24, 1982	President Reagan announces plans to transfer most of the responsibilities of the DOE to the Department of Commerce; Congress does not act on the proposal, however.
January 7, 1983	President Reagan signs the Nuclear Waste Policy Act of 1982, the country's first comprehensive waste legislation. The surface Transportation Assistance Act increases the ethanol subsidy to $0.50 per gallon. It will be increased again to $0.60 per gallon in 1984.
October 25, 1984	National Coal Council is established to advise government and industry on ways to cooperate on coal research, production, transportation, marketing, and use.
April 26, 1986	Two explosions in the Unit 4 reactor of the Chernobyl nuclear plant expose millions of people to radioactive isotopes; now linked to many forms of cancer in Eastern Europe and Russia.
February 18, 1987	DOE report, *America's Coal Commitment*, identifies 37 projects planned for clean coal demonstration facilities.
March 17, 1987	President Reagan's Energy Security Report addresses the continuing growth in the U.S. dependence on foreign oil.
July 29, 1989	President Bush orders the DOE to develop a comprehensive energy policy.
November 9, 1989	The Office of Environmental Restoration and Waste Management is formed at the DOE.
August 1, 1990	Iraq invades Kuwait. Omnibus Budget Reconciliation Act cuts the ethanol subsidy to $0.54 per gallon added to gasoline.
August 15, 1990	DOE director announces plans to increase oil production and decrease consumption to make up for Iraqi-Kuwait war supply losses.
1991–2000	
February 20, 1991	President G. H. Bush presents his administration's new National Energy bill. The bill is presented to Congress on March 1991.
October 24, 1992	President Bush signs the Energy Policy Act of 1992, the nation's first comprehensive energy legislation.

April 21, 1993	President Clinton announces that the United States will stabilize greenhouse gas emission levels by the year 2000.
October 19, 1993	President Clinton and Vice President Gore announce the Climate Change Action Plan, which includes voluntary measures to achieve stabilization.
April 27, 1999	The DOE and Department of the Interior launch the Green Energy Initiative designed to increase use of sustainable energy technology in national parks.
June 3, 1999	President Clinton issues Executive Order 13123, setting new goals for federal energy management. The DOE's Federal Energy Management Program is named program coordinator.
June 21, 1999	The DOE announces the Wind Powering America initiative, designed to significantly increase use of wind power over the next ten years.
September 18, 1999	As part of the Wind Powering America initiative, the world's largest wind power facility is dedicated in Storm Lake, Iowa.
December 10, 1999	The DOE designates the Federal Energy Technology Laboratory as its newest energy research facility; it is co-managed in Morgantown, WV and Pittsburgh, PA, and is the nation's largest fossil energy resource organization.
June–July 2000	The Energy Information Administration revises estimates of oil and other energy prices following a short-lived drop resulting from increased OPEC production.
September 30, 2000	The DOE releases its Strategic Plan, Strength Through Science, Powering the 21st Century.
December 14, 2000	The DOE orders electricity generators and marketers to make power available to power-short California.
2001	
January 17	The DOE announces plans to build the second-largest wind power facility in the United States on 644 acres of the Nevada Test Site.
January 23	Two-week extension of emergency orders requiring energy suppliers to provide natural gas and electricity to California utility companies.
January 29	President G. W. Bush is "deeply concerned" that the economic fallout from the California power crisis will spread to other states. He establishes the Energy Policy Development Group task force to develop a national energy policy to help the private sector and governments at all levels promote dependable, affordable, and environmentally sound energy for the future.

February 6	The DOE offers $95 million in federal matching funds for projects that demonstrate how operators of generating facilities can generate more power and cut emissions.
March 6	Canada, Mexico, and the United States announce formation of the North American Energy Working Group to develop a comprehensive energy policy and facilitate cross-border electricity trade.
March 17	OPEC members cut oil production by 1 million gallons a day.
March 29	The DOE secretary tells companies supplying energy to California that it expects them to minimize chances of a summer supply shortage.
May 1	The DOE establishes an Energy Emergency Task Force within the agency to respond quickly and appropriately to any energy emergencies.
May 17	President Bush releases the National Energy Policy (NEP), with five national goals: modernizing conservation, modernizing the energy infrastructure, increasing energy supplies, accelerating protection and improvement of the environment, and increasing energy security.
May 25	The DOE sees an expanded role for nuclear power in the United States.
June 25	Bonneville Power Administration announces seven wing power projects.
August 2	U.S. House of Representatives passes an energy bill containing major portions of the president's energy plan recommendations. Ethanol subsidy cut to $0.53 per gallon.
August 7	The DOE announces plans to work with the National Governor's Association (NGA) on state and local energy policy issues. The NGA will work cooperatively on energy development and conservation.
September 11	Terrorist attack on the World Trade Center in New York City.
October 25	The DOE announces that its "overarching mission is national security," and lists eight priority objectives: identify new sources of energy, protect critical energy infrastructure, implement the president's energy plan, implement climate change initiative, ensure reliability of nuclear stockpile, address proliferation of nuclear weapons, enhance homeland defense against new terrorist threats, and implement environmental cleanup faster and cheaper.
December 2	Enron Corporation files for bankruptcy.

2002	
February 7	The DOE and the Interior Department call upon the Senate to immediately pass a comprehensive energy bill.
February 14	President Bush announces the Clear Skies Initiative to cut power plant emissions and commits the nation to an aggressive strategy to cut greenhouse gas intensity by 18 percent over the next 10 years. The budget devotes $4.5 billion to addressing climate change.
February 15	The DOE unveils the Nuclear Power 2010 initiative aimed at building new nuclear power plants before the end of the decade. The DOE proposes to invest $38.5 million in FY 2003 in a multiyear program to explore both federal and private sites for new nuclear plants; demonstrate the efficiency and timeliness of key licensing processes to make licensing of new plants more efficient, effective and predictable; and conduct research needed to make the safest and most efficient nuclear plant technologies available.
March 4	The DOE releases a solicitation offering $330 million in federal matching funds for industry-proposed clean coal technology projects, the first competitive stage of a $2 billion, 10-year clean coal technology initiative.
March 6	The DOE tells the House Appropriations Subcommittee on Energy and Water Development that the DOE's FreedomCar program could help create a mass market for hydrogen-fueled vehicles by 2035.
March 8	President Bush signs the economic stimulus bill, which includes several tax credits that benefit energy producers and consumers, including an estimated $1 billion tax credit for power generators producing electricity from renewable sources and about a $150 million tax credit for consumers purchasing electric vehicles.
April 25	The Senate, by a vote of 88 to 11, approves the energy bill. The bill contains $14 billion in tax breaks primarily to promote conservation and renewable energy sources, has no provisions for drilling in the Arctic National Wildlife Refuge but encourages the construction of an Alaskan natural gas pipeline, requires that renewables make up 10 percent of electrical generation by 2020, and leaves setting of fuel efficiency standards to the Bush administration. Ethanol subsidy cut to $0.52 per gallon.
May 8	The DOE releases the National Transmission Grid Study, developed to study the transmission system and identify transmission bottlenecks and measures to eliminate them.
June 6	President Bush proposes a new cabinet-level Homeland Security Department.

June 10	The United States, Canada, and Mexico release *North America—The Energy Picture*, the first report of the North American Energy Working Group.
August 27	The DOE receives 36 proposals for projects valued at more than $5 billion in the first round of President Bush's Clean Coal Power Initiative.
September 16	OPEC decides to hold its production to current levels rather than to increase output to bring down prices, which have been hovering near $30 per barrel.
September 30	The President's Council of Advisors on Science and Technology calls for increased energy efficiency by improving production at coal-fired plants with high-efficiency generation technologies, such as clean coal systems, making the electricity transmission grid more efficient by installing superconducting technologies, increasing use of distributed generation technologies, and making more efficient use of electricity.
November 12	The DOE unveils a National Hydrogen Energy Roadmap to "chart the course to a hydrogen economy" with a transportation system that relies on vehicles powered by fuel cells.
December 12	A Federal Energy Regulatory Commission administrative law judge finds that wholesale power suppliers overcharged California by $1.8 billion from October 2000 to June 2001.
2003	
January 15	The first eight projects chosen by the DOE in President Bush's Clean Coal Power Initiative, valued at more than $1.3 billion, are expected to help pioneer a new generation of innovative power plant technologies that could help meet the Clear Skies and Climate Change initiatives.
January 28	In his State of the Union address, President Bush announces a $1.2 billion Hydrogen Fuel Initiative to reverse the nation's growing dependence on foreign oil by developing the technology needed for commercially viable hydrogen-powered fuel cells. It includes $720 million over five years to develop the technologies and infrastructure to produce, store, and distribute hydrogen for use in fuel cell vehicles and electricity generation.
February 12	The DOE launches the president's Climate VISION (Voluntary Innovative Sector Initiatives: Opportunities Now)—a voluntary, public–private partnership to pursue cost-effective initiatives to reduce projected growth in America's greenhouse gas emissions.

February 27	The DOE announces the formation of a new international effort to advance carbon capture and storage technology as a way to reduce greenhouse gas emissions. The United States will lead a $1 billion, 10-year public-private-international effort to construct the world's first fossil fuel, pollution-free power plant, known as FutureGen; the plant will turn coal into a hydrogen-rich gas, rather than burning it directly.
March 19	The United States and its coalition partners initiate military operations against Iraq. During this month the Department of Energy formed the Office of Electric Transmission and Distribution and the Office of Energy Assurance, which coordinated responses to emergency disruptions and developed ways to harden infrastructure against such disruptions.
May 1	The DOE's Energy Information Administration (EIA) projects that worldwide consumption of commercial energy will grow by 58 percent over the next 2.5 decades. Much of the growth will occur in the developing world, led by China, India, and South Korea, as their consumption increasingly resembles that of the industrialized world.
June 25	The United States, with energy ministers from around the globe, signed the first international charter in support of the Carbon Sequestration Leadership Forum (CSLF), setting the framework for international cooperation in research and development for the separation, capture, transportation, and storage of carbon for reducing greenhouse gas emissions.
August 13	The DOE will provide over $17 million for 187 energy efficiency and renewable energy projects in 48 states, the District of Columbia, and one territory. Funding is provided through the DOE's State Energy Program Special Projects competitive grants. The funds will be used to improve the energy efficiency of schools, homes, and other buildings; promote energy-efficient industrial technologies; and support renewable energy sources such as solar, wind, geothermal, and biomass.
August 14	At 4:10 p.m. EDT, the largest power blackout in North American history swept through eight U.S. States—Ohio, Michigan, Pennsylvania, New York, Vermont, Massachusetts, Connecticut, New Jersey—and the Province of Ontario, Canada, leaving up to 50 million people with no electricity, with cost estimates from $4 to $10 billion in the United States alone. Power was not restored for 4 days in some parts of the United States. Ontario suffers rolling blackouts for more than a week before full power is restored.

2004	
January 9	The United States and Japan sign a joint statement of intent to pursue precompetitive research and development in the field of fuel cell and hydrogen technologies.
January 24	In his State of the Union address, President Bush urges Congress "to pass legislation to modernize our electricity system, promote conservation, and make America less dependent on foreign sources of energy."
January 27	The secretary of the DOE says that he plans no new major initiatives for the DOE as he enters the final year of the Bush administration's first term. Content to fulfill commitments made over the past three years, the secretary says that among his priorities are enacting a comprehensive energy bill, securing management reforms within the DOE, and upgrading and repairing nuclear weapons laboratories and production plants.
February 4	A National Research Council (NRC) report, *The Hydrogen Economy: Opportunities, Costs, Barriers and R&D Needs*, states that a hydrogen energy economy would have dramatic benefits for energy security and the environment, and could fundamentally transform the U.S. energy system, reduce environmental impacts, including CO_2 emissions and pollutants. However, "in the best-case scenario, the transition to a hydrogen economy would take many decades, and any reductions in oil imports and carbon dioxide emissions are likely to be minor during the next 25 years."
February 10	OPEC announces that oil output will be reduced by 1 million barrels to 23.5 million barrels per day effective April 1, 2004.
February 12	The DOE's Office of Science strategic plan sets seven long-term (10–20 year) scientific goals in the areas of: science for energy; harnessing biology for energy and environment; fusion; fundamentals of energy, matter, and time; nuclear physics research from quarks to the stars; computation for the frontiers of science; and building resource foundations for new science.
February 19	The DOE releases a solicitation for the second round of proposals under President Bush's Clean Coal Power Initiative (CCPI). The department plans to provide approximately $280 million in federal funds for demonstrating barrier-breaking technologies that sharply reduce and ultimately eliminate pollution in coal-based power plants.

March 5	The DOE submits to Congress a plan for the FutureGen project, the world's first zero-emission coal facility that will produce both electricity and hydrogen, while sequestering greenhouse emissions. The department will spend $500 million through FY 2018 on construction and operations for a 275-megawatt FutureGen plant and an additional $120 million on carbon sequestration. The plant will be operating at full scale by 2012, with a sequestration system completed by the following year.
April 1	The DOE announces the award of $128.2 million to 30 states and the Navajo Nation to improve the energy efficiency of the homes of low-income families.
April 6	The DOE report, *Wind Power Today and Tomorrow*, states that in 2003 the U.S. wind generating capacity increased by more than 30 percent. Wind power plants of various sizes now operate in 32 states with a total generating capacity of 6,374 MW of power, enough to meet the energy needs of more than 3 million homes.
April 26	President Bush, in a speech in Minneapolis, declares that "we need a different energy strategy than the one we have today, a strategy that uses technology and innovation to diversify our supplies, to make us less dependent on foreign sources of energy."
May 23	The DOE announces it will cooperate with an industry team led by the Tennessee Valley Authority (TVA) to conduct a detailed $4.25 million study of the potential construction of a two-unit Advanced Boiling Water Reactor (ABWR) nuclear plant on the Bellefonte site near Hollywood, Alabama. The plant could produce more than 2600 megawatts of electricity.
June 16	The House of Representatives passes a comprehensive energy bill identical to the bill approved by a House-Senate conference committee in fall 2003.
July 26	The DOE announces that it has received proposals for projects in a new generation of clean coal projects, valued at nearly $6 billion, in the latest phase of the President's Clean Coal Power Initiative (CCPI).
August 9	The DOE will provide $16.3 million for 162 energy efficiency and renewable energy projects in 43 states and the District of Columbia. Funding is being provided through DOE's State Energy Program Special Projects competitive grants. State energy offices will use these funds to improve the energy efficiency of schools, homes, and other buildings; promote energy-efficient industrial technologies; and support renewable energy sources such as solar, wind, and biomass.
September 20	A study by the University of Chicago examining the economic competitiveness of nuclear power indicates that the future cost associated with nuclear power production is comparable with gas and coal-based energy generation.

October 22	President Bush signs the American Jobs Creation Act of 2004. Among its energy-related provisions, the act gives incentives to sponsors of a natural gas pipeline from Alaska to the lower 48 states, renewable energy producers, marginal oil and gas well developers, investor-owned electric utilities, and rural electric cooperatives.
November 17	Resources for the Future, the Washington-based, nonpartisan think tank, announces the release of *New Approaches on Energy and the Environment: Policy Advice for the President*, a collection of 25 analytical prescriptions designed as stand-alone "memos to the president," offering possible policy options for the administration on critical challenges related to energy, the environment, and natural resources.
December 7	The DOE's Energy Information Administration (EIA), in *Electricity Transmission in a Restructured Industry*, states, "the Government does not have the electrical models and data necessary to verify that existing and planned transmission capability is adequate to keep the lights on," adding that collecting the necessary data "would require long-term, coordinated effort across many government organizations."
December 9	The DOE announces awards for five new cost-shared research projects to help meet the nation's growing demand for natural gas.
December 23	The DOE announces 35 research awards to U.S. universities totaling $21 million over three years to engage students and professors in the DOE's major nuclear energy research and development programs, including the Advanced Fuel Cycle Initiative, the Generation IV Nuclear Energy Systems Initiative, and the Nuclear Hydrogen Initiative.
2005	
April	Gulf Gateway Energy Bridge Deepwater Port opens—the first offshore LNG receiving facility and the first new LNG regasification facility to be built in the United States in over 20 years.
August 29	Hurricane Katrina strikes the Gulf Coast of the United States.
September 23	Hurricane Rita strikes the Gulf Coast.
2006	
January 1	Russia attempts to penalize the Ukraine by stopping gas sales; attempt ruled a failure after several days.
July 13	Price of oil hits a record high of $78.40 a barrel on New York Mercantile Exchange.

2007	
March	The European Union (EU) introduces new environmental regulations to reduce greenhouse gases by 20% by 2020.
November 20	Price of oil hits another record high of $99.29 a barrel.
December 19	President George W. Bush signs the Energy Independence and Security Act, which will raise corporate average fuel economy (CAFE) standards for motor vehicles to a combined level of 35 miles per gallon by 2020.
2008	
January 2	Price of oil briefly reaches $100 a barrel for the first time in history.
July 11	Price of crude oil hits peak price of $147.27 a barrel.
November 5	The DOE issues final rules for $25 billion Advanced Technology Vehicles Manufacturing Loan Program (ATVMLP). Loan funds to be used for re-equipping, expanding, and establishing manufacturing facilities to produce advanced technology vehicles and components for such vehicles that provide substantial increases in fuel economy performance.
November 20	Amid fears of a global recession, the price of crude oil drops to below $50 a barrel.
2009	
January 19	Price of crude oil falls to $34 a barrel.
February 17	American Recovery and Reinvestment Act (ARRA) signed into law, providing significant stimulus support for energy-related developments.
March 31	The House of Representatives released the Waxman-Markey Discussion Draft of the American Clean Energy and Security Act of 2009.
May 1	The DOE publishes *20% Wind Energy by 2030: Increasing Wind Energy's Contribution to U.S. Electricity Supply*.
May 21	The American Clean Energy and Security Act of 2009 (H.R. 2454) passed by the House Energy and Commerce Committee.
June	The Khurais oilfield in Saudi Arabi—the largest single oil field development ever—brought onstream.
June 26	The Clean Energy Jobs and American Power Act (S. 1733) is passed by the House of Representatives and sent to the Senate (Senate title and version of House bill H.R. 2454).
September 15	President Obama announces new mandated increases in motor fuel efficiency standards to go into effect with the 2012 model year.

September 30	Democratic Senators Barbara Boxer (CA) and John Kerry (MA) introduced the Clean Energy Jobs and American Power Act (legislation aimed at reducing carbon emissions by 20 percent by 2020 and 80 percent by 2050).
2010	
March 31	President Obama announces plans to open selected sections of the Eastern Seaboard to offshore oil exploration.
April 20	Giant oil spill in the Gulf of Mexico off the coast of Louisiana follows an explosion on the Deepwater Horizon deep ocean drilling rig. The explosion and fire killed 11 workers. It threatens to be much larger than the 1989 Exxon Valdez spill off Alaska—previously considered the worst U.S. ecological disaster ever.
May 10	Senators John F. Kerry (D-Massachusetts) and Joseph I. Lieberman (I-Connecticut) introduce their compromise energy and climate bill—The American Power Act—hoping that public concern over the recent massive oil spill in the Gulf of Mexico will speed passage of the more than 1,000-page bill.

Sources: DOE (U.S. Department of Energy). 2009a. Energy timeline. Washington, DC: Department of Energy, http://www.energy.gov/about/timeline.htm (accessed April 7, 2009); Geo-Help. 2009. History of the world petroleum industry (key dates). Geo-Help, Inc., http://www.geohelp.net/world.html (accessed December 29, 2009).

Appendix C: Energy-Related Acronyms

ACAA: American Coal and Ash Association
ACC: American Coal Council
ACCCE: American Coalition for Clean Coal Electricity
ACE: American Coalition on Ethanol
ACEEE: American Council for an Energy-Efficient Economy
ACORE: American Council on Renewable Energy
ACRI: Air-Conditioning and Refrigeration Institute
AFP: Association of Fundraising Professionals
AGA: American Gas Association
AGI: American Geological Institute
ANWR: Arctic National Wildlife Refuge
APA: American Planning Association
APEC: Asia Pacific Economic Cooperation
APECEWG: Asia Pacific Economic Cooperation Energy Working Group
APERE: Association for the Promotion of Renewable Energies
APGA: American Public Gas Association
API: American Petroleum Institute
ASE: Alliance to Save Energy
ASHRAE: American Society of Heating, Refrigerating and Air-Conditioning Engineers
ASPOG: Association for the Study of Peak Oil and Gas
AWEA: American Wind Energy Association
BERA: Biomass Energy Research Association
BPA: Bonneville Power Administration
BRDI: Biomass Research and Development Initiative
CABO: Council of American Building Officials
CAEF: Committee on America's Energy Future
CARE: Coalition for Affordable and Reliable Energy
CAT: Centre for Alternative Technology
CBTA: Commercial Building Tax Deduction Coalition
CCC: Citizens Coal Council
CCSG: Clean Coal Study Group

CEED: Center for Energy and Economic Development
CEEPR: Center for Energy and Environmental Policy Research (MIT)
CESA: Clean Energy States Alliance
CNTWM: Center for Nuclear and Toxic Waste Management
CREST: Center for Renewable Energy and Sustainable Technology
CRP: Center for Responsive Politics
CSG: Council of State Governments
CTA: Coal Trading Association
DEEP: Diesel Emissions Evaluation Program
DHS: Department of Homeland Security
DNS: Department of Nuclear Safety
DOE: Department of Energy
DPC: Domestic Petroleum Council
DSIRE: Database of State Incentives for Renewable Energy
ECI: Electricity Conservation Institute
ECPA: Energy Consumers and Producers Association
ECS: Energy Charter Secretariat
ECT: Energy Charter Treaty (Europe)
EEI: Edison Electric Institute
EEI: Energy Efficiency Institute
EEN: European Energy Network
EERE: Energy Efficiency and Renewable Energy (DOE)
EETF: Energy and Environment Task Force (Council of State Governments)
EHUB: Ecosustainable Hub
ENS: European Nuclear Society
EPA: Environmental Protection Agency
EPAct: Energy Policy Act (of 2005)
EPIC: Energy Policy Information Center
EPMG: Electric Power Marketing Group
EPRI: Electric Power Research Institute
EPSA: Electric Power Supply Association
ESB: Energy Stability Board (proposed)
Eurostat: European Union Member States
FCEV: Fuel Cell Electric Vehicle
FERC: Federal Energy Regulatory Commission
FOIL: Foreign Oil Independence League
GE: Growth Energy
GES: Greening Earth Society
GGI: Green Government Initiative (National Association of Counties)
GHPC: Geothermal Heat Pump Consortium
GMA: Grocery Manufacturers Association
GPA: Gas Processors Association
GPP: Green Power Partnership
GRI: Gas Research Institute
GSA: Gas Technology Institute
GSA: Geological Society of America
HEI: Health Effects Institute

IAEA: International Atomic Energy Agency
IDEA: International District Energy Association
IEA: International Energy Agency
IEACR: International Energy Agency Coal Research
IECERS: Iowa Energy Center Energy Resource Station
IEF: International Energy Forum
IELE: Institute for Energy, Law, and Enterprise
IGA: International Geothermal Association
IGSHPA: International Ground Source Heat Pump Association
IIEC: International Institute for Energy Conservation
IPPA: Independent Petroleum Association of America
IREC: Interstate Renewable Energy Council
IRENA: International Renewable Energy Agency
ISEO: International Sustainable Energy Association for Renewable Energy and Energy Efficiency
ISES: International Solar Energy Society
JODI: Joint Oil Data Initiative
LBNL: Lawrence Berkeley National Laboratory
LEC: Lignite Energy Council
MCA: Midwest Cogeneration Association
MEA: Midwest Energy Association
MEEA: Midwest Energy Efficiency Alliance
MI: Methanol Institute
NAC: National Association of Counties
NAERC: North American Electric Reliability Council
NAESB: North American Energy Standards Board
NAEWG: North American Energy Working Group
NARO: National Association of Royalty Owners
NARUC: National Association of Regulatory Utility Commissioners
NASEO: National Association of State Energy Officials
NBB: National Biodiesel Board
NBEIA: National BioEnergy Industries Association
NCEP: National Commission on Energy Policy
NCI: Nuclear Control Institute
NEA: Nuclear Energy Agency
NEM: National Energy Marketers Association
NEMA: National Electrical Manufacturers Association
NFRC: National Fuel Cells Research Center
NGEPS: Natural Gas and Electric Power Society
NGSA: Natural Gas Supply Association
NHA: National Hydrogen Association
NHA: National Hydropower Association
NMA: National Mining Association
NNDS: National Nuclear Data Center (Brookhaven National Laboratory)
NPC: National Petroleum Council
NPGA: National Propane Gas Association
NREL: National Renewable Energy Laboratory

NRG: Public Information Service NRG: Nuclear Societies
NRRI: National Regulatory Research Institute
OAPEC: Organization of Arab Petroleum Exporting Countries
OCNEAF: Office of Coal, Nuclear, Electric and Alternate Fuels (U.S. Deptartment of Energy)
OECD: Organization for Economic Cooperation and Development
OLADE: Latin American Energy Organization
OPEC: Organization of Oil Exporting Countries
PATH: Partnership for Advancing the Transition to Hydrogen
PCCMEEP: Public Citizen, Critical Mass Energy and Environment Program
PMMA: Petroleum Marketers Association of America
PVTIP: Photovoltaic Technology Incubator Program
REEEP: Renewable Energy & Energy Efficiency Partnership
RFA: Renewable Fuels Association
RFF: Resources for the Future
RGGI: Regional Greenhouse Gas Initiative
SEFI: Sustainable Energy Finance Institute (UN)
SEIA: Solar Energy Industries Association
SEPA: Solar Electric Power Association
SME: Society for Mining, Metallurgy, and Exploration
TIAP: Tax Incentives Assistance Project
TPCS: Taxpayers for Common Sense
UI: Uranium Institute
UIC: Uranium Information Centre
UNAITFE: United Nations Ad-Hoc Interagency Task Force on Energy
UNDSD: United Nations Division for Sustainable Development
UNSD: United Nations Statistics Division
USC: Union of Concerned Scientists
USDA: United States Department of Agriculture
USDABCAP: U.S. Department of Agriculture Biomass Crop Assistance Program
USEA: United States Energy Association
USEIA: United States Energy Information Administration
USGBC: United States Green Building Council
VEETC: Volumetric Ethanol Excise Tax Credit
WAPA: Western Area Power Administration
WBGEP: World Bank Group Energy Program
WCC: Western Coal Council
WCI: World Coal Institute
WEC: World Energy Council
WEEA: World Energy Efficiency Association
WIM: Women in Mining

Appendix D: Glossary

This glossary is adapted from the U.S. Department of Energy, Energy Information Administration (EIA) glossary at http://www.eia.doe.gov/glossary/index.html.

alternating current (AC): An electric current that reverses its direction at regularly recurring intervals.

anthracite: The highest rank of coal; used primarily for residential and commercial space heating. It is a hard, brittle, and black lustrous coal, often referred to as hard coal, containing a high percentage of fixed carbon and a low percentage of volatile matter. The heat content of anthracite ranges from 22 to 28 million Btu per ton on a moist, mineral-matter-free basis.

anthropogenic: Made or generated by a human or caused by human activity. The term is used in the context of global climate change to refer to gaseous emissions that are the result of human activities, as well as other potentially climate-altering activities, such as deforestation.

ash: Impurities consisting of silica, iron, alumina, and other noncombustible matter contained in coal. Ash increases the weight of coal, adds to the cost of handling, and can affect its burning characteristics.

base load capacity: The generating equipment normally operated to serve loads on an around-the-clock basis.

base load plant: An electricity generating plant that is normally operated to take all or part of the minimum load of a system, and that consequently produces electricity at an essentially constant rate and runs continuously. These units are operated to maximize system mechanical and thermal efficiency and minimize system operating costs.

bbl: The abbreviation for barrels.

bbl/d: The abbreviation for barrel(s) per day.

biodiesel: A fuel typically made from soybean, canola, or other vegetable oils; animal fats; and recycled grease. It can serve as a substitute for petroleum-derived diesel or distillate fuel.

biofuels: Liquid fuels and blending components produced from biomass feedstocks; used primarily for transportation.

biomass: Organic nonfossil material of biological origin constituting a renewable energy source.

bituminous coal: A dense coal, usually black, sometimes dark brown, often with well-defined bands of bright and dull material, used primarily as fuel in steam-electric power generation, with substantial quantities also used for heat and power applications in manufacturing and to make coke. Bituminous coal is the most abundant coal in active U.S. mining regions. The heat content of bituminous coal ranges from 21 to 30 million Btu per ton on a moist, mineral-matter-free basis.

CAFE: Corporate average fuel economy.

capacity factor: The ratio of the electrical energy produced by a generating unit for the period of time compared to the electrical energy that could have been produced at continuous full power operation during the same period.

capacity utilization: Capacity utilization is computed by dividing production by productive capacity and multiplying by 100.

capital cost: The cost of field development and plant construction and the equipment required for industry operations.

capital stock: Property, plant, and equipment used in the production, processing, and distribution of energy resources.

carbon dioxide (CO_2): A colorless, odorless, nonpoisonous gas that is a normal part of Earth's atmosphere. Carbon dioxide is a product of fossil-fuel combustion as well as other processes. It is considered a greenhouse gas because it traps heat (infrared energy) radiated by the Earth into the atmosphere and thereby contributes to the potential for global warming. The global warming potential (GWP) of other greenhouse gases is measured in relation to that of carbon dioxide, which by international scientific convention is assigned a value of one (1).

carbon dioxide equivalent: The amount of carbon dioxide by weight emitted into the atmosphere that would produce the same estimated radiative forcing as a given weight of another radiatively active gas. Carbon dioxide equivalents are computed by multiplying the weight of the gas being measured (for example, methane) by its estimated global warming potential (which is 21 for methane). *Carbon equivalent units* are defined as carbon dioxide equivalents multiplied by the carbon content of carbon dioxide (i.e., 12/44).

carbon sequestration: The fixation of atmospheric carbon dioxide in a carbon sinks through biological or physical processes.

carbon sink: A reservoir that absorbs or takes up released carbon from another part of the carbon cycle. The four sinks, which are regions of the Earth within which carbon behaves in a systematic manner, are the atmosphere, terrestrial biosphere, including freshwater systems, oceans, and sediments, which include fossil fuels.

Clean Development Mechanism (CDM): A Kyoto Protocol program that enables industrialized countries to finance emissions-avoiding projects in developing countries and receive credit for reductions achieved against their own emissions limitation targets.

climate change: A term used to refer to all forms of climatic inconsistency, but especially to change from one prevailing climatic condition to another. Scientists generally prefer this term to *global warming* since the impacts of anthropogenic climate change vary across the globe.

cogeneration: The production of electrical energy and another form of useful energy (such as heat or steam) through the sequential use of energy.

combined cycle: An electric generating technology in which electricity is produced from otherwise lost waste heat exiting from one or more gas (combustion) turbines. The exiting heat is routed to a conventional boiler or to a heat recovery steam generator for utilization by a steam turbine in the production of electricity. This process increases the generating unit's efficiency.

conservation: (1) Reduced consumption of energy services, which leads to reduced use of and demand for primary or secondary energy (contrast with *energy efficiency*); (2) an ethic of resource use that aims to protect the natural environment.

criteria pollutant: A pollutant determined to be hazardous to human health and regulated under the EPA's National Ambient Air Quality Standards. The 1970 amendments to the Clean Air Act require the EPA to describe the health and welfare impacts of a pollutant as the "criteria" for inclusion in the regulatory regime.

crude oil: A mixture of hydrocarbons that exists in liquid phase in natural underground reservoirs and remains liquid at atmospheric pressure after passing through surface separating facilities.

current (electric): A flow of electrons in an electrical conductor. The strength or rate of movement of the electricity is measured in amperes.

DC: Direct current; a flow of electrical charge in one direction.

demand-side management (DSM): The planning, implementation, and monitoring of utility activities designed to encourage consumers to modify patterns of electricity usage, including the timing and level of electricity demand. It refers to only energy and load-shape modifying activities that are undertaken in response to utility-administered programs.

depletion allowance: An income tax deduction that is related to the exhaustion of mineral reserves. Depletion is included as one of the elements of amortization of the mineral asset.

electric hybrid vehicle: An electric vehicle that either (1) operates solely on electricity, but contains an internal combustion motor that generates additional electricity (series hybrid); or (2) contains an electric system and an internal combustion system and is capable of operating on either system (parallel hybrid).

electric power: The rate at which electric energy is transferred. Electric power is measured by capacity and is commonly expressed in megawatts (MW).

electric power grid: A system of synchronized power providers and consumers connected by transmission and distribution lines and operated by one or more control centers. In the continental United States, the electric power grid consists of three systems: the Eastern Interconnect, the Western Interconnect, and the Texas Interconnect. In Alaska and Hawaii, several systems encompass areas smaller than the state (e.g., the interconnect serving Anchorage, Fairbanks, and the Kenai Peninsula; individual islands).

electric utility: A corporation or other legal entity aligned with distribution facilities for delivery of electric energy for use primarily by the public. Included are investor-owned electric utilities, municipal and state utilities, federal electric utilities, and rural electric cooperatives. A few entities that are tariff based and corporately aligned with companies that own distribution facilities are also included.

electricity: A form of energy characterized by the presence and motion of elementary charged particles generated by friction, induction, or chemical change.

energy: There is no completely satisfactory definition. A common one is that energy is the capacity for doing work. Smil (2008, 13) suggests it is better described as "the ability to transform a system." Energy has several forms (e.g., electrical, chemical, thermal, nuclear), some of which are easily convertible and can be changed to another form useful for work. Most of the world's convertible energy comes from fossil fuels that are burned to produce heat that is then used as a transfer medium to mechanical or other means in order to accomplish tasks. Electrical energy is usually measured in kilowatt-hours, while heat energy is usually measured in British thermal units (Btu).

energy density: Energy per unit of volume of a fuel or power source; typically measured in joules/meters^3.

energy efficiency: (1) Use of improved technology and systems to reduce the amount of energy used by devices and systems (e.g., lighting and heating), while maintaining a given level of energy services; (2) improving the rate of energy output of a conversion process.

energy reserves: Estimated quantities of energy sources that are demonstrated to exist with reasonable certainty on the basis of geologic and engineering data (proved reserves) or that can reasonably be expected to exist on the basis of geologic evidence that supports projections from proved reserves (probable/indicated reserves). Knowledge of the location, quantity, and grade of probable/indicated reserves is generally incomplete or much less certain than it is for proved energy reserves.

ethanol (CH_3-CH_2OH): A clear, colorless, flammable oxygenated hydrocarbon. Ethanol is typically produced chemically from ethylene, or biologically from fermentation of various sugars from carbohydrates found in agricultural crops and cellulosic residues from crops or wood. It is used in the United States as a gasoline octane enhancer and oxygenate (blended up to a 10 percent concentration).

externalities: Benefits or costs generated as a by-product of an economic activity that do not accrue to the parties involved in the activity. Environmental externalities are benefits or costs that manifest themselves through changes in the physical or biological environment.

facility: An existing or planned location or site at which prime movers, electric generators, and/or equipment for converting mechanical, chemical, and/or nuclear energy into electric energy are resituated or will be situated.

firm power: Power or power-producing capacity, intended to be available at all times during the period covered by a guaranteed commitment to deliver, even under adverse conditions.

First Law of Thermodynamics: In a conversion process, energy can never be created or destroyed.

fission: The process whereby an atomic nucleus of appropriate type, after capturing a neutron, splits into (generally) two nuclei of lighter elements, with the release of substantial amounts of energy and two or more neutrons.

futures market: A trade center for quoting prices on contracts for the delivery of a specified quantity of a commodity at a specified time and place in the future.

gigawatt (GW): One billion watts or one thousand megawatts.

global warming potential (GWP): An index used to compare the relative radiative forcing of different gases without directly calculating the changes in atmospheric concentrations. GWPs are calculated as the ratio of the radiative forcing that would result from the emission of one kilogram of a greenhouse gas to that from the emission of one kilogram of carbon dioxide over a fixed period of time, such as 100 years.

greenhouse gases: Those gases, such as water vapor, carbon dioxide, nitrous oxide, methane, hydrofluorocarbons (HFCs), perfluorocarbons (PFCs), and sulfur hexafluoride, that are transparent to solar (short-wave) radiation but opaque to long-wave (infrared) radiation, thus preventing long-wave radiant energy from leaving Earth's atmosphere. The net effect is a trapping of absorbed radiation and a tendency to warm the planet's surface.

investor-owned utility (IOU): A privately owned electric utility whose stock is publicly traded. It is rate regulated and authorized to achieve an allowed rate of return.

kilowatt (kW): One thousand watts.

kinetic energy: Energy available as a result of motion that varies directly in proportion to an object's mass and the square of its velocity.

levelized cost: The present value of the total cost of building and operating a generating plant over its economic life, converted to equal annual payments. Costs are levelized in real dollars (i.e., adjusted to remove the impact of inflation).

LIHEAP: Low-Income Home Energy Assistance Program.

liquefied natural gas (LNG): Natural gas (primarily methane) that has been liquefied by reducing its temperature to −260 degrees Fahrenheit at atmospheric pressure.

megawatt (MW): One million watts of electricity.

methane: A colorless, flammable, odorless hydrocarbon gas (CH_4) that is the major component of natural gas. It is also an important source of hydrogen in various industrial processes. Methane is a greenhouse gas.

metric ton: A unit of weight equal to 2,204.6 pounds.

mill: A monetary cost and billing unit used by utilities; it is equal to 1/1000 of the U.S. dollar (equivalent to 1/10 of 1 cent).

N_2O: Nitrous oxide.

natural gas: A gaseous mixture of hydrocarbon compounds, the primary one being methane.

natural gas hydrates: Solid, crystalline, waxlike substances composed of water, methane, and usually a small amount of other gases, with the gases being trapped in the interstices of a water-ice lattice. They form beneath permafrost and on the ocean floor under conditions of moderately high pressure and at temperatures near the freezing point of water.

nonfirm power: Power or power-producing capacity supplied or available under a commitment having limited or no assured availability.

nonspinning reserve: The generating capacity not currently running but capable of being connected to the bus and load within a specified time.

NYMEX: New York Mercantile Exchange.

oil shale: A sedimentary rock containing kerogen, a solid organic material.

OPEC (Organization of Oil Exporting Countries): An intergovernmental organization created in 1960 whose stated objective is to "coordinate and unify the petroleum policies of member countries." Current members include Algeria, Angola, Ecuador, Iran, Iraq, Kuwait, Libya, Nigeria, Qatar, Saudi Arabia, United Arab Emirates, and Venezuela.

ozone: A molecule made up of three atoms of oxygen. Occurs naturally in the stratosphere and provides a protective layer shielding the Earth from harmful ultraviolet radiation. In the troposphere, it is a chemical oxidant, a greenhouse gas, and a major component of photochemical smog.

particulate: A small, discrete mass of solid or liquid matter that remains individually dispersed in gas or liquid emissions. Particulates take the form of aerosol, dust, fume, mist, smoke, or spray. Each of these forms has different properties.

peak load: The maximum load during a specified period of time.

petroleum: A broadly defined class of liquid hydrocarbon mixtures that includes crude oil, lease condensate, unfinished oils, refined products obtained from the processing of crude oil, and natural gas plant liquids.

photovoltaic cell (PVC): An electronic device consisting of layers of semiconductor materials fabricated to form a junction (adjacent layers of materials with different electronic characteristics) and electrical contacts; capable of converting incident light directly into electricity (direct current).

plant: A term commonly used either as a synonym for an industrial establishment or a generating facility or to refer to a particular process within an establishment.

power: The rate at which work is performed or energy is converted; the rate of flow of energy. In electrical systems power is measured in watts.

power density: Rate of energy flow per unit of surface area (Smil 2008, 20).

primary energy: Energy in the form that it is first accounted for in a statistical energy balance, before any transformation to secondary or tertiary forms of energy. For example, coal can be converted to synthetic gas, which can be converted to electricity; in this example, coal is the primary energy, synthetic gas is the secondary energy, and electricity is the tertiary energy.

prime mover: The engine, turbine, water wheel, or similar machine that drives an electric generator; or, for reporting purposes, a device that converts energy to electricity directly (e.g., photovoltaic solar and fuel cells).

proved energy reserves: Estimated quantities of energy sources that analysis of geologic and engineering data demonstrates with reasonable certainty are recoverable under existing economic and operating conditions. The location, quantity, and grade of the energy source are usually considered to be well established in such reserves.

public utility: Enterprise providing essential public services, such as electric, gas, telephone, water, and sewer under legally established monopoly conditions.

quadrillion: The quantity 1,000,000,000,000,000(10^{15}).

radiation: The transfer of heat through matter or space by means of electromagnetic waves.

radiative forcing: A change in average net radiation at the top of the troposphere (known as the tropopause) because of a change in either incoming solar or exiting infrared radiation. A positive radiative forcing tends on average to warm the Earth's surface; a negative radiative forcing on average tends to cool the Earth's surface. Greenhouse gases, when emitted into the atmosphere, trap infrared energy radiated from the Earth's surface and therefore tend to produce positive radiative forcing.

radioactivity: The spontaneous emission of radiation from the nucleus of an atom. Radionuclides lose particles and energy through this process.

Rankine cycle: The thermodynamic cycle that is an ideal standard for comparing performance of heat-engines, steam power plants, steam turbines, and heat pump systems that use a condensable vapor as the working fluid. Efficiency is measured as work done divided by sensible heat supplied.

rate base (electric): The value of property upon which a utility is permitted to earn a specified rate of return as established by a regulatory authority.

rate case: A proceeding, usually before a regulatory commission, involving the rates to be charged for a public utility service.

renewable energy resources: Energy resources that are naturally replenishing but flow-limited. They are virtually inexhaustible in duration but limited in the amount of energy that is available per unit of time. Renewable energy resources include biomass, hydro, geothermal, solar, wind, ocean thermal, wave action, and tidal action.

royalty: An amount paid to the owner of a resource (often a government entity) per unit of material extracted.

Second Law of Thermodynamics: Processes in energy systems increase the entropy (disorder) of the universe. The key concept is that systems do not spontaneously become more ordered. Another formulation is that heat does not flow from a colder to a warmer material.

solar radiation: A general term for the visible and near-visible (ultraviolet and near-infrared) electromagnetic radiation that is emitted by the sun. It has a spectral or wavelength distribution that corresponds to different energy levels; short wavelength radiation has a higher energy than long-wavelength radiation.

solar thermal energy (as used at electric utilities): Energy radiated by the sun as electromagnetic waves (electromagnetic radiation) that is converted at electric utilities into electricity by means of solar (photovoltaic) cells or concentrating (focusing) collectors.

spinning reserve: That reserve generating capacity running at a zero load and synchronized to the electric system.

spot purchases: A single shipment of fuel or volumes of fuel purchased for delivery within one year. Spot purchases are often made by a user to fulfill a certain portion of energy requirements, to meet unanticipated energy needs, or to take advantage of low-fuel prices.

stocks: Inventories of fuel stored for future use.

strip mine: An open cut in which the overburden is removed from a coal bed prior to the removal of coal.

tariff: A levy applied to imports.

thermal: A term used to identify a type of electric generating station, capacity, capability, or output in which the source of energy for the prime mover is heat.

time-of-day rate: The rate charged by an electric utility for service to various classes of customers. The rate reflects the different costs of providing the service at different times of the day.

transmission (electric): The movement or transfer of electric energy over an interconnected group of lines and associated equipment between points of supply and points at which it is transformed for delivery to consumers or is delivered to other electric systems. Transmission is considered to end when the energy is transformed for distribution to the consumer.

unbundling: Separating vertically integrated monopoly functions into their component parts for the purpose of separate service offerings.

voltage: The difference in electrical potential between any two conductors or between a conductor and ground. It is a measure of the electric energy per electron that electrons can acquire and/or give up as they move between the two conductors.

watt (W): Unit of power; rate of doing work at 1 joule/second.

wind energy: Kinetic energy present in wind motion that can be converted to mechanical energy for driving pumps, mills, and electric power generators.

wind turbine: Wind energy conversion device that produces electricity; typically three blades rotating about a horizontal axis and positioned upwind of the supporting tower.

References

Aalto, P. 2008. Conclusion: Prospects for a Pan-European energy policy. In *The EU-Russian dialogue: Europe's future energy security*, ed. Pami Aalto, 193–208. Aldershot, U.K.: Ashgate.

Aaron, H. 2009. *There is no entitlement crisis*. The Brookings Institution (February 23), http://www.brookings.edu/opinions/2009/0223_entitlements_aaron.aspx (accessed February 23, 2010).

Abbasi, S., and N. Abbasi. 2000. The likely adverse environmental impacts of renewable energy sources. *Applied Energy* 65 (1–4): 121–144.

Abbot, A. M. 2003. Leadership from within: A succession management leadership development program. *NPPA Bulletin* 57 (December): 13–14.

Achenbach, J. 2010. The 21st century grid. *National Geographic* 218 (1): 118–139.

Adam, S. and H. Kriesi. 2007. The network approach. In P. A. Sabatier, ed. *Theories of the Policy Process*. Boulder, CO: Westview Press, pp. 129–154.

Adeyeye, A., J. Barrett, J. Diamond, L. Goldman, J. Pendergrass, and D. Schramm. 2009. *Estimating U.S. government subsidies to energy sources: 2002–2008*. Washington, DC: Environmental Law Institute.

AFDC (Alternative Fuels and Advanced Vehicles Data Center). 2008. *Energy Independence and Security Act of 2007, Public Law 110–140*. U.S. Department of Energy, Energy Efficiency and Renewable Energy, Alternative Fuels and Advanced Vehicles Data Center, http://www.afdc.energy.gov/afdc/incentives_laws_security.htm (accessed April 9, 2009).

Ageworks.com. 2010. *Demographics of an aging population: Module 2*. Ageworks.com, http://www.ageworks.com/course_demo/200/module2/module2.htm (accessed June 26, 2010).

Agranoff, R., and M. McGuire. 2001. American federalism and the search for models of management. *Public Administration Review* 61 (6): 671–681.

AICPA (American Institute of CPAs). 1999. Statement of governmental accounting standards No. 34: Basic financial statements and management's discussion and analysis for state and local government. *Journal of Accountancy* 188 (October): 112–131.

Allen, F. L. 1940. *Since yesterday*. New York: Bantam Books.

Allison, G. 2006. Emergence of schools of public policy. In *The Oxford handbook of public policy*, ed. M. Moran, M. Rein, and R. E. Goodin, 58–79. Cambridge, U.K.: Oxford University Press.

Alvarado, F. and R. Rajaraman. 2003. The 2003 blackout: did the system operator have enough power? http://www.hks.harvard.edu/hepg/Standard_Mkt_dsgn/Blackout_Investigation_lrca.pdf (accessed August 18, 2010).

Anadon, L., M. Bunn, C. Jones, and V. Narayanamurti. 2010. *U.S. public energy innovation institutions and mechanisms: Status and deficiencies*. Cambridge, MA: Science, Technology and Public Policy Program, Belfer Center for Science and International Affairs, Harvard Kennedy School (January).

Andreen, W. 2009. Federal climate change legislation and preemption. *Environmental and Energy Law and Policy Journal* 3 (2): 261–303.

ANWR.com (Arctic national Wildlife Refuge). Congressman Don Young introduces American Energy Independence and Price Reduction Act. http://www.anwr.org/Headlines/Congressman-Don-Young-introduces-American-Energy-Independence-and-Price-Reduction-Act.php (accessed January 17, 2011).

APEC (Asia-Pacific Economic Cooperation). 2009. APEC Energy Working Group. http://www.apec.org/content/apec_groups/som_committee_on_economic/working_groups/energy.html (accessed ______).

API (American Petroleum Institute). 2010. Motor fuel taxes. American Petroleum Institute, http://www. api.org/statistics/fueltaxes/ (accessed August 20, 2010).

Aringhoff, R., G. Brakmann, M. Geyer, S. Teske. 2005. Concentrated Solar Thermal Power-Now! Amsterdam: Greenpeace International. http://www.greenpeace.org/raw/content/international/press/reports/Concentrated-Solar-Thermal-Power.pdf (accessed January 25, 2010).

Armstrong, J. R. 1985. The evolution of state level energy institutions. In *State energy policy: Current issues, future directions*, ed. Stephen W. Sawyer and John R. Armstrong, 283–302. Boulder, CO: Westview.

Arnold, T., and L. Evans. 2001. Governance in the New Zealand electricity market: A law and economics perspective on enforcing obligations in markets based on a multilateral contract. *The Antitrust Bulletin* 46 (fall): 611–641.

Associated Press. 2010. Fate of cape wind decided. *Cape Cod Times Online* (April 27), http://www.capecodonline.com/apps/pbcs.dll/article?AID=/20100427/NEWS11/100429794 (accessed June 22, 2010).

AWEA (American Wind Energy Association). 2009a. *Achieving 20% wind energy by 2030: An overview.* American Wind Energy Association, http://www.20percentwind.org/AWEA20percentReportCard09.pdf (accessed June 22, 2009).

AWEA (American Wind Energy Association). 2009b. *Summary of the American Recovery and Reinvestment Act* (AARA) *of 2009: Provisions of interest to the wind industry*. American Wind Energy Association, http://www.awea.org/legislative/pdf/ARRA_Provisions_of_Interest-to-the-Wind-Energy_Industry.pdf (accessed June 22, 2009).

AWEA (American Wind Energy Association). 2010. *U.S. wind industry annual market report year ending 2009.* American Wind Energy Association, http://www.awea.org/reports/Annual_Market_Report_Press_Release_Teaser.pdf (accessed June 22, 2010).

Bailey, R. 2009. Energy futures: A quick guide to alternative energy. *Reason* (June), http://reason.com/news/printer/133235.html (accessed June 7, 2009).

Ballesteros, A., S. Nakhooda, J. Werksman, K. Hurlburt, and S. Kumar. 2009. Power, responsibility and accountability: Re-thinking the legitimacy of institutions for climate finance. WRI Working Paper. World Resources Institute, http://www.wri.org/publication/power-responsibility-accountability (accessed June 4, 2010).

Bamberger, R. 2006. *Energy policy framework and continuing issues.* (May 11). Washington, DC: Congressional Research Service (CSR), Library of Congress.

Barackobama.com. 2008. *Barack Obama and Joe Biden: New energy for America.* Organizing for America, http://www.barackobama.com/pdg/factsheet_energy_speech_080308.pdf (accessed January 31, 2009).

Bardach, E. 2008. Policy analysis. In *The Oxford handbook of public policy*, ed. M. Moran, M. Rein, and R. E. Goodin, 336–366. Oxford, U.K.: Oxford University Press.

Barrett, S. 2009. Rethinking global climate change governance. Special Issue of *Economics: The Open-Access, Open Assessment E-Journal. Global Governance—Challenges and Proposals for Reform* 3 (5). Np. http://www.economics-ejournal.org/economics/journalarticles/2009-5 (accessed January 16, 2011).

Barrionuevo, A. 2007. Brazil discovers an oil field can be a political tool. *New York Times* (November 19). A3, NYTimes.com, http://www.nytimes.com/2007/11/19/world/americas/19braziloil.html (accessed June 21, 2010).

Barry, D. 2001. *Dave Barry Hits Below the Beltway*. New York: Ballantine Books.

Barstow, D., L. Dodd, J. Glanz, S. Saul, and I. Urbina. 2010. Regulators failed to address risks in oil rig fail-safe device. *New York Times* (June 21): A1, NYTimes.com, http://www.nytimes.com/2010/06/21/us/21blowout.html?hp (accessed June 21, 2010).

Bartis, J. T. 2009. Research priorities for fossil fuels. Testimony before the Senate Energy and Natural Resources Committee, March 5, 2009. Washington, DC: The Rand Corporation.

Bartsch, U., B. Müller, and A. Aaheim 2000. *Fossil fuels in a changing climate: Impacts of the Kyoto Protocol and developing country participation.* Oxford, U.K.: Oxford University Press.

Bary, A. 2009. The long and binding road. *Barron's* (May 11). http://online.barrons.com/article/SB124183159872002803.html#articleTabs_panel_article%3D1 (accessed May 29, 2010).

Barzelay, M. 2001. *The new public management.* Berkeley: University of California Press.

Batrkovick, B. R. 1989. *Regulatory interventionism in the utility industry.* New York: Quorum Books.

Bauer, J. H. 1925. Effective regulation of public utilities. New York: Macmillan.

Beder, S. 2003. Power play: The fight to control the world's electricity. New York: The New Press.

Berg, E. 2008. The Baltic gateway: A corridor leading three different directions? In *The EU-Russian dialogue: Europe's future energy security*, ed. Pami Aalto, 145–162. Aldershot, U.K.: Ashgate.

Berrie, T. W. 1993. *Electricity economics and planning.* London, U.K.: Peter Peregrinus.

Betsill, M. 2010. International climate change policy: Toward the multilevel governance of global warming. In *The global environment: Institutions, law and policy,* 3rd. ed., ed. R. Axelrod, S. Vandeveer, and D. L. Downie, 111–131. Washington, DC: CQ Press.

Bettelheim, E. C., and G. D'Origny. 2002. Carbon sinks and emissions trading under the Kyoto Protocol: A legal analysis. *Philosophical transactions: Mathematical, physical and engineering sciences* 360 (1797): 1827–1851.

Betts, R. A., M. Collins, D. L. Hemming, C. D. Jones, J. A. Lowe, and M. Sanderson. 2009. *When could global warming reach 4°C?* Hadley Center Technical Note 80 (December), http://www.metoffice.gov.uk/publications/HCTN/HCTN_80.pdf (accessed January 27, 2010).

Bezdek, R. H., and R. M. Wendling. 2004. The case against gas dependence: Greater reliance on gas-fired power implies serious economic, technical, and national security risks. *Public Utilities Fortnightly* 142 (April): 43–47.

Bezdek, R. H., and R. M. Wendling. 2007. A half century of U.S. federal government energy incentives: Value, distribution, and policy implications. *International Journal of Global Energy Issues* 27 (1): 42–60.

Bird, C. 1966. *The invisible scar.* New York: Donald McKay.

Birger, J. 2009. Field general. *Fortune* 160 (1): 86–90.

Bishop, B. 2008. *The big sort.* New York: Houghton Mifflin.

Bittle, S., J. Rochkind, and A. Ott. 2009. *The energy learning curve.* Public Agenda, http://www.publicagenda.org/files/pdf/energy_learning_curve.pdf (accessed February 17, 2009).

Blackford, M. G. 1988. *The rise of modern business in Great Britain, the United States, and Japan.* Chapel Hill: University of North Carolina Press.

Blackford, M. G., and K. A. Kerr. 1990. *Business enterprise in American history.* 2nd ed. Boston: Houghton Mifflin.

Blackman, A. 2010. Will REDD really be cheap? Washington, DC: Resources for the Future (February 5), http://www.rff.org/RFF/Documents/Resources_174_Will_REDD_Be_Cheap.pdf (accessed May 7, 2010).

BLM (U.S. Bureau of Land Management). 2008. Inventory of onshore federal oil and natural gas resources and restrictions to their development. In *Phase III inventory—Onshore United States, executive summary.* Washington, DC: U.S. Bureau of Land Management, http://www.blm.gov/pgdata/etc/medialib/blm/wo/MINERALS__REALTY__AND_RESOURCE_PROTECTION_/energy/EPCA_Text_PDF.Par.18155.File.dat/Executive%20Summary%20text.pdf (accessed April 20 2010).

BLM [U.S. Bureau of Land Management] 2010. BrightSource Energy Ivanpah Solar Electric Generating System Factsheet. http://www.blm.gov/pgdata/etc/medialib/blm/ca/pdf/pa/energy/factsheets.Par.81531.File.dat/Ivanpah-Fact-Sheet.pdf (accessed January 25, 2010).

Blood, M. R. 2010. Endangered tortoises snarl solar-energy plans. Associated Press (January 2), http://theguzzler.blogspot.com/2010/01/endangered-tortoises-snarl-solar-energy.html

Bobrow, D. B. 2006. Social and cultural factors: Constraining and enabling. In *The Oxford handbook of public policy*, M. Moran, M. Rein, and G. E. Goodin, 572–586. Oxford, U.K.: Oxford University Press.

BOEMRE (Bureau of Ocean Energy Management, Regulation, and Enforcement). Offshore Production, Development & Resources. Washington, DC: U.S. Department of the Interior. http://www.boemre.gov/omm/pacific/production-development-resources/prod-dev-resource-eval.htm (accessed January 16, 2011).

Bolet, A. M., ed. 1985. *Forecasting U.S. electricity demand: Trends and methodologies.* Boulder, CO: Westview Press.

Bolinger, M., R. Wiser, and N. Darghouth. 2010. Preliminary evaluation of the impact of the Section 1603 treasury grant program on renewable energy deployment in 2009. Berkeley, CA: Lawrence Berkeley National Laboratory, http://eetd.lbl.gov/ea/emp/reports/lbnl-3188e.pdf

Bonvillian, C. and W. Weiss. 2009. *Structuring an Energy Technology Revolution.* Cambridge, MA: MIT University Press.

Bozeman, B., and J. D. Straussman. 1991. *Public management strategies.* San Francisco: Jossey-Bass.

BP (British Petroleum). 2009. *Statistical review of world energy, June 2009.* British Petroleum http://www.bp.com/statisticalreview (accessed June 1, 2010).

BP (British Petroleum). 2010. *Statistical review of world energy, June 2010.* British Petroleum, http://www.bp.com/liveassets/bp_internet/globalbp/globalbp_uk_english/reports_and_publications/statistical_energy_review_2008/STAGING/local_assets/2010_downloads/statistical_review_of_world_energy_full_report_2010.pdf

Brennan, T. J., K. L. Palmer, and S. A. Martinez. 2002. *Alternating currents.* Washington, DC: Resources for the Future.

Brinkley, A. 1995. *The end of reform.* New York: Knopf.

Broder, D., and H. Johnson. 1996. *The system.* New York: Little, Brown.

Brown, A. C. 2002. The power-line problem starts with the states. *Newsday* (August 18), http://www.newsday.com/news/opinion (accessed February 1, 2009).

Brown, G. 1937. *The romance of city light.* Seattle: Star Publishing Co.

Brown, M. A. 2007. Energy myth one: Today's energy crisis is "hype." In *Energy and American society: Thirteen myths*, ed. Benjamin K. Sovacool and Marilyn A. Brown, 23–50. Dordrecht, The Netherlands: Springer.

Brown, M. H., and R. P. Sedano. 2003. *A comprehensive view of U.S. electric restructuring with policy options for the future.* Washington, DC: National Council on Electricity Policy.

Bruchley, S. 1990. *Enterprise: The dynamic economy of a free people.* Cambridge, MA: Harvard University Press.

Bryant, K. L., Jr., and H. C. Dethloff. 1990. *A history of American business.* 2nd ed. Englewood Cliffs, NJ: Prentice Hall.

Bryce, R. 2008. *Gusher of lies: The dangerous delusions of "energy independence."* New York: PublicAffairs.

Bryson, J. M. 1988. *Strategic planning for public and nonprofit organizations.* San Francisco: Jossey-Bass.

BTS (Bureau of Transportation Statistics). 2003. *Vehicle Ownership and Availability.* Washington, DC: Bureau of Transportation Statistics, U.S. Department of Transportation. http://www.bts.gpov/publications/transportation-statistics-annual-report/2003/html/chapter-02/vehicle-ownership-and-availability-html (accessed January 11, 2011).

Bumpus, A. G., and D. M. Liverman. 2008. Accumulation by decarbonization and the governance of carbon offsets. *Economic Geography* 84 (2): 127–155.

Bureau of Economic Analysis. 2009. Table 3.3ES. Historical-cost net stock of private fixed assets by industry. U.S. Department of Commerce, http://www.bea.gov/national/FA2004/SelectTable.asp#S3 (accessed December 21, 2009).

Burgelman, R. A., C. M. Christensen, and S. C. Wheelwright. 2004. *Strategic management of technology and innovation*, 4th ed. Boston: McGraw-Hill Irwin.

Burstein, P. 2003. The impact of public opinion on public policy: A review and an agenda. *Political Research Quarterly* 56 (1): 29–40.

Burton, B. 2007. Battle tanks: How think tanks shape the public agenda. Center for Media and Democracy, PR Watch.org, http://www.prwatch.org/prwissues/2005Q4/battletanks (accessed March 20, 2010).

Busch, P-O., H. Jörgens, and K. Tews. 2005. The global diffusion of regulatory instruments: The making of a new international environmental regime. *Annals of the American Academy of Political and Social Science* 598 (March): 146–167.

California Public Utilities Commission. 2009. *Status of energy utility service disconnections in California.* Sacramento, CA: Division of Ratepayer Advocates, http://www.dra.ca.gov/NR/rdonlyres/2A0C5457-56FC-4821-8C4D-457F4CF204D1/0/20091119_DRAdisconnectionstatusreport.pdf (accessed January 4, 2010).

Caperton, R., and S. Gandhi. *America's hidden power bill: Examining federal energy tax expenditures.* Washington, DC: Center for American Progress.

Carpini, M. X. D., and S. Keeter. 1996. *What Americans Know about Politics and Why it Matters*. New Haven: Yale University Press.

Caruson, K., and V. A. Farar-Myers. 2007. Promoting the president's foreign policy agenda. *Political Research Quarterly* 60 (4): 631–644.

Cascio, J. 2005. The 2000 watt society. WorldChanging (June 2), http://www.worldchanging.com/archives/002829.html (accessed June 22, 2010).

Castern, T. R., and R. U. Ayres. 2007. Energy myth eight: Worldwide power systems are economically and environmental optimal. In *Energy and American society: Thirteen myths*, ed. Benjamin K. Sovacool and Marilyn A. Brown, 201–238. Dordrecht, The Netherlands: Springer.

CBO (Congressional Budget Office). 2003. *Cost estimate S.14 Energy Policy Act of 2003* (May 7). Washington, DC: Congressional Budget Office, http://www.cbo.gov/ftpdocs/42xx/doc4206/s14.pdf (accessed May 20, 2010).

CBO (Congressional Budget Office). 2008a. *Nuclear power's role in generating electricity*. Pub. 2986 (May). Washington, DC: Congressional Budget Office, http://www.cbo.gov/ftpdocs/91xx/doc9133/05-02-Nuclear.pdf (accessed May 11, 2010).

CBO (Congressional Budget Office). 2008b. *Policy options for reducing CO_2 emissions*. Washington, DC: Congressional Budget Office, http://www.cbo.gov/ftpdocs/89xx/doc8934/02-12-Carbon.pdf (accessed March 19, 2010).

CBO (Congressional Budget Office). 2009. *Budget options, volume 2* (August). Washington, DC: Congressional Budget Office.

CBO (Congressional Budget Office). 2010. *Federal climate change programs: Funding history and policy issues*. Pub. No. 4025 (March). Washington, DC: Congressional Budget Office, http://www.cbo.gov/ftpdocs/112xx/doc11224/03-26-ClimateChange.pdf (accessed May 17, 2010).

CBS. 2009. Obama lays out new energy plan, CBS News.com (January 26), http://www.cbsnews.com/stories/2009/01/26/politics/100days/main4753112.shtml (accessed June 20, 2009).

CESA (Clean Energy States Alliance). 2005. *Overview of Energy Policy Act of 2005*. Clean Energy States Alliance, http://www.cleanenergystates.org/library/Reports/CESA_Analysis_Energy%20Bill_%202005_8.4.05.pdf (accessed April 8, 2009).

Chandler, J. 2009. Trendy solutions: Why do states adopt sustainable energy portfolio standards? *Energy Policy* 37 (2009): 3274–3281.

Charles, D. 2009. Leaping the efficiency gap. *Science* 325 (August 14): 804–811.

Chasek, P., D. Downie, and J. W. Brown. 2010. *Global environmental politics*. 5th ed. Boulder, CO: Westview Press.

Cheney, D. 2001. *Report of the National Energy Policy Development Group*. Washington, DC: U.S. Government Printing Office.

Chertow, M. R., and D. C. Esty, eds. 1997. *Thinking ecologically: The next generation of environmental policy*. New Haven, CT: Yale University Press.

Chestney, N. 2010. Hopes for $2 trillion global carbon market fade. Reuters (March 3), http://www.reuters.com/article/idUSTRE6223KZ20100303 (accessed May 29, 2010).

Choma, R. 2010. Renewable energy money still going abroad, despite criticism from Congress. Investigative Reporting Workshop (February 8), http://investigativereportingworkshop.org/investigations/wind-energy-funds-going-overseas/story/renewable-energy-money-still-going-abroad/

Christensen, T., and P. Lægreid, eds. 2002. *New public management*. Aldershot, U.K.: Ashgate Publishing.

Chu, S. 2010. *Meeting the energy and climate challenge*. Presentation at Energy Information Administration Conference, Washington, DC, April 6, http://www.eia.doe.gov/conference/2010/plenary/chu.pdf (accessed May 19, 2010).

Chupka, M., R. Earle, P. Fox-Penner, and R. Hledik. 2008. *Transforming America's power industry: The investment challenge 2010–2030*. Washington, DC: The Edison Foundation, http://www.eei.org/ourissues/finance/Documents/Transforming_Americas_Power_Industry.pdf (accessed April 14, 2010).

Ciolek, M., W. Jones, and W. Wilson. 2003. Utility ratemaking and ROE: Rethinking the tools of the trade. *Public Utilities Fortnightly* 141 (October 15): 24–29.

Clark, J. G. 1990. *The political economy of world energy*. Chapel Hill, NC: University of North Carolina Press.

Claussen, E., and J. Peace. 2007. Energy myth twelve: Climate policy will bankrupt the U.S. economy. In *Energy and American society: Thirteen myths*, ed. Benjamin K. Sovacool and Marilyn A. Brown, 311–340. Dordrecht, The Netherlands: Springer.

Clayton, M. 2004. The coal rush. *The Seattle Times* (February 27): A3.

Cleetus, R. 2008. *We Need a Well-Designed Cap-and-Trade Program to Fight Global Warming*. Union of Concerned Scientists Fact Sheet. http://www.ucsusa.org/global_warming/solutions/big_picture_solutions/cap-and-trade.html (accessed January 16, 2011).

Cleveland, C. 2008. Ten fundamental principles of net energy. In *Encyclopedia of Earth*, ed. C. Cleveland (last revised September 23). Washington, DC: Environmental Information Coalition, National Council for Science and the Environment, http://www.eoearth.org/article/Ten_fundamental_principles_of_net_energy (accessed June 22, 2010).

Coburn, T., and J. McCain. 2009. Out of gas: Congress raids the highway trust fund for pet projects while bridges and roads crumble. American Trails, http://www.americantrails.org/resources/fedfund/Out-of-gas-Senator-Coburn-McCain-2009.html (accessed October 31, 2010).

Cochran, C. E., L. C. Mayer, T. R. Carr, and N. J. Cayer. 1996. *American public policy.* 5th ed. New York: St. Martin's Press.

Cochran, T. C. 1972. *American business in the twentieth century.* Cambridge, MA: Harvard University Press.

Coe, G. 1985. California's experience in promoting renewable energy development. In *State energy policy: Current issues, future directions*, ed. Stephen W. Sawyer and John R. Armstrong, 193–212. Boulder, CO: Westview.

Cohen, M. D., J. G. March, and J. P. Olsen, 1972. A garbage can model of organizational choice. *Administrative Science Quarterly* 17 (1): 1–25.

Committee on America's Energy Future. 2009. *America's energy future: Technology and transformation*. National Academy of Sciences, National Academy of Engineering, National Research Council of the National Academies. Washington, DC: National Academy of Sciences.

Cooper, H., and J. Broder. 2010. Obama vows end to "cozy" oversight of oil industry. *New York Times* (May 15): A13.

Cory, K., and T. Couture 2009. *State Clean Energy Policies Analysis (SCEPA) Project: An Analysis of Renewable Energy Feed-in Tariffs in the United States*, National Renewable Energy Laboratory (May). http://www.nrel.gov/docs/fy09osti/45551.pdf (accessed January 16, 2011).

Cory, K. S. and B. Swezey. (2007). Renewable portfolio standards in the states: balancing goals and rules (NREL Report No. JA-640-41876). *Electricity Journal*, 40 (4): 21–32.

Costello, K. W. 2003. *Exploratory questions and issues pertaining to the future role of liquefied natural gas in the U.S. market.* Columbus, OH: National Regulatory Research Institute.

Costello, K. W., and R. Burns. 2003. Era of low gas prices may be behind us. *The NRRI Networker* (summer): 1.

Costello, K. W., and D. J. Duann. 1996. Turning up the heat in the natural gas industry. *Regulation: The Cato Review of Business and Government.* 19 (1).

Craddock, J., and B. Hogue. 2004. Natural gas storage: Now more than ever. *Public Utilities Fortnightly* 142 (July): 60–66.

Craig, P., A. Gadgil, and J. Koomey. 2002. What can history teach us? A retrospective examination of long-term energy forecasts for the United States. *Annual Review of Energy and the Environment* 27 (2002): 119–158.

Crandall, R. W. 1992. Policy watch: Corporate average fuel economy standards. *Journal of Economic Perspectives* 6 (2): 171–180.

Crandall, R. W., and J. D. Graham. 1989. The effect of fuel economy standards on automobile safety. *Journal of Law and Economics* 32 (1): 97–118.

Crane, K., A. Goldthau, M. Toman, T. Light, S. E. Johnson, A. Nader, A. Ragasa, and H. Dogo. 2009. *Imported oil and U.S. national security.* Santa Monica, CA: RAND Corporation, http://www.rand.org/pubs/monographs/2009/RAND_MG838.pdf (accessed June 12, 2010).

Crew, M. A., ed. 1985. *Analyzing the impact of regulatory change in public utilities.* Lexington, MA: Lexington Books.

Crew, M. A., and P. R. Kleindorfer 1986. *The economics of public utility regulation.* Cambridge, MA: MIT Press.

Creyts, J., A. Derkach, S. Nyquist, K. Ostrowski, and J. Stephenson. 2007. *Reducing U.S. greenhouse gas emissions: How much at what cost? U.S. greenhouse gas initiative executive report.* Chicago: McKinsey and Company.

Crosbie, T. 2008. Household energy consumption and consumer electronics: The case of television. *Energy Policy* 36 (2008): 2191–2199.

Cross, P. S. 2003. A survey of recent PUC hearings. *Public Utilities Fortnightly* 141 (November 15): 32–36.

CRS (Congressional Research Service). 2005. *Energy tax policy: An economic analysis.* RL30406. Washington, DC: Congressional Research Service, http://ncseonline.org/NLE/CRSreports/05jun/RL30406.pdf (accessed April 19, 2010).

CRS (Congressional Research Service). 2006. *The federal excise tax on gasoline and the highway trust fund: A short history.* RL30304. Washington, DC: Congressional Research Service, http://ncseonline.org/NLE/CRSreports/06May/RL30304.pdf (accessed May 25, 2010).

CRS (Congressional Research Service). 2009a. *Carbon tax and greenhouse gas control: Options and considerations for congress.* R40242. Washington, DC: Congressional Research Service, http://www.fas.org/sgp/crs/misc/R40242.pdf (accessed May 9, 2010).

CRS (Congressional Research Service). 2009b. *The role of federal gasoline excise taxes in public policy.* R40808. Washington, DC: Congressional Research Service, http://assets.opencrs.com/rpts/R40808_20090911.pdf (accessed May 25, 2010).

CRS (Congressional Research Service). 2010a. *Federal research and development funding*: FY2011. R41098. Washington, DC: Congressional Research Service, http://www.ntis.gov/search/product.aspx?ABBR=ADA516413 (accessed March 17, 2010).

Cummings, C. M., and D. A. Chase. 1992. Survival for utilities: A question of people. *Public Utilities Fortnightly* 129 (June 1): 24–27.

Danielsen, A. L., and D. R. Kamerswchen. 1983. *Current issues in public-utility economics.* Lexington, MA: Lexington Books.

Dasgupta, S. M. Huq, M. Khaliquzzaman, K. Pandey, and D. Wheeler. 2006. Who suffers from indoor air pollution? Evidence from Bangladesh. *Health Policy and Planning*, 21 (6) 444–458.

Davis, D. H. 1982. *Energy Politics.* 3rd ed. New York: St. Martin's Press.

Deffeyes, K. 2005. *Hubbert's peak: The impending world oil shortage.* Princeton, NJ: Princeton University Press.

de Gorter, H., and D. Just. 2010. The social costs and benefits of biofuels: The intersection of environmental, energy and agricultural policy. *Applied Economic Perspectives and Policy* 32 (1): 4–32.

DeLeon, P. 2006. The historical roots of the field. In *The Oxford handbook of public policy*, ed. M. Moran, M. Rein, and R. E. Goodin, 39–57. Cambridge, U.K.: Oxford University Press.

Delli Carpini, M., and S. Keeter. 1996. *What Americans know about politics and why it matters.* New Haven, CT: Yale University Press.

Denhardt, K. G. 1989. The management of ideals: A political perspective on ethics. *Public Administration Review* 49 (January): 187–193.

Depledge, J. 2000. *Tracing the origins of the Kyoto Protocol: An article-by-article textual history.* U.N. Framework Convention on Climate Change, http://unfccc.int/resource/docs/tp/tp0200.pdf (accessed February 22, 2010).

Dimock, M. E. 1935. *Business and government.* New York: Henry Holt.

Dixit, A. 2002. Incentives and organizations in the public sector: An interpretive review. *Journal of Human Resources* 37 (fall): 697–727.

DOD (U.S. Department of Defense). 2008. *National defense strategy.* Washington, DC: U.S. Department of Defense (June), http://permanent.access.gpo.gov/lps103291/2008%20national%20defense%20strategy.pdf (accessed June 22, 2010).

DOE (U.S. Department of Energy). 2000. Resource planning approval criteria. Department of Energy Rules and Regulation. Washington, DC: *Federal Register* 65 (March 20): 16788–17802.

DOE (U.S. Department of Energy). 2007. *Renewable portfolio standards in the states: Balancing goals and implementation strategies.* NREL/TP-670-41409. Golden, CO: National Renewable Energy Laboratory, http://www.nrel.gov/docs/fy08osti/41409.pdf (accessed June 1, 2010).

DOE (U.S. Department of Energy). 2008. *Analysis of crude oil production in the Arctic National Wildlife Refuge* SR/OIAF/2008-03 (May). Washington, DC: Energy Information Administration, http://www.eia.doe.gov/oiaf/servicerpt/anwr/pdf/sroiaf(2008)03.pdf (accessed June 19, 2010).

DOE (U.S. Department of Energy). 2009a. Energy timeline. Washington, DC: Department of Energy, http://www.energy.gov/about/timeline.htm (accessed April 7, 2009).

DOE (U.S. Department of Energy). 2009b. *Feed-in tariff policy: Design, implementation and RPS policy interactions.* NRL/TP-6A2-45549. Golden, CO: National Renewable Energy Laboratory, http://www.nrel.gov/docs/fy09osti/45549.pdf (accessed June 1, 2010).

DOE (U.S. Department of Energy). 2009c. *Strategic petroleum reserve—Quick facts and frequently asked questions.* Washington, DC: Department of Energy, http://www.fossil.energy.gov/programs/reserves/spr/spr-facts.html (accessed May 12, 2009).

DOE (U.S. Department of Energy). 2009d. *FY 2010 budget by organization.* Washington, DC: Department of Energy, http://www.cfo.doe.gov/budget/10budget/content/ApprSum.pdf (accessed June 29, 2009).

DOE (U.S. Department of Energy). 2009e. *Office of electricity delivery and energy reliability program specific recovery plan.* Washington, DC: Department of Energy, http://www.energy.gov/recovery/documents/Office_of_Electric_Delivery-Energy_Reliability_Recovery_Program_Plan.pdf (accessed December 28, 2009).

DOE (U.S. Department of Energy). 2009f. *Carbon capture and storage R&D overview.* Washington, DC: Department of Energy, http://www.fossil.energy.gov/sequestration/overview.html (accessed January 25, 2010).

DOE (U.S. Department of Energy). 2009g. *2008 renewable energy data book.* Golden, CO: National Renewable Energy Laboratory, http://www.nrel.gov/docs/fy09osti/45654.pdf (accessed November 24, 2009).

DOE (U.S. Department of Energy). 2009h. Crude oil proved reserves, reserves changes, and production. U.S. Energy Information Administration, http://www.eia.doe.gov/dnav/pet/pet_crd_pres_dcu_NUS_a.htm (accessed June 21, 2010).

DOE. 2010. DOE *Announces More than $5 Million to Support Wind Energy Development: Funds to Enhance Short-Term Wind Forecasting and Accelerate Midsize Wind Turbine Development.*" from http://www.energy.gov/9479.htm (accessed January 15, 2011).

DOE/NETL (U.S. Department of Energy/National Energy Technology Laboratory). 2007. *Peaking of world oil production: Recent forecasts.* Washington, DC: Department of Energy/National Energy Technology Laboratory, http://www.netl.doe.gov/energy-analyses/pubs/Peaking%20of%20World%20Oil%20Production%20-%20Recent%20Forecasts%20-%20NETL%20Re.pdf (accessed February 5, 2010).

DOI (U.S. Department of the Interior). 1974. *Federal energy administration project independence blueprint. Final task force report on coal.* Washington, DC: U.S. Department of the Interior. Report No. FE 1.18:C63.

Donaldson, T., and T. W. Dunfee. 1999. *Ties that bind: A social contracts approach to business ethics.* Boston: Harvard Business School Press.

DOT. 2009. CARS Will Put Safer, Cleaner, More Fuel Efficient Vehicles on Road. (June 24, 2010). Washington, DC: U.S. Department of Transportation. http://www.dot.gov/affairs/2009/dot8709.htm (accessed January 15, 2011).

Douglass, E. 2004. PUC promises to keep power plants in check. *Los Angeles Times,* May 7: C2.

Dulles, E. L. 1936. *Depression and reconstruction: A study of causes and controls.* Philadelphia: University of Pennsylvania Press.

Eadie, D. C. 1999. Putting a powerful tool to practical use. In *Public sector performance: Management, motivation, and measurement,* ed. Richard C. Kearney and Evan M. Berman, 133–147. Boulder, CO: Westview Press.

EBR (*Energy Business Review*). 2009. How dependent are we on foreign oil? *Energy Business Daily* (September), http://energybusinessdaily.com/oil/how-dependent-are-we-on-foreign-oil (accessed June 18 2010).

Economist, The. 2009a. Face value: The alternative choice. *Economist* 392 (8638): 64.

Economist, The. 2009b. Touch wood: Climate change and forests. *Economist* 393 (8662): 112.

Economist, The. 2009c. The EIA puts a date on peak oil production. *Economist* 393 (8661): 82.

Economist, The. 2009d. The other kind of solar power. *Economist* 391 (8634): 22.

Economist, The. 2010. The wrong sort of recycling. *Economist* 394 (March 27): 84.

Economist, The. 2010a. Fed up: Germany's solar subsidies. *Economist* 394 (8664): 66.

Edison Electric Institute. 2004. *Key facts about the electric power industry.* Edison Electric Institute, http://http://www.cewd.org/toolkits/teacher/eeipub_keyfacts_electric_industry.pdf (accessed November 15, 2004).

EERE (Energy Efficiency and Renewable Energy, U.S. Department of Energy). 2008. *State policy: Efficiency and renewable energy.* Washington, DC: U.S. Department of Energy, http://apps1.eere.energy.gov/states/state_policy.cfm?print (accessed January 31, 2009).

EERE (Energy Efficiency and Renewable Energy, U.S. Department of Energy). 2009. *Research funding summary by program ($US thousands).* Washington, DC: U.S. Department of Energy, http://www.eere.energy.gov/ba/pba/pdfs.FY2009-budget-brief.pdf (accessed June 8, 2010).

Ehrhardt, M. K., and J. Laitner. 2008 . *The size of the U.S. energy efficiency market: Generating a more complete picture.* Report Number E083 (May). Washington, DC: American Council for an Energy-Efficient Economy, http://www.aceee.org/research-report/e083 (accessed June 19, 2010).

EIA (U.S. Department of Energy, Energy Information Administration). 1997. *U.S. electric utility demand-side management 1996.* Washington, DC: U.S. Department of Energy, Office of Coal, Nuclear, Electric and Alternative Fuels, www.osti.gov/bridge/servlets/purl/563214-Rl8DxB/webviewable/

EIA (U.S. Department of Energy, Energy Information Administration). 2000. *The changing structure of the electric power industry 2000: An update* (October). Washington, DC: U.S. Department of Energy, Office of Coal, Nuclear, Electric and Alternate Fuels. No. DOE/EIA-0562 00.

EIA (U.S. Department of Energy, Energy Information Administration). 2002. *Petroleum chronology of events 1970–2000.* Washington, DC: Energy Information Administration, U.S. Department of Energy.

EIA (U.S. Department of Energy, Energy Information Administration). 2003. *Status of state electric industry restructuring activity as of February 2003* (last modified December 2, 2003). Energy Information Administration, http://www.eia.doe.gov/electricity/chg_str/restructure.pdf (accessed 5 May 2004).

EIA (U.S. Department of Energy, Energy Information Administration). 2004a. *Annual energy outlook: Issues in focus.* Washington, DC: Department of Energy, Energy Information Administration, http://www.eia.doe.gov/oiaf/archive/aeo04/pdf/0383(2004).pdf

EIA (U.S. Department of Energy, Energy Information Administration). 2004b. *Annual energy outlook 2004 with projections to 2025.* Washington, DC: Department of Energy. Report No. DOE/EIA-0383 (January), http://www.eia.doe.gov/oiaf/aeo/pdf/aeotab_21.pdf (accessed June 12, 2009).

EIA (U.S. Department of Energy, Energy Information Administration). 2004c. *Status of State Electric Industry Restructuring Activity—As of February 2003* from http://www.eia.doe.gov/cneaf/electricity/chg_str/regmap.html . (accessed June 4 2004).

EIA (U.S. Department of Energy, Energy Information Administration). 2005. *Annual energy outlook 2005, natural gas supply and disposition.* Washington, DC: U.S. Department of Energy, http://tonto.eia.doe.gov/ftproot/forecasting/0383(2005).pdf

EIA (U.S. Department of Energy, Energy Information Administration). 2008a. *International energy outlook 2008.* Washington, DC: U.S. Department of Energy, http://www.eia.doe.gov/oiaf/ieo/pdf/0484(2008).pdf (accessed December 26, 2009).

EIA (U.S. Department of Energy, Energy Information Administration). 2008b. *Federal financial interventions and subsidies in energy markets 2007.* Washington, DC: U.S. Department of Energy Office of Coal, Nuclear, Electric and Alternate Fuels.

EIA (U.S. Department of Energy, Energy Information Administration). 2008c. *Annual energy outlook 2008 with projections to 2030.* Washington, DC: Energy Information Administration, U.S. Department of Energy.

EIA (U.S. Department of Energy, Energy Information Administration). 2008d. *Analysis of crude oil production in the Arctic National Wildlife Refuge.* Washington, DC: Energy Information Administration, U.S. Department of Energy, http://www.eia.doe.gov/oiaf/servicerpt/anwr/pdf/sroiaf(2008)03.pdf (accessed June 19, 2010).

EIA (U.S. Department of Energy, Energy Information Administration). 2009a. *How dependent are we on foreign oil?* Washington, DC: U.S. Department of Energy, Energy Information Administration (April), http://tonto.eia.doe.gov/cfapps/energy_in_brief/foreign_oil_dependence.cfm?featureclicked=3 (accessed April 27, 2009).

EIA (U.S. Department of Energy, Energy Information Administration). 2009b. *Short-term energy outlook* (May). Washington, DC: U.S. Department of Energy, Energy Information Administration, http://www.eia.doe.gov/emeu/steo/pub/archives/may09.pdf

EIA (U.S. Department of Energy, Energy Information Administration). 2009c. *Annual energy outlook 2009 with projections to 2030: Electricity demand.* Global Oil Watch, http://www.globaloilwatch.com/reports/EIAEO09.pdf (accessed June 28, 2009).

EIA (U.S. Department of Energy, Energy Information Administration). 2009d. *U.S. carbon dioxide emissions from energy sources: 2008 flash estimate.* Washington, DC: U.S. Department of Energy, Energy Information Administration (May), http://www.eia.doe.gov/oiaf/1605/flash/pdf/flash.pdf (accessed February 5, 2010).

EIA (U.S. Department of Energy, Energy Information Administration). 2009e. *Emissions of greenhouse gases in the United States 2008.* Washington, DC: U.S. Department of Energy, Energy Information Administration, http://www.eia.doe.gov/oiaf/1605/ggrpt/pdf/0573(2008).pdf (accessed March 30 2010).

EIA (U.S. Department of Energy, Energy Information Administration). 2009f. Table 2.1.A, Coal: Consumption for electricity generation by sector, 1995 through September 2009. Washington, DC: U.S. Department of Energy, http://www.eia.doe.gov/cneaf/electricity/epm/table2_1_a.html (accessed December 21, 2009).

EIA (U.S. Department of Energy, Energy Information Administration). 2009g. *Annual energy review 2008.* Washington, DC: U.S. Department of Energy, http://people.virginia.edu/~gdc4k/phys111/fall09/important_documents/aer_2008.pdf (accessed December 29, 2009).

EIA (US Department of Energy, Energy Information Administration). 2009h. International Energy Outlook 2010: Electricity. Washington, DC: U.S. Department of Energy. http://www.eia.doe.gov/oiaf/ieo/electricity.html (accessed January 16, 2011).

EIA (US Department of Energy, Energy Information Administration). 2009i. *U.S. Crude Oil, Natural Gas, and Natural Gas Liquids Proved Reserves, 2009.* http://www.eia.doe.gov/natural_gas/data_publications/crude_oil_natural_gas_reserves/cr.html (accessed January 16, 2011).

EIA (U.S. Department of Energy, Energy Information Administration). 2010a. *Electricity explained: Your guide to understanding energy.* Washington, DC: U.S. Department of Energy, Energy Information Administration, http://www.eia.doe.gov/energyexplained/print.cfm?page=electricity_in_the_united-states.pdf (accessed March 31, 2010).

EIA (U.S. Department of Energy, Energy Information Administration). 2010b. *U.S. [crude oil] imports by country of origin.* Washington, DC: Energy Information Administration, U.S. Department of Energy.

EIA (U.S. Department of Energy, Energy Information Administration). 2010c. *Annual energy review 2009.* Washington, DC: U.S. Department of Energy, http://www.eia.doe.gov/aer/pdf/aer.pdf

EIA (U.S. Department of Energy, Energy Information Administration). 2010d. *Annual coal report 2008.* Washington, DC: U.S. Department of Energy, Energy Information Administration, http://www.eia.doe.gov/cneaf/coal/page/acr/acr.pdf (accessed August 31, 2010).

EIA (U.S. Department of Energy, Energy Information Administration). 2010e. *Electric power annual 2008.* Washington, DC: U.S. Department of Energy, Energy Information Administration, http://www.eia.doe.gov/cneaf/electricity/epa/epa.pdf (accessed August 31, 2010).

EIA (US. Department of Energy, Energy Information Administration). 2010f. *Natural Gas Annual 2009.* Washington, DC: U.S. Department of Energy, Energy Information Administration, http://www.eia.gov/pub/oil_gas/natural_gas/data_publications/natural_gas_annual/current/pdf/nga09.pdf (accessed January 19, 2011).

EIA (US. Department of Energy, Energy Information Administration). 2010g. *Energy in brief: how dependent are we on foreign oil?* http://www.eia.doe.gov/energy_in_brief/foreign_oil_dependence.cfm (accessed January 21, 2010).

Eilperin, J. 2010. Frank Luntz's climate advice for environmentalists. *Washington Post* blog (January 21), http://views.washingtonpost.com/climate-change/post-carbon/2010/01/luntzs_climate_advice_for_environmentalists.html (accessed June 15, 2010).

Einstein, A. 2007. Albert Einstein quoted by U.S. Internal Revenue Service (IRS), Tax Quotes. http://www.irs.gov/newsroom/article/0,,id=110483,00.html (accessed January 15, 2011).

Eland, Ivan. 2008. U.S. dependence on foreign oil: Why we shouldn't be alarmed. Washington, DC: Independent Institute (September 1), http://www.independent.org/newsroom/article.asp?id=2306 (accessed June 21, 2010).

Ely, R. T. 1910. *Monopolies and trusts.* New York: Macmillan.

Emerson, S. M. 2002. California's electric deregulation and its implications. *Public Works Management and Policy* 7 (July): 19–31.

Energy Charter. *About the charter.* Energy Charter website, http://www.encharter.org/index.php?id=7&L=0 (accessed August 20, 2010).

Engel, K. 2006. State and local climate change initiatives: What is motivating state and local governments to address a global problem and what does this say about federalism and environmental law? *Urban Lawyer* 38 (4): 1015–1027.

Engels, A. 2005. The science-policy interface. *Bridging Sciences and Policy* 5 (1): 7–26.

ENS. 2010. Nuclear Power Plants, World-wide. Brussels, Belgium: European Nuclear Society. http://www.euronuclear.org/info/encyclopedia/n/nuclear-power-plant-world-wide.htm (accessed January 15, 2011).

EPA (U.S. Environmental Protection Agency). 2005. *Energy Policy Act of 2005: Public Law* 109-58 (August 8, 2005). U.S. Environmental Protection Agency, http://www.epa.gov/oust/fedlaws/publ_109-058.pdf (accessed April 9, 2009).

EPA. (U.S. Environmental Protection Agency). 2008. *Advanced Notice of Proposed Rulemaking.* http://www.epa.gov/climatechange/emissions/downloads/ANPRPreamble5.pdf (accessed January 15, 2011).

EPA (U.S. Environmental Protection Agency). 2009a. *EPA analysis of the American Clean Energy and Security Act of 2009, H.R. 2454 in the 111th Congress.* U.S. Environmental Protection Agency, Office of Atmospheric Programs, http://www.epa.gov/climatechange/economics/pdfs/HR2454_Analysis.pdf

EPA. (U.S. Environmental Protection Agency). 2009b. *Inventory of U.S. greenhouse gas emissions and sinks*: 1990–2007 (April 15). http://www.epa.gov/climatechange/emissions/downloads09/InventoryUSGhG1990-2007.pdf (accessed January 5, 2010).

EPA (U.S. Environmental Protection Agency). 2010. *Energy Star® overview of 2009 achievements.* Washington, DC: U.S. Environmental Protection Agency, https://www.EnergyStar.gov/ia/partners/annualreports/2009_achievements.pdf (accessed May 20, 2010).

European Environment Agency. 1997. *Life cycle assessment: A guide to approaches, experiences and information sources.* Copenhagen: European Environment Agency, http://www.eea.europa.eu/publications/GH-07-97-595.../Issue-report-No-6.pdf (accessed June 21, 2010).

Ewers, J. 2008. The myth of energy independence. *U.S. News and World Report* (March 17), http://www.usnews.com/articles/news/2008/03/17/the-myth-of-energy-independence_print (accessed March 29, 2009).

Farris, M. T., and R. J. Sampson. 1973. *Public utilities: Regulation, management, and ownership.* Boston: Houghton Mifflin.

Fattouh, B. 2006. OPEC's discount on heavy crude oil: Is a new policy instrument taking shape? *Oxford Energy Comment* (June 2006), Oxford Institute for Energy Studies, http://www.oxfordenergy.org/pdfs/comment_0606-3.pdf (accessed March 30, 2010).

Fattouh, B. 2009a. *Oil price dynamics and the role of inventories: Evidence for the crude oil market.*Oxford, U.K.: Oxford University, Oxford Institute for Energy Studies.

Fattouh, B. 2009b. *Light-duty diesel vehicles: Market issues and potential emissions impacts.* Washington, DC: U.S. Department of Energy (January), http://www.eia.doe.gov/oiaf/servicerpt/lightduty/pdf/sroiaf(2009)02.pdf (accessed May 26, 2010).

Federal Register. 2009. Endangerment and cause or contribute findings for greenhouse gases under Section 202(a) of the Clean Air Act. 2009. Final Rule. 74 *Federal Register* (December 15): 239.

Felder, F. A., and R. Haut. 2008. Balancing alternatives and avoiding false dichotomies to make informed U.S. electricity policy. *Policy Sciences* 41 (5): 165–180.

Felix, E. and D. R. Tilley. 2008. Integrated energy, environmental and financial analysis of ethanol production from cellulosic switchgrass. *Energy*, 34 (4): 410–436.

FHA (U.S. Deparatment of Transportation, Federal Highway Administration) 2008. *Highway Statistics 2007.* Table VM-1, Annual Vehicle Distance Traveled in Miles and Related Data – 2007. http://www.fhwa.dot.gov/policyinformation/statistics/2007/vm1.cfm. (accessed January 4, 2010).

Field, C. B., J. E. Campbell, and D. B. Lobell. 2008. Biomass energy: The scale of the potential resource. *Trends in Ecology and Evolution*, 23(2): 65–72.

Finon, D., T. A. Johnsen, and A. Midttun. 2004. Challenges when electricity markets face the investment phase. *Energy Policy* 32 (12): 1355–1362.

Flannery, Tim. 2005. *The Weather Makers: How Man is changing the Climate and What it Means to Life on Earth*. New York: Atlantic Monthly Press.

Folmer, H., and T. Tietenberg, eds. 1999. *International yearbook of environmental and resource economics 1999/2000: A survey of current issues.* Cheltenham, U.K.: Edward Elgar.

Fox-Penner, P., and G. Basheda. 2001. A short honeymoon for utility deregulation. *Issues in Science and Technology* 17 (spring): 51–57.

Frankel, J. 2009. Addressing the leakage/competitiveness issue in climate change policy proposals. In *Climate change, trade and investment: Is a collision inevitable?*, ed. L. Brainard. Washington, DC: Brookings Institution Press.

Frankena, M. W. 2001. Geographic market delineation for electric utility mergers. *The Antitrust Bulletin* 46 (summer): 357–402.

Freed J., A. Zevin, and J. Jenkins. 2009. *Jumpstarting a clean energy revolution with a National Institutes of Energy*. Breakthrough Institute, http://thebreakthrough.org/blog/Jumpstarting_Clean_Energy_Sept_09.pdf (accessed May 20, 2010).

Freeman, J., and C. Kolstad, eds. 2007. *Moving to markets in environmental regulation*. New York: Oxford University Press.

Frenzel, C. W. 1999. *Management of information technology.* Cambridge, MA: Course Technology ITP.

Friedman, T. 2008. *Hot, flat and crowded.* New York: Farrar, Straus and Giroux.

Frohock, F. M. 1979. *Public policy: Scope and logic.* Englewood Cliffs, NJ: Prentice Hall.

Frondel, M., N. Ritter, and C. Schmidt, 2008. Germany's solar cell promotion: Dark clouds on the horizon. *Energy Policy* 36 (2008): 4198–4204.

Fuglestvedt, J. S., and B. Romstad, 2006. Who's to blame for global warming? *Cicerone* (Oslo, Norway: Center for International Climate and Environmental Research), 2006 (2). http://www.cicero.uio.no/fulltext/index_e.aspx?id=4218 (accessed January 11, 2011).

Fung, I. Y., S. C. Doney, K. Lindsay, and J. John. 2005. Evolution of carbon sinks in a changing climate. *Proceedings of the National Academy of Sciences of the United States of America* 102 (32): 11201–11206.

Furman, J., J. Bordoff, M. Deshpande, and P. Noel. 2007. *An economic strategy to address climate change and promote energy security*. Washington, DC: Brookings Institution, http://www.brookings.edu/~/media/Files/rc/papers/2007/10climatechange_furman/10_climatechange_furman.pdf (accessed May 28, 2010).

Furrey, L., S. Nadel, and J. Laitner. 2009. *Laying the foundation for implementing a federal energy efficiency resource standard.* Report E091 (March). Washington, DC: American Council for an Energy-Efficient Economy, http://www.aceee.org/pubs/e091.htm (accessed April 19, 2010).

Galston, W. A. 2006. Politics and feasibility: Interests and power. In *The Oxford handbook of public policy*, ed. M. Moran, M. Rein, and G. E. Goodin, 543–556. Oxford, U.K.: Oxford University Press.

GAO (U.S. Government Accountability Office). 2005. *National energy policy: Inventory of major federal energy programs and status of policy recommendations.* GAO-05-379. Washington, DC: U.S. Government Accountability Office.

GAO (U.S. Government Accountability Office). 2006. *Department of energy: Key challenges remain for developing and deploying advanced energy technologies to meet future needs.* GAO-07-106. Washington, DC: U.S. Government Accountability Office.

GAO (U.S. Government Accountability Office). 2007a. *Energy efficiency: Long-standing problems with DOE's program for setting efficiency standards continue to result in forgone energy savings.* Report GAO-07-42. Washington, DC: U.S. Government Accountability Office.

GAO (U.S. Government Accountability Office). 2007b. *International energy: International forums contribute to energy cooperation within constraints.* GAO-07-170. Washington, DC: U.S. Government Accountability Office.

GAO (U.S. Government Accountability Office). 2007c. *Passenger vehicle fuel economy: Preliminary observations on corporate average fuel economy standards.* GAO-07-551T. Washington, DC: U.S. Government Accountability Office.

GAO (U.S. Government Accountability Office). 2007d. *Vehicle fuel economy: Reforming fuel economy standards could help reduce oil consumption by cars and light trucks, and other options could complement these standards.* GAO-07-921. Washington, DC: U.S. Government Accountability Office.

GAO (U.S. Government Accountability Office). 2008a. *Carbon offsets: The U.S. voluntary market is growing, but quality assurance poses challenges for market participants.* Report GAO-08-1048. Washington, DC: U.S. Government Accountability Office.

GAO (U.S. Government Accountability Office). 2008b. *Department of energy: Key challenges remain for developing and deploying advanced energy technologies to meet future needs.* GAO-07-106. Washington, DC: U.S. Government Accountability Office.

GAO (U.S. Government Accountability Office). 2008c. *Advanced energy technologies: Budget trends and challenges for DOE's energy R&D program.* GAO-08-556T. Washington, DC: U.S. Government Accountability Office.

GAO (U.S. Government Accountability Office). 2008d. *Department of Energy: New loan guarantee program should complete activities necessary for effective and accountable program management.* GAO-08-750. Washington, DC: U.S. Government Accountability Office.

GAO (U.S. Government Accountability Office). 2009a. *Green affordable housing: HUD has made progress in promoting green building, but expanding efforts could help reduce energy costs and benefit tenants.* GAO-09-46. Washington, DC: U.S. Government Accountability Office.

GAO (U.S. Government Accountability Office). 2009b. *Clean coal: DOE's decision to restructure FutureGen should be based on a comprehensive analysis of costs, benefits, and risks.* GAO-09-248. Washington, DC: U.S. Government Accountability Office.

GAO (U.S. Government Accountability Office). 2009c. *International climate change programs: Lessons learned from the European Union's emissions trading scheme and the Kyoto Protocol's clean development mechanism.* GAO-09-151. Washington, DC: U.S. Government Accountability Office.

GAO (U.S. Government Accountability Office). 2009d. *Climate change policy: Preliminary observations on options for distributing emissions allowances and revenue under a cap-and-trade program.* GAO-09-950T. Washington, DC: U.S. Government Accountability Office.

Gaskell, G., and B. Joerges, eds. 1987. *Public policies and private actions.* Aldershot, U.K.: Gower.

Gaudiano, N. 2010. Vermont's delegation in Washington calls to end tax breaks for oil companies. *Burlington Free Press* (June 16): A1, http://www.burlingtonfreepress.com/article/20100616/NEWS03/100615014/Vermont-s-delegation-in-Washington-calls-to-end-tax-breaks-for-oil-companies#ixzz0r1wKOtg0 (accessed June 18, 2010).

Gaul, D., and L. W. Young. 2003. *U.S. LNG markets and uses.* Washington, DC: Energy Information Administration, Office of Oil and Gas.

Gecan, R., R. Johansson, and K. FitzGerald. (April 2009). *The impact of ethanol use on food prices and greenhouse-gas emissions.* Washington, DC: Congressional Budget Office.

Gentemann, K. M., ed. 1981. *Social and political perspectives on energy policy.* New York: Praeger.

Geo-Help. 2009. History of the world petroleum industry (key dates). Geo-Help, Inc., http://www.geohelp.net/world.html (accessed December 29, 2009).

Ghahramani, B. 2003. A telecommunication's lean management information system for the utility industry. *International Journal of Information Technology & Decision Making* 2 (4): 693–715.

Gillingham, K., R. G. Newell, and K. Palmer. 2004. *Retrospective examination of demand-side energy efficiency policies.* Discussion paper RFF-DP-04. Resources for the Future, http://www.rff.org/Documents/RFF-DP-04-19rev.pdf (accessed May 14, 2010).

Gillingham, K., R. G. Newell, and K. Palmer. 2009. *Energy efficiency economics and policy.* Resources for the Future, http://www.rff.org/documents/RFF-DP-09-13.pdf (accessed May 20, 2010).

Gipe, P. 2010. Germany to raise solar target for 2010 and adjust tariffs. Renewable Energy World.com (June 2), http://www.renewableenergyworld.com/rea/news/article/2010/06/germany-to-raise-solar-target-for-2010-adjust-tariffs (accessed June 2, 2010).

Glaeser, M. G. 1957. *Public utilities in American capitalism.* New York: Macmillan.

Globalwarmingart.com. Gallery of Data Related to Carbon Dioxide. Global Warming Art, http://www.globalwarmingart.com/wiki/Carbon_Dioxide_Gallery (accessed February 5, 2010).

Glover, L. 2006. From love-in to logos: Charting the demise of renewable energy as a social movement. In *Transforming power: Energy, environment and society in conflict*, ed. J. Byrne, N. Toly, and L. Glover. New Brunswick, NJ: Transaction Publishers.

Goffman, E. 1974. Frame analysis: An essay on the organization of experience. New York: Harper and Row.

Goldenbemberg, J. T., B. Johansson, A. K. N. Reddy, and R. H. Williams. 2001. Energy for the new millennium. *Royal Swedish Academy of Sciences* 30 (6): 330–337.

Goodale, C. L., M. J. Apps, R. A. Birdsey, C. B. Field, L. S. Heath, R. A. Houghton, J. C. Jenkins, G. H. Kohlmaier, W. Kurz, S. Liu, G-T. Nabuurs, S. Nilsson, and A. Z. Shvidenko. 2002. Forest carbon sinks in the Northern Hemisphere. *Ecological Applications* 12 (3): 891–899.

Goode, D. 2010. Are climate meetings too big to succeed? *National Journal* (January 9): 12.

Goodell, J. 2006. *Big coal*. New York: Houghton Mifflin.

Goodin, R. E., M. Rein, and M. Moran. 2006. The public and its policies. In *The Oxford handbook of public policy*, ed. Michael Moran, Martin Rein, and Robert E. Goodin, 3–35. Oxford, U.K.: Oxford University Press.

Gordon, R. L. 2001. Don't restructure electricity; deregulate. *CATO Journal* 20 (winter): 327–359.

Gormley, W. T. 1983. *The politics of public utility regulation*. Pittsburgh, PA: University of Pittsburgh Press.

Goss, R. P. 1996. A distinct public administration ethics? *Journal of Public Administration Research & Theory* 6 (October): 573–598.

Gottron, F. 2001. *Energy efficiency and the rebound effect: Does increasing efficiency decrease demand?* Washington, DC: Congressional Research Service.

Grace, J. 2004. Understanding and managing the global carbon cycle. *Journal of Ecology* 92 (2): 189–202.

Granade, H., J. Creyts, A. Derkach, P. Farese, S. Nyquist, and K. Ostrowski. 2009. *Unlocking energy efficiency in the U.S. economy*. Washington, DC: McKinsey Global Energy and Materials.

Green, K., S. Hayward, and K. Hassett. 2007. Climate change: Caps vs. taxes. *Environmental Policy Outlook* 2 (June). Washington, DC: American Enterprise Institute, http://www.aei.org/outlook/26286 (accessed June 8, 2010).

Griffin, J. M., and H. B. Steele. 1980. *Energy economics and policy*. New York: Academic Press.

Grubb, M., C. Vrolijk, and D. Brack. 1999. *The Kyoto Protocol: A guide and assessment*. London: Royal Institute of International Affairs and Earth Scan.

Grubb, M., T. Brewer, B. Müller, J. Drexhage, K. Hamilton, T. Sugiyama, T. Aiba, A. Sharma, A. Michaelowa, C. Azar, and J. Karas. 2003. *A strategic assessment of the Kyoto-Marrakech System: Synthesis report*. London: Royal Institute of International Affairs Briefing Paper (June 6).

Grueber, M., and T. Studt. 2009. Re-emerging U.S. R&D. *R&D*, http://www.rdmag.com/Featured-Articles/2009/12/Policy-and-Industry-Re-Emerging-U-S-R-D/ (accessed May 5, 2010).

Gruening, E. 1931. *The public pays: A study of power propaganda*. New York: Vanguard Press.

Grunewald, D., and H. L. Bass, eds. 1966. *Public policy and the modern corporation*. New York: Meredith Publishing.

Gstalder, C. 2009. Only one in ten Americans are very knowledgeable about sources of electricity. *The Harris Poll* 85 (July 30), http://www.istockanalyst.com/article/viewiStockNews/articleid/3381083

Gumbel, P. 2009. Meet Shell's new CEO. *Fortune* 160 (2): 136.

Gupta, J. 2001. *Our simmering planet: What to do about global warming?* London: Zed Books.

Gupta, J., and R. Tol. 2003. Why reduce greenhouse gas emissions? Reasons, issue-linkages and dilemmas. In *Issues in international climate policy: Theory and policy*, ed. E. van Ierland, J. Gupta and M. Kok. Cheltenham, U.K.: Edward Elgar.

Gurko, L. 1947. *The angry decade*. New York: Dodd, Mead.

Haber, S. 1964. *Efficiency and uplift: Scientific management in the progressive era, 1890–1920*. Chicago: University of Chicago Press.

Hain, P. 2003. Why bigger is better. *New Economy: Journal of the Institute for Public Policy Research* 10 (June): 95–100.

Hajer, M. 2003. A frame in the fields: Policymaking and the reinvention of politics. In *Deliberative policy analysis: Understanding governance in the network society*, ed. M. A. Hajer, and H. Wagenaar. New York: Cambridge University Press.

Hajer, M., and D. Laws. 2006. Ordering through discourse. In *The Oxford handbook of public policy*, ed. M. Moran, M. Rein, and R. E. Goodin, 251–268. Oxford: Oxford University Press.

Hakes, J. 2008. *A declaration of energy independence.* Hoboken, NJ: John Wiley.
Hawley, E. W. 1966. *The New Deal and the problem of monopoly.* Princeton, NJ: Princeton University Press.
Hays, S. W., and R. C. Kearney. 2001. Anticipated changes in human resource management: Views from the field. *Public Administration Review* 61 (September): 585–596.
Heiman, M. K., and B. D. Solomon. 2004. Power to the people: Electric utility restructuring and the commitment to renewable energy. *Annals of the Association of American Geographers* 94 (1): 94–116.
Helgesson, C. F. 1999. *Making a natural monopoly.* Stockholm, Sweden: Economic Research Institute of the Stockholm School of Economics.
Helm, D, ed. 2005. *Climate change policy.* New York: Oxford University Press.
Helman, C., C. R. Schoenberger, and R. Wherry. 2005. The silence of the nuke protesters. *Forbes* 175 (January 31): 84–92.
Herzog, H., and D. Golomb. 2004. Carbon capture and storage from fossil fuel use. In *Encyclopedia of Energy,* ed. C. Cleveland, 277–287. San Diego, CA: Elsevier Science.
Heywood, A. 2000. *Key concepts in politics.* Houndmills, U.K.: Palgrave.
HHS (U.S. Department of Health and Human Services) 2009. *LIHEAP Home Energy Notebook for Fiscal Year 2007.* Washington, D.C.: U.S. Department of Health and Human Services, Administration for Children and Families, http://www.acf.hhs.gov/programs/ocs/liheap/publications/notebook2007.pdf (accessed December 30, 2009).
Hicks, J. D. 1960. *Normalcy and reaction, 1921–1933.* Washington, DC: Service Center for Teachers of History.
Hirst, E. 1985. Improving energy efficiency of existing homes: The residential conservation service. In *State energy policy: Current issues, future directions,* ed. Stephen W. Sawyer and John R. Armstrong, 85–106. Boulder, CO: Westview.
Hodgkinson, C. 1983. *The philosophy of leadership.* New York: St. Martin's Press.
Hogan, W. W. 2003. Electricity is a federal issue. *Wall Street Journal* (August 18). Available at Harvard Kennedy School of Government, http://www.hks.harvard.edu/fs/whogan/Hogan_WSJ_081803.pdf (accessed October 31, 2010).
Holden, M., Jr. 2006. Reflections on how political scientists (and others) might think about energy and policy. In *The Oxford handbook of public policy,* ed. Michael Moran, Martin Rein, and Robert E. Goodin, 874–891. Oxford, U.K.: Oxford University Press.
Hornsby, D. J., A. J. S. Summerlee, and K. B. Woodside. 2007. NAFTA's shadow hangs over Kyoto's implementation. *Canadian Public Policy/Analyse de Politiques* 33 (3): 285–297.
House of Representatives. 2009. *The American Clean Energy and Security Act of 2009 (ACES) discussion draft.* Energy and Commerce committees, http://energycommerce.house.gov/Press_111/20090331/acesa_discussiondraft.pdf (accessed April 12, 2009).
Hughes, L. 2009. The four R's of energy security. *Energy Policy* 37 (2009): 2459–2461.
Hughes, O. E. 2003. *Public management & administration.* Houndmills, U.K.: Palgrave Macmillan.
Hultman, N. 2010. *International climate governance: Will redefining "insiders" enable global progress?* Brookings, http://www.brookings.edu/opinions/2010/0430_climate_governance_hultman.aspx (accessed June 8, 2010).
Hyman, D., J. Bridger, J. Shingler, and M. Van Loon. 2001. Paradigms, policies, and people exploring the linkages between normative beliefs, public policies and utility consumer payment problems. *Policy Studies Review* 18 (summer): 89–122.
ICAO (International Civil Aviation Organization) 2009. *Annual Report of the Council.* Table 4. Regional distribution of scheduled traffic — 2008. http://www.icao.int/icaonet/dcs/9921/9921_en.pdf (accessed December 27, 2009).
IEA (International Energy Agency). 2007. *World energy assessment 2007.* Paris: Organization for Economic Cooperation and Development.
IEA (International Energy Agency). 2007b. *Energy Policies of IEA Countries Germany 2007 Review.* Paris: Organization for Economic Cooperation and Development.
IEA (International Energy Agency). 2009a. *Annual energy outlook 2009 with projections to 2030: Electricity demand.* Paris: Organization for Economic Cooperation and Development.

IEA (International Energy Agency). 2009b. *Towards a More Energy Efficient Future: Applying indicators to enhance energy policy*. http://www.iea.org/papers/2009/indicators_brochure2009.pdf (accessed January 18, 2011).

IEA (International Energy Agency). 2009c. *Key world energy statistics*. Washington, DC. U. S. Department of Energy, http://www.iea.org/textbase/nppdf/free/2009/key_stats_2009.pdf

IEA (International Energy Agency). 2010. *International energy agency: About the IEA*. Paris: Organization for Economic Cooperation and Development, http://www.iea.org/about/index.asp (accessed February 2, 2010).

IEF (International Energy Forum). 2010. International Energy Forum homepage, http://www.ief.org (accessed February 7, 2010).

Ilic, M. 2001. Understanding demand: The missing link in efficient electricity markets. Unpublished Working Paper No. MIT EL 01-014WP. Cambridge, MA: MIT Energy Laboratory.

Immergut, E. M. 2006. Institutional constraints on policy. In *The Oxford handbook of public policy*, ed. M. Moran, M. Rein, and G. E. Goodin, 557–571. Oxford, U.K.: Oxford University Press.

Ingram, H., A. L. Schneider, and P. deLeon (2007). Social construction and policy design. In *Theories of the policy process*, 2nd ed., ed. P. Sabatier, 93–128. Boulder, CO: Westview Press.

Institute for Energy, Law, and Enterprise. 2003. *Introduction to LNG*. Houston, TX: University of Houston Law Center, https://www.piersystem.com/clients/crisis_569/UniversityofHouston,IntituteofEnergy-INTRODUCTIONTOLNG.pdf

Interagency Working Group on Social Cost of Carbon, U.S. Government. 2010. Appendix 16A. Social cost of carbon for regulatory impact analysis under executive order 12866. In *Residential heating products final rule technical support document*, U.S. Department of Energy, http://www1.eere.energy.gov/buildings/appliance_standards/residential/printable_versions/heating_products_fr_tsd.html (accessed June 19, 2010).

Ivanovich, D. 2006. Oil lease blunder hidden 6 years. *Houston Chronicle* (September 14), http://www.chron.com/disp/story.mpl//4185660.html (accessed June 21, 2010).

Jackson, Peter. 2009. "The future of global oil supply: understanding the building blocks." Cambridge, MA: IHS CERA. http://www.cera.com/aspx/cda/client/report/report.aspx?KID=5&CID=10720 (accessed January 23, 2010).

Jackson, P. J. 2006. *The Federal Excise Tax on Gasoline and the Highway Trust Fund: A Short History*. U.S. Library of Congress, Congressional Research Service (CRS). http://www.cnie.org/nle/crsreports/06may/rl30304.pdf (accessed January 15, 2011).

Jacobson, C. D. 2000. *Ties that bind: Economic and political dilemmas of urban utility networks, 1800–1990*. Pittsburgh: University of Pittsburgh Press.

Jacobson, M., and M. Delucchi. 2009. A path to sustainable energy by 2030. *Scientific American* (November): 58–65.

Jacobson, S., A. Bergek, D. Finon, V. Lauber, C. Mitchell. D. Toke, and A. Verbruggen. 2009. EU renewable energy support policy: Faith or facts? *Energy Policy* 37 (2009): 2143–2146.

Jochem, E., D. Favrat. K. Hungerbühler, P. R. von Rohr, D. Spreng, A. Wokaun, and M. Zimmerman. 2002. *Steps toward a 2000-watt society*. Zurich: Novatlantis, http://www.efficientpowersupplies.org/pages/Steps_towards_a_2000_WattSociety.pdf (accessed June 21, 2010).

Jones, C. O. 1984. *An introduction to the study of public policy*. 3rd ed. Belmont, CA: Wadsworth.

Jones, J. 2010. Oil spill alters views on environmental protection. Gallop.com (May 27), http://www.gallup.com/poll/137882/oil-spill-alters-views-environmental-protection.aspx (accessed June 18, 2010).

John Deere Co. 2010. Jobs and economic impact. *Biopower Facts*. http://biopowerfacts.com/jobs/ (accessed January 11, 2011).

Joseph, S., J. Schultz, and M. Castan. 2000. *The international covenant on civil and political rights*. Oxford, U.K.: Oxford University Press.

Joskow, P. L. 1983. *Markets for power*. Cambridge, MA: MIT Press.

Joskow, P. L. 2003. *The difficult transition to competitive electricity markets in the U.S.* Paper prepared for the April 4, 2003 conference Electricity Deregulation: Where From Here? at Texas A&M University. Washington, DC: AEI-Brookings Institute Joint Center for Regulatory Studies working papers, http://tisiphone.mit.edu/RePEc/mee/wpaper/2003-008.pdf

Joyner, C. 2005. Rethinking international environmental regimes: What role for partnership coalitions? *Journal of International Law and International Relations* 1 (1/2): 89–119.
Juhasz, A. 2008. *The tyranny of oil.* New York: Harper Collins.
Kalicki, J. H., and D. L. Goldwin, eds. 2005. *Energy and security.* Baltimore, MD: Johns Hopkins University Press.
Kamarck, E. 2004. Government innovation around the world. KSG Working Paper No. RWP04-010, Social Science Research Network, http://ssrn.com/abstract=517666 (accessed March 4, 2004).
Kammen, D. M., and G. F. Nemer. 2007. Energy myth eleven: Energy R&D investment takes decades to reach the market. In *Energy and American society: Thirteen myths*, ed. Benjamin K. Sovacool and Marilyn A. Brown, 289–310. Dordrecht, The Netherlands: Springer.
Kamrany, N., ed. 1981. *Energy independence for the United States: Alternative policy proposals.* Santa Monica, CA: Fundamental Books.
Kanellos, B., and B. Prior. 2010. Are solar thermal plants doomed? (October 18) Greentech Solar.com, http://www.greentechmedia.com/articles/read/is-CSP-doomed/ (accessed January 24, 2010).
Karl, T. 1999. The perils of the petro-state: Reflections on the paradox of plenty. *Journal of International Affairs* 53 (1): 31–48.
Karl, T., Jerry M. Melillo, and T. Peterson, eds. 2009. *Global climate change impacts in the United States.* New York: Cambridge University Press, http://downloads.globalchange.gov/usimpacts/pdfs/climate-impacts-report.pdf (accessed June 15, 2010).
Karns, M., and K. Mingst. 2004. *International organizations: The politics and processes of global governance.* Boulder, CO: Lynne Rienner.
Katz, J. L. 2001. A web of interests: Stalemate on the disposal of spent nuclear fuel. *Policy Studies Journal* 29 (3): 456–477.
Kearney, R. C., and E. M. Berman, eds. 1999. *Public sector performance: Management, motivation, and measurement.* Boulder, CO: Westview Press.
Kennedy, M. E. 2004. Using customer relationship management to increase profits. *Strategic Finance* 85 (9): 37–42.
Kent, C. A. 1993. *Utility Holding Company Act of 1935: 1935–1992.* Washington, DC: Energy Information Administration.
Kenworthy, T. 2009. Reverse Bush-Era Pillage of the West. *Center for American Progress*, http://www.americanprogress.org/issues/2009/03/kenworthy_pillage.html (accessed January 15, 2011).
Keohane, R. O., and J. Nye. 2000. Introduction. In *Governance in a globalizing world*, ed. Robert O. Keohane and John D. Donahue, 1-41. Washington, DC: Brookings Institution Press.
King, C. S., and C Stivers. 1998. *Government is Us.* Thousand Oaks, CA: Sage.
Kingdon, J. 2003. *Agendas, alternatives, public policy.* 2nd ed. New York: Longman.
Kingsley, G. A. 1992. U.S. energy conservation policy: Themes and trends. *Policy Studies Journal* 20 (1): 114–123.
Klass, D. L. 1991. Human resources and the gas industry. *Public Utilities Fortnightly* 128 (July 1): 19–24.
Komor, P. 2004. *Renewable energy policy.* New York: iUniverse.
Koontz, H. D. 1941. *Government control of business.* Boston: Houghton Mifflin.
Koplow, D. 2006a. Biofuels—*At what cost? Government support for ethanol and biodiesel in the United States.* Cambridge, MA: Earth Track, Inc., http://www.earthtrack.net/files/legacy_library/biofuels_subsidies_us.pdf (accessed July 1, 2008).
Koplow, D. 2006b. *Subsidies in the U.S. energy sector: Magnitude, causes, and options for reform.* Cambridge, MA: Earth Track, Inc., http://www.earthtrack.net/files/legacy_library/SubsidyReformOptions.pdf (accessed July 1, 2008).
Koplow, D. 2008. Tax subsidies. *Encyclopedia of Earth*, http://www.eoearth.org/article/Tax_subsidies (accessed May 20, 2010).
Koplow, D. 2009. State and federal subsidies to biofuels: Magnitude and options for redirection. *International Journal of Biotechnology* 11 (1–2): 92–106.
Koteen, J. 1997. *Strategic management in public and nonprofit organizations.* 2nd ed. Westport, CT: Praeger.
Kraft, M. E. 2000. Policy design and the acceptability of environmental risks: Nuclear waste disposal in Canada and the United States. *Policy Studies Journal* 28 (1): 206–218.

Kraft, Michael E. 2010. *Environmental Policy and Politics*, 5th ed. New York: Pearson Longman.

Krasner, S. 1982. Structural causes and regime consequences: Regimes as intervening variables. *International Organization* 36 (2): 185–205.

Krauthhammer, C. 2007. Energy independence? *Washington Post*, January 26, A21, from http://www.washingtonpost.com/wp-dyn/content/article/2007/01/25/AR2007012501547.html (accessed March 29, 2009).

Krosnick, J. 2010. The climate majority. *New York Times* (June 9): A25, NYTimes.com, http://www.nytimes.com/2010/06/09/opinion/09krosnick.html?adxnnl=1&ref=opinion&adxnnlx=1276873360-Fm2cRjKyVCgN4X/6WQPkLw (accessed June 18, 2010).

Kruyt, B., D. P. van Vuuren, H. J. M. deVries, and H. Groenenberg. 2009. Indicators for energy security. *Energy Policy* 37 (2009): 2166–2181.

Kwoka, J. E., Jr. 2002. Governance alternatives and pricing in the U.S. electric power industry. *Journal of Law, Economics, & Organization* 18 (April): 278–294.

LaFollette, M. C. 1994. The politics of research misconduct: Congressional oversight, universities, and science. *Journal of Higher Education* 65 (May/June): 261–265.

Lakoff, G. 2004. *Don't think of an elephant.* White River Junction, VT: Chelsea Green.

Lakoff, G. 2008. *The political mind.* New York: Viking.

Lakoff, G., and the Rockridge Institute. 2006. *Thinking points: Communicating our American values and vision: A progressive's handbook.* New York: Farrar, Straus and Giroux.

Lane, J. E. 2000. *New public management.* London: Routledge.

Lashgari, M. 2004. Corporate governance: Theory and practice. *Journal of American Academy of Business* 5 (September): 46–51.

Lassa, T. 2010. Comparing America's top 10 best-selling vehicles, 2009 vs. 2000. *Motor Trend.com* (February), http://www.motortrend.com/features/auto_news/2010/112_1004_america_top_10_best_selling_vehicle_comparison_2009_2000/index.html (accessed May 26, 2010).

Lazzari, S. 2005. *Energy Tax Policy: History and Current Issues.* U.S. Library of Congress, Congressional Research Service (CSR). http://www.fas.org/sgp/crs/misc/RL33578.pdf (accessed January 15, 2011).

LBNL (Lawrence Berkeley National Laboratory). 2009. Reactions on the Obama Chu announcement from colleagues and notables. U.S. Department of Energy, Berkeley Lab, http://www.lbl.gov/Publications/Director/chu-reactions.html# (accessed September 7, 2009).

Leiserowitz, A., E. Maibach, and C. Roser-Renouf. 2010. *Global warming's six Americas*, January 2010. Yale University and George Mason University. New Haven, CT: Yale Project on Climate Change, http://www.climatechangecommunication.org/images/files/SixAmericasJan2010(1).pdf (accessed June 22, 2010).

Leiserowitz, A., E. Maibach, C. Roser-Renouf, and N. Smith. 2010. *Climate change in the American mind: Americans' global warming beliefs and attitudes in June 2010.* Yale University and George Mason University. New Haven, CT: Yale Project on Climate Change Communication, http://environment.yale.edu/climate/files/ClimateBeliefsJune2010.pdf (accessed June 23, 2010).

LeQuéré, C., M. R. Raupach, J. G. Canadell, and G. Marland. 2009. Trends in the sources and sinks of carbon dioxide. *Nature Geoscience*, (17 December 2009). http://www.globalcarbonproject.org/global/pdf/LeQuere-2009-Trends%20sources%20&%20sinks%20CO2.NatureGeo.pdf (accessed January 11, 2011).

Leuchtenburg, W. E. 1963. *Franklin D. Roosevelt and the New Deal.* New York: Harper & Row.

Levy, P. F. 1985. The role of public utility commissions. In *State energy policy: Current issues, future directions*, ed. Stephen W. Sawyer and John R. Armstrong, 49–266. Boulder, CO: Westview.

Lewis, P. 2003. Texas utility-billing company told to quit doing business in state. *The Seattle Times* (September 4): B3.

Lillenthal, D. E. 1944. *TVA: Democracy on the march.* Chicago: Quadrangle Books.

Linder, J. 2008. Congressman John Linder co-sponsors American energy act, JohnLinder.com, http://www.johnlinder.com/news_details.asp?id=39 (accessed April 12 2010).

Liner, C. 2010. CO_2-reactions. *Seismos*, http://seismosblog.blogspot.com/2010/04/CO_2-reactions.html (accessed August 28, 2010).

LLNL (Lawrence Livermore National Laboratory). 2009. *Estimated U.S. energy use in 2008.* U.S. Department of Energy, Lawrence Livermore National Laboratory, https://publicaffairs.llnl.gov/news/energy/energy.html (accessed December 8, 2009).

Lovins, A. B. 1976. Energy strategy: The road not taken? *Foreign Affairs* 55 (6): 65–96.

Lovins, A. B. 2007. Energy myth nine: Energy efficiency improvements have already reached their potential. In *Energy and American society: Thirteen myths*, ed. Benjamin K. Sovacool and Marilyn A. Brown, 239–264. Dordrecht, The Netherlands: Springer.

Lowry, William. 2008. Disentangling energy policy from environmental policy. *Social Science Quarterly* 89 (5): 1195–1211.

Luger, Stan. 1995. Market ideology and administrative fiat: The rollback of automobile fuel economy standards. *Environmental History Review* 19 (1): 77–93.

Lynd, L. R., M. S. Laser, J. Mcbride, K. Podkaminer, and J. Hannon. 2007. Energy myth three: High land requirements and an unfavorable energy balance preclude biomass ethanol from playing a large role in providing energy services. In *Energy and American society: Thirteen myths*, ed. Benjamin K. Sovacool and Marilyn A. Brown, 75–102. Dordrecht, The Netherlands: Springer.

MacAvoy, P. W. 1983. *Energy policy: An economic analysis.* New York: W. W. Norton.

MacAvoy, P. W. 2000. *The natural gas market: Sixty years of regulation and deregulation.* New Haven, CT: Yale University Press.

Majone, G. 2006. Agenda setting. In *The Oxford handbook of public policy*, ed. M. Moran, M. Rein, and R. E. Goodin, 228–250. Cambridge, U.K.: Oxford University Press.

Mankiw, N. G. 2006. Raise the gas tax. *Wall Street Journal* (October 20): A12.

Mann, A. 1963. *The Progressive Era: Liberal renaissance or liberal failure?* New York: Holt, Rinehart and Winston.

Mann, T., and N. Ornstein. 2006. *The broken branch.* New York: Oxford.

Manning, A. W. 2003. The changing workforce—the latest information. What's going to happen and what to do about it. Address given before the AMSA Winter Conference, February 4, in Santa Fe, New Mexico.

Mariner-Volpe, B. 2007. *Status of natural gas residential choice programs by state as of December 2007.* http://http://www.eia.doe.gov/conf_pdfs/Tuesday/mariner-volpe_eia.pdf

Marufu, L., B. Taubman, B. Bloomer et al. 2004. The 2003 North American electrical blackout: An accidental experiment in atmospheric chemistry. *Geophysical Research Letters* 31, L13106, doi:10.1029/2004GL019771.

Mattoon, R. 2002. The electricity system at the crossroads—policy choices and pitfalls. *Economic Perspectives* 26 (1): 2–18.

McCann, J. C. 2004. *Industry profiles: Electric utilities.* Standard & Poor's NetAdvantage, http://www.netadvantage.standardandpoors.com/docs/indusr////elu_0204/elu30204.htm (accessed May 13, 2004).

McCarthy, R. W., and J. Ogden. 2005. *Assessing reliability in transportation energy supply pathways: A hydrogen case study.* University of California, Davis, Institute of Transportation Studies, eScholarship, http://www.escholarship.org/uc/item/8zs63172 (accessed September 24, 2008).

McElroy, M. B. 2010. *Energy: Perspectives, problems, and prospects.* Oxford, U.K.: Oxford University Press.

McElvaine, R. S. 1984. *The Great Depression.* New York: Times Books.

McKeown, R. 2007. Energy myth two: The public is well informed about energy. In *Energy and American society: Thirteen myths*, ed. Benjamin K. Sovacool and Marilyn A. Brown, 51–74. Dordrecht, The Netherlands: Springer.

McKinsey & Co. 2007. *Pathways to a Low Carbon Economy.* http://www.mckinsey.com/clientservice/sustainability/pathways_low_carbon_economy.asp (accessed January 15, 2011).

McKinsey & Co. 2009. *American Recovery and Reinvestment Act (ARRA).* March 13, 2009 Briefing presentation at the Hawaii state capitol (March 13, 2009), http://ploneadmin.hawaii.gov/gov/recovery/cleanenergy/ARRABriefing.pdf

McKinsey Global Institute. 2009. *Averting the next energy crisis: The demand challenge.* San Francisco: McKinsey Global Institute, http://www.mckinsey.com/mgi/reports/pdfs/next_energy_crisis/MGI_next_energy_crisis_full_report.pdf (accessed March 10, 2010).

McNabb, D. E. 2005. *Public utilities: Management challenges for the twenty-first century.* Northampton, MA: Edward Elgar.

Menz, F., and S. Vachon. 2006. The effectiveness of different policy regimes for promoting wind power: Experiences from the states. *Energy Policy* 34 (14): 1786–1796.

Metcalf, G. E. 2009. *Investment in Energy Infrastructure and the Tax Code*. Cambridge: Massachusetts Institute of Technology, Center for Energy and Environmental Policy Research (CEEPR). http://web.mit.edu/ceepr/www/publications/workingpapers/2009-020.pdf (accessed January 15, 2011).

Michaels, R. 2008. A federal renewable energy requirement: What's not to like? *Policy Analysis* 36 (627), CATO Institute, http://www.cato.org/pubs/pas/pa-627.pdf (accessed April 19, 2010).

Milken, M. 2010. Remarks on America's energy future. Milken Institute Global Conference, April 26.

Miller, A. D. 2010. The false religion of Mideast peace. *Foreign Policy* (May/June): 50–57.

Miller, H., and C. Fox. 2007. *Postmodern public administration, revised edition*. Armonk, NY: ME Sharpe.

MMS (Minerals Management Service). 2006. *Report to Congress: Comprehensive Inventory of U.S. OCS Oil and Natural Gas Resources*. Washington, DC: U.S. Department of the Interior. http://www.boemre.gov/revaldiv/PDFs/FinalInventoryReportDeliveredToCongress-corrected3-6-06.pdf (accessed January 16, 2011).

Moore, K. 2009. Energy plant may bring jobs. *Mason County Journal*. (September 10).A1, 7. Shelton, WA

Morgenthau, H. J. 1978. *Politics among nations: The struggle for power and peace*. 5th ed. New York: Knopf.

Moriarty, P., and D. Honnery. 2009. What energy levels can the Earth sustain? *Energy Policy* 37 (2009): 2469–2474.

Morse, E. L. 2009. Low and behold: Making the most of cheap oil. *Foreign Affairs* 88 (5): 36–52.

Mott, J. 2011. *American Government and Politics*. "The policy process." http://www.thisnation.com/textbook/processes-policyprocess.html (accessed February 16, 2010).

Mouawad, J. 2009. Estimate places natural gas reserves 35% higher. *New York Times* (June 18), NYTimes.com, http://www.nytimes.com/2009/06/18/business/energy-environment/18gas.html?_r_1&pageswanted=print (accessed May 22, 2010).

Mowry, G. E. 1972. *The Progressive Era, 1900–1920: The reform persuasion*. Washington, DC: American Historical Society.

Mumford, L. 1937. *Technics and civilization*. New York: Harcourt, Brace.

Munasinghe, M., and P. Meier. 1993. *Energy policy and modeling*. Cambridge, U.K.: Cambridge University Press.

Munoz, M, V. Oschmann, and J. Tabara. 2007. Harmonization of renewable electricity feed-in laws in the European Union. *Energy Policy* 35 (2007): 3104–3114.

Nadal, S. 2010. *Energy Efficiency Opportunities and Policies* 101. Presented June 1, 2010 at the Washington, DC, Energy Foundation Conference. http://theenergyfoundationconference2010.com/?page_id=30 (accessed January 15, 2011).

National Energy Policy Development Group. 2001. *Report of the U.S. National Energy Policy Development Group*. Washington, D.C.: U.S. Government Printing Office.

NCEP (National Commission on Energy Policy). 2003. *Reviving the electricity sector: Findings of the National Commission on Energy Policy*. Washington, DC: National Commission on Energy Policy, http://www.whrc.org/resources/publications/pdf/HoldrenetalNatlCommEnergyPol.08.03.pdf

New York Times/CBS News. 2010. Poll on the Gulf of Mexico oil spill, June 16–20. *New York Times* (June 21). NYTimes.com, http://documents.nytimes.com/new-york-timescbs-news-poll-on-the-gulf-of-mexico-oil-spill?scp=2&sq=Poll%20on%20the%20Gulf%20of%20Mexico%20oil%20spill&st=cse (accessed June 21, 2010).

Newell, R., A. Jaffe, and R. Stavins, 1999. The induced innovation hypothesis and energy-saving technological change. *Quarterly Journal of Economics* 114 (3): 941–975.

Newport, F. 2010. Americans leery of too much gov't regulation of business (February 2). Gallup, http://www.gallup.com/poll/125468/americans-leery-govt-regulation-business.aspx (accessed May 6, 2010).

NHTSA (National Highway Traffic Safety Administration). 2009. *Consumer Assistance to Recycle and Save Act of 2009*. Report to Congress. Washington, DC: U.S. Department of Transportation, National Highway Traffic Safety Administration. Car Allowance Rebate System (CATS), http://www.cars.gov/files/official-information/CARS-Report-to-Congress.pdf (accessed May 12, 2010).

Nivola, P. S. 2009. *The long and winding road: Automotive fuel economy and American politics*. Washington, DC: The Brookings Institute. Governance Studies at Brookings, http://www.brookings.edu/~/media/Files/rc/papers/2009/0225_cafe_nivola/0225_cafe_nivola.pdf (accessed March 12, 2010).

Nordhaus, W. 2007. To tax or not to tax: Alternative approaches to slowing global warming. *Review of Environmental Economics and Policy* 1 (1): 26–44.
Nordhaus, W. 2008. *A question of balance*. New Haven, CT: Yale University Press.
Nowotny, K., D. B. Smith, and H. M. Tebing, eds. 1989. *Public utility regulation.* Boston: Kluwer Academic Publishers.
NRRI (National Regulatory Research Institute). 2005. *Summary of the Energy Policy Act of 2005*, 21–31. Silver Spring, MD: National Regulatory Research Institute.
Numark, N. J., and M. O. Terry. 2003. New nuclear construction: Still on hold. *Public Utilities Fortnightly* 141 (December): 32–38.
Nye, D. 1998. *Consuming power*. Boston: MIT Press.
Obama, B. 2011. Address before a joint session of the Congress on the state of the union. Washington, DC: The White House. http://www.whitehouse.gov/the-press-office/2011/01/25/remarks-president-state-union-address (accessed January 26, 2011).
Obama, President Barak. 2010a. Remarks by the president on energy security at Andrews Air Force Base, March 31, 2010. White House Press Office. http://www.whitehouse.gov/the-press-office/remarks-president-energy-security-andrews-air-force-base-3312010
Obama, President Barak. 2010b. Remarks by the president on the economy at Carnegie Mellon University, June 2, 2010. http://www.whitehouse.gov/the-press-office/remarks-president-economy-carnegie-mellon-university
Oberthür, S., and H. Ott. 1999. *The Kyoto Protocol: International climate policy for the 21st century.* Berlin: Springer Verlag.
Ogden, P., J. Podesta, and J. Deutch. 2008. A new strategy to spur energy innovation. *Issues in Science and Technology (Online)* (winter): 35–44, http://www.issues.org/24.2/ogden.html (accessed November 1, 2010).
Olmstead, S., and R. Stavins. 2009. An expanded three-part architecture for post-2012 international climate policy. Discussion Paper 2009-29, Cambridge, MA: Harvard Project on International Climate Agreements (September). Belfer Center, Harvard Kennedy School, http://belfercenter.ksg.harvard.edu/files/Stavins_Olmstead_Final%20_2.pdf (accessed January 29, 2010).
OMB (U.S. Office of Management and Budget). 2010. Table 15.5: Total government expenditures by major category of expenditure as percentages of GDP: 1948–2008. *Economic Report of the President: Historical Tables*. 331. http://www.gpoaccess.gov/usbudget/fy10/pdf/hist.pdf (accessed January 11, 2011).
Osborne, D., and P. Plastrik. 1992. *Banishing bureaucracy.* Reading, MA: Addison-Wesley.
Owens, B. 2003. Combined-cycle profitability as a market barometer. *Public Utilities Fortnightly* 141 (1): 13.
Ōyama, K. 1998. The policymaking process behind petroleum industry regulatory reform. In *Is Japan really changing its ways?*, ed. L. E. Carlile and M. C. Tilton, 142–162. Washington, DC: Brookings Institution Press.
Pacala, S., and R. Socolow. 2004. Stabilization wedges: Solving the climate problem for the next 50 years with current technologies. *Science* 305 (13): 968–972.
Page, E. G. 2006. The origins of policy. In *The Oxford handbook of public policy*, M. Moran, M. Rein and R. E. Goodin, 207–227. Cambridge, U.K.: Oxford University Press.
Parrington, V. L. 1963. The Progressive Era: A liberal renaissance. In *The Progressive Era*, ed. Arthur Mann, 6–12. New York: Holt, Rinehart and Winston.
Parry, I. 2009a. How much should highway fuels be taxed? Resources for the Future discussion paper 09-52 (December). Washington, DC: RFD.
Parry, I. 2009b. Raise $100 billion from a $20 CO_2 tax. *Tax Notes* 123 (2): 243–247.
Parry, I., and K. Small. 2009. Should Urban Transit Subsidies Be Reduced? *American Economic Review*, 99 (3): 700-724.
Pearson, C. M., and J. A. Clair. 1998. Reframing crisis management. *The Academy of Management Review* 22 (January): 59–76.
Pelast, G., J. Oppenheim, and T. MacGregor 2003. *Democracy and regulation.* London: Pluto Press.
Perloff, R. M. 1998. *Political communication: Politics, press, and public opinion in America.* Mahwah, NJ: Lawrence Erlbaum Associates.
Peterson, K. S. 2001. Would I lie to you? *USA TODAY* (5 July): 8D.
Petrick, J. A., and J. F. Quinn. 1997. *Management ethics: Integrity at work.* Newbury Park, CA: Sage.
Petroleum Economist. 2005. Energy policy. London: Euromoney Institutional Investor PLC (September): 1.

Pew Center on Global Climate Change. 2006. *Agenda for climate action.* Pew Center for Global Climate Change, http://www.pewclimate.org/docUploads/PCC_Agenda_2.08.pdf (accessed June 22, 2010).

Pew Center on Global Climate Change. 2010a. *History of Kyoto Protocol.* Pew Center for Global Climate Change, http://www.pewclimate.org/history_of_kyoto.cfm (accessed February 22, 2010).

Pew Center on Global Climate Change. 2010b. *Carbon market design and oversight: A short overview* (February). Pew Center for Global Climate Change, http://www.pewclimate.org/economics/brief/carbon-market-design-oversight (accessed May 29, 2010).

Pielke Jr., R. 2010. *The climate fix.* New York: Basic Books.

Pindyck, R. S. 2001. Energy policy and the American economy. In *Energy independence for the United States: Alternative policy proposals,* ed. N. Kamrany, 67–76. Santa Monica, CA: Fundamental Books.

Pirog, R. 2009. *The Role of Federal Gasoline Excise Taxes in Public Policy.* Congressional Research Service (CRS) Report R-40808. http://fpc.state.gov/documents/organization/130217.pdf (accessed January 15, 2011).

Pizer, W. 1997. *Choosing Price Or Quantity Controls For Greenhouse Gases.* Climate Issues Brief No. 17. Washington, DC: Resources for the Future. http://www.rff.org/rff/Documents/RFF-CCIB-17.pdf (accessed January 15, 2011).

Pollin, R. H., G. Peltier, J. Heintz, and H. Scharber. 2008. *Green recovery: A program to create good jobs and start building a low-carbon economy.* Washington, DC: Center for American Progress, http://www.americanprogress.org/issues/2008/09/pdf/green_recovery.pdf

Polovinkin, V. N. 2004. Energy saving: The energy of the future—energy from conservation. *Metallurgist* 48 (3–4): 101–103.

Pope, Daniel. 2008. "A Northwest distaste for nuclear power." *Seattle Times,* July 31st. http://seattletimes.nwsource.com/html/opinion/2008082460_nukeop31.html (accessed January 22, 2010).

Portney, P. R. 1990. *Public policies for environmental protection.* Washington, DC: Resources for the Future.

Portney, P. R., I. W. H. Parry, H. K. Gruenspecht, and W. Harrington. 2003. Policy watch: The economics of fuel economy standards. *Journal of Economic Perspectives* 17 (4): 203–217.

Prins, G., and S. Raymer. 2007. Time to ditch Kyoto. *Nature* 449 (October 25): 973–975.

Prins, G., I. Galiana, C. Green, R. Grundmann, M. Hulme, A. Korhola, F. Laird, T. Nordhaus, R. Pielke Jr., S. Rayner, D. Sarewitz, M. Shellenberger, N. Stehr and H. Tezuka. 2010. *The Hartwell Paper: a New Direction for Climate Policy After the Crash of 2009.* Oxford, UK: Oxford University. http://sciencepolicy.colorado.edu/admin/publication_files/resource-2821-2010.15.pdf (accessed January 16, 2011).

Public Citizen. 2004. Price-Anderson Act: The billion dollar bailout for nuclear power mishaps. *Public Citizen,* http://www.citizen.org/documents/Price%20Anderson%20Factsheet.pdf (accessed May 10, 2010).

Qualter, T. H. 1965. *Propaganda and psychological warfare.* New York: Random House.

Rabe, B. 2006. *Race to the top: The expanding role of U.S. state renewable portfolio standards.* Arlington, VA: Pew Center on Global Climate Change, http://www.pewcenteronthestates.org/uploadedFiles/Race%20to%20the%20Top.pdf

Raloff, J. 2000. Liquid assets. *Science News* 157 (January 29): 72–75.

Ramseur, J. L. 2010. *Oil Spills in U.S. Coastal Waters: Background, Governance and Issues for Congress.* Testimony before the U. S. Senate Committee on Energy and Natural Resources (May 25, 2010). http://energy.senate.gov/public/_files/RamseurTestimony5252010.pdf (accessed January 15, 2011).

Ramseur, J. L., and L. Parker. 2009. *Carbon Tax and Greenhouse Gas Control: Options and Considerations for Congress.* U.S. Library of Congress, Congressional Research Service (CSR). http://leahy.senate.gov/imo/media/doc/CRSCarbonTax.pdf (accessed January 15, 2011).

Randolf, J. 1985. Inventory of current state energy activities. In *State energy policy: Current issues, future directions,* ed. Stephen W. Sawyer and John R. Armstrong, 27–58. Boulder, CO: Westview.

Reed, W. L. 2003. Competitive electricity markets and innovative technologies: Hourly pricing can pave the way for the introduction of technology and innovation. Rand Corporation working papers, http://www.rand.org/content/dam/rand/pubs/conf_proceedings/CF170z1-1/CF170.1.reed.pdf

Reese, S.D. 2001. Prologue—Framing Public Life: A Bridging Model for Media Research. In S. D. Reese, O. H. Gandy, Jr. and A. E. Grant, eds. *Framing Public Life.* Mahwah, NJ: Lawrence Erlbaum. 1–24.

Reese, S. D., O. H. Gandy Jr., and A. E. Grant. 2001. *Framing public life: Perspectives on media and our understanding of the social world.* Mahwah, NJ: Erlbaum.

Reicher, D. 2009. Testimony before the Senate Committee on Environment and Public Works Legislative Hearing on S. 1733, Clean Energy Jobs and American Power Act (October 28, 2009). http://epw.senate.gov/public/index.cfm?FuseAction=Files.View&FileStore_id=8d3195f8-9107-4fb0-86ca-51d9c5fbca46 (accessed January 15, 2011).
Rein, M. 2006. Reframing problematic policies. In *The Oxford handbook of public policy*, ed. M. Moran, M. Rein, and R. E. Goodin, 389–405. Cambridge, U.K.: Oxford University Press.
Reinhardt, M. 1991. Attacking utility fraud. *Public Utilities Fortnightly* 128 (August): 20–22.
REN21. 2009. *Renewables global status report: 2009 update.* Paris: REN21 Secretariat, http://www.ren21.net/pdf/RE_GSR_2009_Update.pdf (accessed June 4, 2010).
Revkin, A. 2010. Kerry, Edison and the energy quest. *New York Times* (May 5), Dot Earth Blog, NYTimes.com, http://dotearth.blogs.nytimes.com/2010/05/05/kerry-edison-and-the-energy-quest/ (accessed May 19, 2010).
Rhodes, R. A. W. 2006. Policy network analysis. In *The Oxford handbook of public policy*, M. Moran, M. Rein, and R. E. Goodin, 425–447. Cambridge, U.K.: Oxford University Press.
Rich, F. 2010. Clean the Gulf, clean house, clean their clock. *New York Times* (June 20): WK8. NYTimes.com, http://www.nytimes.com/2010/06/20/opinion/20rich.html?src=me&ref=homepage (accessed June 21, 2010).
Rittel, H., and M. Webber. 1973. Dilemmas in a general theory of planning. *Policy Sciences* 4: 155–169.
Robinson, C., ed. 2002. *Utility regulation and competitive policy.* Northampton, MA: Edward Elgar.
Rogelj, J., B. Hare, J. Nabel, K. Macey, M. Schaeffer, K. Markmann, and M. Meinshausen. 2009. Halfway to Copenhagen, no way to 2°C. *Nature Reports Climate Change* (June 11), http://www.nature.com/climate/2009/0907/full/climate.2009.57.html (accessed June 4, 2010).
Rohr, J. A. 1998. *Public service, ethics, and constitutional practice.* Lawrence: University of Kansas Press.
Romanova, T. 2008. Energy dialogue from strategic partnership to the regional level of the northern dimension. In *The EU-Russian dialogue: Europe's future energy security*, ed. Pami Aalto, 63–92. Aldershot, U.K.: Ashgate.
Romm, J. 2007. Energy myth four: The hydrogen economy is a panacea to the nation's energy problems. In *Energy and American society: Thirteen myths*, ed. Benjamin K. Sovacool and Marilyn A. Brown, 103–124. Dordrecht, The Netherlands: Springer.
Romm, Joseph. 2008. "Cleaning up on carbon." *Nature Reports Climate Change* (19 June). http://www.nature.com/climate/2008/0807/full/climate.2008.59.html (accessed January 25, 2010).
Rosenbaum, W. S. 1981. *Energy, politics and public policy.* Washington, DC: Congressional Quarterly Press.
Rosenberg, M., and D. R. Wolcott. 1985. State government's role in the development of energy service contracting. In *State energy policy: Current issues, future directions*, ed. Stephen W. Sawyer and John R. Armstrong, 231–248. Boulder, CO: Westview.
Rosenberg, R. G. 2003. The dividend bust? *Public Utilities Fortnightly* 141 (October 15): 45–48.
Rosner, K. 2010. The evolution of energy security. *Journal of Energy Security* (May 2010 Issue). http://www.ensec.org/index.php?option=com_content&view=article&id=249:the-evolution-of-energy-security&catid=105:editorenergysecurity0510&Itemid=361 (accessed January 12, 2011).
Rosenthal, E. 2009. Soot from third-world stoves is new target in climate fight. *New York Times* (April 16): A1, A12.
Rothard, M. N. 1963. *America's great depression.* Princeton, NJ: C. Van Nostrand.
Russell, D. 2009. *North American energy relationships.* Draft paper for discussion. The Pembina Institute (Alberta, Canada) and the International Institute for Sustainable Development (Winnipeg, Canada).
Russo, T. N. 2004. Making hydro sustainable. *Public Utilities Fortnightly* 133 (January 1): 14–20.
Ruth, J. S. 1999. The new wave of public-private partnerships. *Business News New Jersey* 12 (27): 18.
Saad, L. 2010. Banking reform sells better when "Wall Street" is mentioned. (April 20). Gallup, http://www.gallup.com/poll/127448/banking-reform-sells-better-wall-street-mentioned.aspx (accessed May 6, 2010).
Sabatier, P. A, ed. 2007. *Theories of the policy process.* 2nd ed. Cambridge, MA: Westview Press.
Said, C. 2004. Utility lawyer's tenacity leads to smoking gun in Enron case. *Seattle Post-Intelligencer* (June 26): 1+.
Salamon, L., ed. 2002. *The tools of government: A guide to the new governance.* New York: Oxford.
Sanders, R. 2008. Obama chooses Nobelist Steven Chu as secretary of energy. University of California Berkeley News, http://berkeley.edu/news/media/releases/2008/12/15_obama.shtml (accessed September 7, 2009).

Sandia National Laboratories. 2009. *Perspectives on energy policy: Security, economics, and the environment.* University of California, San Diego, Sandia National Laboratories, http://www.sandia.gov/news/publications/white-papers/PerspectivesReportFinal.pdf (accessed May 20, 2010).

Sawyer, S. W. 1985a. Energy and the states. In *State energy policy: Current issues, future directions*, ed. Stephen W. Sawyer and John R. Armstrong, 5–26. Boulder, CO: Westview.

Sawyer, S. W. 1985b. State renewable energy policy: Program characteristics, projections, needs. *State & Local Government Review* 17 (1): 147–154.

Sawyer, S. W. 1985c. Unmet energy needs: The state's perspective. In *State energy policy: Current issues, future directions*, ed. Stephen W. Sawyer and John R. Armstrong, 213–230. Boulder, CO: Westview.

Sawyer, S. W., and J. R. Armstrong, eds. 1985. *State energy policy: Current issues, future directions.* Boulder, CO: Westview.

Sawyer, S. W., and R. Lancaster. 1985. Renewable energy tax incentives: Status, evaluation attempts, continuing issues. In *State energy policy: Current issues, future directions*, ed. Stephen W. Sawyer and John R. Armstrong, 171–192. Boulder, CO: Westview.

Saylor Foundation. 2010. Iron Triangle (U.S. Politics). http://www.saylor.org/site/wp-content/uploads/2010/12/IronTriangle.pdf (accessed January 24, 2011).

Scalingi, P. L., and M. Morrison. 2003. Power to the people. *Security Management* 47 (December): 93–101.

ScienceDaily. 2010. Americans favor conservation, but few practice it. *Sciencedaily.com*, (February 24) http://www.sciencedaily.com/releases/2010/02/100216113559.htm (accessed June 21, 2010).

Schlissel, D., M. Mullett, and R. Alvarez. 2009. Nuclear loan guarantees: Another taxpayer bailout ahead? Union of Concerned Scientists, http://www.ucsusa.org/nuclear_power/nuclear_power_and_global_warming/nuclear-loan-guarantees.html (accessed May 12, 2010).

Schmidt, C. W. 2009. Carbon offsets: Growing pains in a growing market. *Environmental Health Perspectives* 117 (2): 62–68.

Schneider, S. 2009. The worst case scenario. *Nature* 458 (7242): 1077–1212.

Schneider, S., and K. Kuntz-Duriseti. 2002. Uncertainty and climate change policy. In *Climate change policy: A survey*, ed. S. Schneider, A. Rosencranz, and J. Niles. Washington, DC: Island Press.

Schneyer, J. 2009. Buy or sell: Should heavy crude cost as much as light? Thompson Reuters (August 12), http://www.reuters.com/article/idUSTRE57B5FG20090812

Schön, D. A., and M. Rein. 1994. *Frame reflection: Toward the resolution of intractable policy controversies.* New York: Basic Books.

Schwabe, P., K. Cory, and J. Newcomb. 2009. *Renewable energy project financing: Impacts of the financial crisis and federal legislation.* NREL Report No. TP-6A2-44930. National Renewable Energy Laboratory, http://www.nrel.gov/docs/fy09osti/44930.pdf (accessed May 20, 2010).

Seattle City Light. 1976. *Energy 1990—Final Report.* Seattle, Washington: Seattle City Light Office of Environmental Affairs.

Seib, G. 2010. Senate woes flag wider disease. *Wall Street Journal* (February 15): A2.

Selin, H., and S. D. VanDeveer. 2009. *Changing Climates in North American Politics.* Boston: MIT Press.

Shaffer, B., D. Rode, and S. Dean. 2010. Best among equals? Choosing tax incentives for wind projects. *Renewable Energy World North America* 1 (2), http://www.renewableenergyworld.com/rea/news/article/2010/01/best-among-equals-choosing-between-tax-incentives-for-wind-projects (accessed May 20, 2010).

Sharpe, V. A. 2008. "Clean" nuclear energy? Global warming, public health, and justice. *The Hastings Center Report* 38 (4): 16–18.

Shere, C. 2004. *Natural gas.* Standard & Poor's Industry Surveys. (May 13), http://www.netadvantage.standardandpoors.com/docs/indsur///ngd_0504/ngd4054.htm (accessed March 20, 2009).

Sherman, M. 1985. Shifting winds, changing tides: Navigating low-income conservation programs through uncharted waters. In *State energy policy: Current issues, future directions*, ed. Stephen W. Sawyer and John R. Armstrong, 107–130. Boulder, CO: Westview.

Simanaitis, D. 2010. Will the first hydrogen highway signs be in German? Japanese? Or 'Merican? *Road & Track* (July): 104.

Simon, S. 2010. Even Boulder finds it isn't easy going green. *Wall Street Journal* (February 13): A1.

Slavin, M. 2009. The federal energy subsidy scorecard: How renewables stack up. Renewal Energy World.com, http://www.renewableenergyworld.com/rea/news/article/ 2009/11/the-federal-energy-subsidy-scorecard-how-renewables-stack-up (accessed April 19, 2010).

Smeloff, E., and P. Asmus. 1997. *Reinventing electric utilities.* Washington, DC: Island Press.

Smil, V. 1999. *Energies.* Cambridge, MA: MIT Press.

Smil, V. 2003. *Energy at the crossroads.* Cambridge, MA: MIT Press.

Smil, V. 2006. *Transforming the Twentieth Century: Technical Innovations and Their Consequences.* Oxford University Press, New York.

Smil, V. 2008. *Energy in nature and society.* Cambridge, MA: MIT Press.

Smil, V. 2009. U.S. energy policy: The need for radical departures. *Issues in Science and Technology* (summer): 47–50.

Smith, M., and K. Hargroves. 2008. Seeing the wood for the trees. *Ecos* 140 (December–January): 12–14.

Snyder, J. 2010. Top 5 hurdles climate bill must clear. The Hill.com (March 31), http://thehill.com/business-a-lobbying/89899-top-five-hurdles-climate-bill-must-clear-to-get-senate-ok (accessed April 13, 2010).

Sobin, R. 2007. Energy myth seven: Renewable energy systems could never meet growing electricity demands in America. In *Energy and American society: Thirteen myths*, ed. Benjamin K. Sovacool and Marilyn A. Brown, 171–200. Dordrecht, The Netherlands: Springer.

Sorrel, S. 2007. *The rebound effect: An assessment of the evidence for economy-wide energy savings from improved energy efficiency.* London: U.K. Energy Research Center.

Sovacool, B. 2006. The power production paradox. Unpublished Ph.D. diss. Blacksburg, VA: Virginia Polytechnic Institute and State University.

Sovacool, B. 2007. Coal and nuclear technologies: Creating a false dichotomy for American energy policy. *Policy Sciences* 40 (2): 101–122.

Sovacool, B. 2008a. *The dirty energy dilemma: What's blocking clean power in the United States.* Westport, CT: Praeger.

Sovacool, B. 2008b. The problem with the "portfolio approach" in American energy policy. *Policy Sciences* 41 (3): 245–261.

Sovacool, B., and M. A. Brown, eds. 2007. *Energy and American society: Thirteen myths.* Dordrecht, The Netherlands: Springer.

Sovacool, B., and R. F. Hirsh. 2007. Energy myth six: The barriers to new and innovative energy technologies are primarily technical: The case of distributed generation (DG). In *Energy and American society: Thirteen myths*, ed. Benjamin K. Sovacool and Marilyn A. Brown, 145–170. Dordrecht, The Netherlands: Springer.

Sovacool, B., and K. E. Sovacool. 2009. Identifying future electricity-water tradeoffs in the United States. *Energy Policy* 37 (2009): 2763–2773.

Speck, S. 2008. The design of carbon and broad-based energy taxes in European countries. *Vermont Journal of Environmental Law* 10, http://www.vjel.org/journal/pdf/VJEL10074.pdf (accessed June 9, 2010).

Starling, G. 1998. *Managing the public sector.* Ft. Worth, TX: Harcourt Brace.

State of California, Public Utilities Commission. 2009. *Renewables portfolio standard quarterly report Q4 2009.* Public Utilities Commission, State of California Sacramento: California Energy Commission. http://www.cpuc.ca.gov/NR/rdonlyres/52BFA25E-0D2E-48C0-950C-9C82BFEEF54C/0/FourthQuarter2009RPSLegislativeReportFINAL.pdf (accessed March 23, 2010).

State of Ohio, Office of the Governor. 2007. Coordinating Ohio energy policy and state energy utilization. http://governor.ohio.gov/Portals/0/Executive%20Orders/Executive%20Order%202007-02S.pdf (accessed January 31, 2009).

State of Washington. 1998. *Washington State Electricity System Study.* Olympia, WA: Washington Utilities and Transportation Commission.

Stavins, R. 2003. *Market-based environmental policies: What can we learn from U.S. experience and related research?* RWP03-031. Boston: John F. Kennedy School of Government, Harvard University Faculty Research Working Papers Series.

Stavins, Robert, 2007. *A U.S. Cap-and-Trade System to Address Global Climate Change*. Working Paper Series rwp07-052, Cambridge, MA: Harvard University, John F. Kennedy School of Government. http://ksgnotes1.harvard.edu/Research/wpaper.nsf/rwp/RWP07-052/$File/rwp_07_052_stavins.pdf (accessed January 16, 2011).

Stavros, R. 2003. Is the recovery here? *Public Utilities Fortnightly* 141 (October 15): 31–38.

Stephenson, J. 2009. Climate change: Observations on the potential role of carbon offsets in climate change legislation. March 5, 2009 testimony before the U.S. House of Representatives subcommittee on energy and environment. Report GAO-O9-456T. http://www.gao.gov/new.items/d09456t.pdf

Stern, N. 2009. *The global deal*. New York: Public Affairs.

Stillman, R. J. 2003. Twenty-first century United States governance: Statecraft and the peculiar governing paradox it perpetuates. *Public Administration* 81 (1): 19–40.

Stoft, S. 2002. *Power system economics*. Piscataway, NJ: IEEE Press.

Strauss, M. J. 2007. Efficiency replaces conservation as the goal of energy saving policies (October 30). *New York Times*, http://www.nytimes.com/2007/10/30/business/worldbusiness/30iht-reneff.1.8110766.html (accessed June 19, 2009).

Swanson, D., A. Brown, and M. Smith. 2009. American Recovery and Reinvestment Act of 2009: Summary of energy-related provisions. Dorsey and Whitney, LLP, http://www.dorsey.com/files/Publication/d6cc0bd0-3574-45d5-8a1e-3bd04a2e4caa/Presentation/PublicationAttachment/7cc8b851-efcf-495a-8ea8-489cd9a537ee/American_Recovery_Reinvestment_Act%20(2).pdf (accessed March 10, 2010).

Switzer, J. V. 2001. *Environmental politics*. 3rd ed. Boston, MA: Bedford/St. Martin's Press.

Taylor, J., and P. Van Doren. 2007. Energy myth five: Price signals are insufficient to induce energy investments. In *Energy and American society: Thirteen myths*, ed. Benjamin K. Sovacool and Marilyn A. Brown, 125–144. Dordrecht, The Netherlands: Springer.

Teske, P. 2003. State regulation: Captured Victorian-era anachronism or "reinforcing" autonomous structure? *Perspectives on Politics* 1 (June): 291–306.

Thaler, R. H., and C. R. Sunstein. 2009. *Nudge: Improving Decisions about Health, Wealth and Happiness*. New Haven, CT: Yale University Press.

Thompson, C. D. 1932. *Confessions of the power trust*. New York: E. P. Dutton.

Thurow, L. C. 2001. Solving the energy problem. In *Energy independence for the United States: Alternative policy proposals*, ed. Nake Kamrany, 23–27. Santa Monica, CA: Fundamental Books.

Tietenberg, T. H. 2006. *Emissions Trading Principles and Practice*, 2nd ed. Washington, DC: Resources for the Future.

Tietenberg, T., and L. Lewis. 2009. *Environmental and natural resource economics*. 8th ed. New York: Pearson-Addison Wesley.

Tkachenko, S. 2008. Actors in Russia's energy policy towards the EU. In *The EU-Russian dialogue: Europe's future energy security*, ed. Pami Aalto, 163–192. Aldershot, U.K.: Ashgate.

Tol, R. 2008a. The social cost of carbon: Trends, outliers and catastrophes. *Economics: The Open-Access, Open-Assessment E-Journal* 2 (2008): 25, http://www.economics-ejournal.org/economics/journalarticles/2008-25 (accessed January 29, 2010).

Tol, R. 2008b. Why worry about climate change? A research agenda. *Environmental Values* 17 (4): 437–470.

Tomain, J., and R. Cudahy. 2004. *Energy law in a nutshell*. St. Paul, MN: Thomson-West.

Toman, M., J. Griffin, and R. J. Lempert. 2008. *Impacts on U.S. energy expenditures and greenhouse-gas emissions of increased renewable-energy use*. Santa Monica, CA: Rand Corporation.

Trapmann, W. 2004. *The natural gas industry and markets in 2002*. Washington, DC: Energy Information Administration, U.S. Department of Energy.

Trenberth, K. 2010. More knowledge, less certainty. *Nature Reports Climate Change*, 4 (February). http://www.cgd.ucar.edu/cas/Trenberth/trenberth.papers/climate.2010.06.pdf (accessed January 11, 2011).

Turban, E., E. R. Mclean, and J. C. Wetherbe. 1999. *Information technology for management: Making connections for strategic advantage*. New York: John Wiley.

Twentieth Century Fund. 1948. *Electric power and government policy*. New York: TCF.

UCS (Union of Concerned Scientists). 2009. *Nuclear Power: A Resurgence We Can't Afford*. http://www.ucsusa.org/nuclear_power/nuclear_power_and_global_warming/nuclear-power-resurgence.html (accessed January 15, 2011).

UN (United Nations). 2000. *Emissions scenarios*. Geneva, Switzerland: United Nations Intergovernmental Panel on Climate Change, http://www.ipcc.ch/ipccreports/sres/emission/index.php?idp=0 (accessed February 2, 2010).

UN (United Nations). 2007. Synthesis report. Contribution of Working Groups I, II and III to the Fourth Assessment Report of the Intergovernmental Panel on Climate Change [Core writing team: R. K. Pachauri and A. Reisinger, A. (eds.)]. Geneva, Switzerland: Intergovernmental Panel on Climate Change, http://www.ipcc.ch/publications_and_data/ar4/syr/en/contents.html (accessed January 7, 2010).

UN (United Nations). 2007. *Climate Change 2007*: Synthesis Report. Geneva, Switzerland: United Nations Intergovernmental Panel on Climate Change (IPCC).

UN (United Nations). 2008. *Issue paper: Definition of primary and secondary energy*. United Nations Statistics Division: Energy Statistics, http://unstats.un.org/UNSD/envaccounting/londongroup/meeting13/LG13_12a.pdf (accessed December 28, 2009).

UN (United Nations). 2009. Energy balances and electricity profiles—concepts and definitions. United Nations Statistics Division: Energy Statistics, http://unstats.un.org/unsd/energy/balance/concepts.htm (accessed December 26, 2009).

UNCED (United Nations Conference on Environment and Development). 1992. *Report of the United Nations Conference on Environment and Development*. Rio de Janeiro, June 3–14. United Nations, http:www.un.org/documents/ga/conf151/aconf15126-1annex1.htm (accessed February 20, 2010).

UNDP (United Nations Development Program). 2009. *Decarbonizing growth: Some countries are doing better than others*. United Nations Development Program, Human Development Reports, http://hdr.undp.org/en/statistics/data/climatechange/growth/ (accessed April 1, 2010).

UNFCCC (United Nations Framework Convention on Climate Change). 2002. *Report of the Conference of the parties on its seventh session*. Presentation in Marrakesh, October 29–November 10. United Nations Framework Convention on Climate Change, http://unfccc.int/resource/docs/cop7/13a02.pdf (accessed February 20, 2010).

UNFCCC (United Nations Framework Convention on Climate Change). 2005. Emissions trading. United Nations Framework Convention on Climate Change, http://unfccc.int/kyoto_protocol/mechanisms/emissions_trading/items/2731.php (accessed February 20, 2010).

U.S. Census Bureau. 2008. *American community survey*, Table B25045. U.S. Census Bureau, http://www.census.gov/acs/www/index.html (accessed January 4, 2010).

U.S. Census Bureau. 2009. *Highlights of annual 2008 characteristics of new housing*. Washington, DC: U.S. Department of Commerce. U.S. Census Bureau, http://www.census.gov/const/www/highanncharac2008.html (accessed December 27, 2009).

U.S. Congress, Joint Committee on Taxation. 2010. *Estimates of federal tax expenditures for fiscal years 2009–2013*. Report No. JCS-1-10 (January 11). Joint Committee on Taxation, http://www.jct.gov/publications.html?func=startdown&id=3642 (accessed May 17, 2010).

U.S. Congress, Senate Committee on Energy and Natural Resources. 2009. Hearing to receive testimony on current national security challenges. 111th Cong, 1st Session. Washington, DC: Government Printing Office.

USDA (U.S. Department of Agriculture). 2006. 2007 *Farm Bill theme paper: Energy and agriculture*. Washington, DC: U.S. Department of Agriculture.

U.S. Primary Energy Overview. (From EIA, U.S. Department of Energy, Energy Information Administration). 2010c. *Annual energy review 2009*. Washington, DC: U.S. Department of Energy.

U.S. Senate. 2008. S.2758, a bill to amend the Agricultural Research, Extension, and Education Reform Act of 1998. Washington, DC: US Government Printing Office. http://origin.www.gpo.gov/fdsys/pkg/BILLS-111s2758is/pdf/BILLS-111s2758is.pdf (accessed April 9, 2009).

Veigel, J. M., and S. Lakoff. 1985. Stimulating innovation through alternate institutions. In *State energy policy: Current issues, future directions*, ed. Stephen W. Sawyer and John R. Armstrong, 267–282. Boulder, CO: Westview.

Verrastro, F. A., J. A. Placke, and A. S. Hegburg. 2004. Securing U.S. energy in a changing world. *Middle East Policy* 11 (4): 1–25.

Verschoor, C. C. 1999. Corporate performance is closely linked to a strong ethical commitment. *Business and Society* 104 (winter): 407–415.

Victor, D. G., and L. Yueh. 2010. The new energy order. *Foreign Affairs* 89 (1): 61–73.

Vietor, R. H. K. 1984. *Energy policy in America since 1945.* Cambridge, U.K.: Cambridge University Press.

Vietor, R. H. K. 1987. *Energy Policy in America Since 1945: A Study of Business-Government Relations.* Cambridge, UK: Cambridge University Press.

Vine, E., M. Kushler, and D. York. 2007. Energy myth ten: Energy efficiency measures are unreliable, unpredictable and unenforceable. In *Energy and American society: Thirteen myths*, Benjamin K. Sovacool and Marilyn A. Brown, 265–288. Dordrecht, The Netherlands: Springer.

Viscidi, L. 2010. The twilight of the western oil majors. *Foreign Policy.com.* (April 27), http://www.foreignpolicy.com/articles/2010/04/26/the_twilight_of_the_western_oil_majors (accessed June 21, 2010).

Voorhees, J. 2009. White House rolls out details of auto fuel economy, emissions standard. NYTimes.com, http://www.nytimes.com/gwire/2009/09/15/15greenwire-white-house-rolls-out-details-of-auto-fuel-eco-13342.html (accessed March 11, 2010).

Wald, M. 2010. Selling cape wind's future wares. *New York Times.* Green Blog (May 7), NYTimes.com, http://green.blogs.nytimes.com/2010/05/07/selling-cape-winds-future-wares/ (accessed June 22, 2010).

Wald, M., and L. Kaufman. 2010. U.S. tightens requirements for Energy Star certification. *New York Times* (April 14), http://www.nytimes.com/2010/04/15/business/energy-environment/15star.html (accessed May 8, 2010).

Walker, F. D. 1999. Corporate character and ethics—a competitive difference? *Business and Society* 104 (winter): 439–458.

Ward, N. 1992. Patrons of husbandry. *American Heritage* 43 (8): 28–9.

Warkentin. D. 1996. *Energy marketing handbook.* Tulsa, OK: PennWell Books.

Warren, C. 1928. *The Supreme Court in United States history.* Boston: Little, Brown.

Warren, H. G. 1959. *Herbert Hoover and the Great Depression.* New York: Oxford University Press.

Watkins, S. 2003. Former Enron vice president Sherron Watkins on the Enron collapse. *Academy of Management Executive* 17 (November): 119–125.

Watkins, T. H. 1999. *The hungry years.* New York: Henry Holt.

Watson, D. J., and W. L. Hassett. 2002. Capital-intensive privatization: Return to public ownership. *Public Works Management and Policy* 7 (October): 115–123.

Webber, R. E. 1985. Evaluating energy programs in Michigan. In *State energy policy: Current issues, future directions*, ed. Stephen W. Sawyer and John R. Armstrong, 59–84. Boulder, CO: Westview.

Weinberger, C. W. 2005. An energy bill—at last. *Forbes* 176 (6): 35.

Weiss, C., and W. Bonvillian. 2009. *Structuring an energy technology revolution.* Cambridge, MA: MIT Press.

Weiss, C. H. 1998. *Evaluation.* 2nd ed. Upper Saddle River, NJ: Prentice Hall.

Weiss, D., and A. Wingate. 2007. Big oil's favorite representatives. Center for American Progress Action Fund, http://www.americanprogressaction.org/issues/2007/house_oil.html (accessed March 17, 2010).

Weiss, R. 2004. Data quality law is nemesis of regulation. *Washington Post* (August 16): A1.

Weitzman, M. 2009a. Some basic economics of extreme climate change. Harvard University, http://www.economics.harvard.edu/faculty/weitzman/files/Cournot%2528Weitzman%2529.pdf (accessed January 29, 2010).

Weitzman, M. 2009b. Some basic economics of extreme climate change. In *Changing climate, changing economy?*, ed. Jean-Philippe Touffut. Northampton, MA: Edward Elgar.

Wellstone, P. D., and B. M. Gasper. 1985. Politics and policy: The Minnesota community energy program. In *State energy policy: Current issues, future directions*, ed. Stephen W. Sawyer and John R. Armstrong, eds. 131–146. Boulder, CO: Westview.

Whalen, J. J. 1991. The utility industry's changing face. *Public Utilities Fortnightly* 128 (November 1): 22–23.

Whellen, T. L., and J. D. Hunger. 2002. *Strategic management and business policy.* 8th ed. Upper Saddle River, NJ: Prentice Hall.

White House, Office of the Press Secretary. 2009. North American leaders' declaration of climate change and clean energy (August 10). White House, Office of the Press Secretary. http://www.whitehouse.gov/the_press_office/North-American-Leaders-Declaration-on-Climate-Change-and-Clean-Energy/ (accessed February 7, 2010).

White House. Office of the Press Secretary. 2009. Statement of U.S. energy policy goals. White House. Office of the Press Secretary. http://www.isa.org/filestore/af/Energy-Policy-Proclamation.pdf (accessed May 22, 2010).Wijen, F., and K. Zoeteman. 2004. *Final report of the study* Past and future of the Kyoto Protocol. Tilburg, the Netherlands: Tilburg University Globus Institute for Globalization and Sustainable Development.

Wilbanks, T. J. 1981. Introduction. In *Social and political perspectives on energy policy*, ed. Karen M. Gentemann, xv–xxxvi. New York: Praeger.

Wilbanks, T. J. 2007. Energy myth thirteen: Developing countries are not doing their part in responding to concern about climate change. In *Energy and American society: Thirteen myths*, ed. Benjamin K. Sovacool and Marilyn A. Brown, 341–350. Dordrecht, The Netherlands: Springer.

Wildavsky, A. 1988. *Searching for safety*. Piscataway, NJ: Transaction Publishers.

Wilks, D. 2006. Testimony before U.S. Senate Committee on Energy and Natural Resources, hearing on coal-based generation reliability (May 25). http://energy.senate.gov/public/index.cfm?FuseAction=Hearings.Testimony&Hearing_ID=1560&Witness_ID=4410

Wilson, J. A. 1985. Efficiency standards in California's energy policy. In *State energy policy: Current issues, future directions*, ed. Stephen W. Sawyer and John R. Armstrong, 147–170. Boulder, CO: Westview.

Wilson, J. Q. 1980. *Politics of regulation*. New York: Basic Books.

Wilson, J. Q. 1989. *Bureaucracy*. New York: Basic Books.

Winebrake, J., and D. Sakva. 2006. An evaluation of errors in U.S. energy forecasts: 1982–2003. *Energy Policy* 34 (2006): 3475–3483.

Wirth, T. E., C. B. Gray, and J. D. Podesta. 2003. The future of energy policy. *Foreign Affairs* 82 (4): 132–155.

Wiser, R., C. Namovicz, M. Gielecki, and R. Smith. 2007. *Renewable portfolio standards: A factual introduction to experience from the United States*. LBNL-62569. Berkeley, CA: Lawrence Berkeley National Laboratory.

Wiser R., G. Barbose. 2008. *Renewables Portfolio Standards in the United States: A Status Report with Data Through 2007*. Lawrence Berkeley National Laboratory, Report LBNL 154-E. http://eetd.lbl.gov/ea/ems/reports/lbnl-154e.pdf (accessed January 16, 2011).

Wood, D. J. 1986. *Strategic uses of public policy: Business and government in the Progressive Era*. Marshfield, MA: Pitman Publishing.

World Bank. 2008. *Rising food and fuel prices: Addressing the risks to future generations*. World Bank, http://siteresources.worldbank.org/DEVCOMMEXT/Resources/Food-Fuel.pdf (accessed January 4, 2010).

WTO (World Trade Organization). 2006. *World Trade Report 2006*. Geneva, Switzerland: World Trade Organization, http://www.wto.org/english/res_e/booksp_e/anrep_e/world_trade_report06_e.pdf (accessed May 1, 2010).

Yergin, D. 1993. *The prize: The epic quest for oil, money, & power*. New York: Free Press.

Yergin, D. 2006. Ensuring energy security. *Foreign Affairs* (March/April), http://www.foreignaffairs.com/articles/61510/daniel-yergin/ensuring-energy-security (accessed June 12, 2010).

Young, O. 2010. The effectiveness of international environmental regimes. In *Human footprints on the global environment*, ed. E. A. Rosa, A. Diekman, T. Dietz, and C. A. Jaeger, 165–202. Cambridge, MA: MIT Press.

Zamani, M. 2005. Energy conservation: An alternative for investment in the oil sector for OPEC member countries. *OPEC Review: Energy Economist and Related Issues* 29 (2): 107–114.

Index

C

D

E

G

H

I

R

S

T

U

V

W

9781439841891